Research on Intelligent Manufacturing

Research on Intelligent Manufacturing (RIM) publishes the latest developments and applications of research in intelligent manufacturing—rapidly, informally and in high quality. It combines theory and practice to analyse related cases in fields including but not limited to:

Intelligent design theory and technologies
Intelligent manufacturing equipment and technologies
Intelligent sensing and control technologies
Intelligent manufacturing systems and services

This book series aims to address hot technological spots and solve challenging problems in the field of intelligent manufacturing. It brings together scientists and engineers working in all related branches from both East and West, under the support of national strategies like Industry 4.0 and Made in China 2025. With its wide coverage in all related branches, such as Industrial Internet of Things (IoT), Cloud Computing, 3D Printing and Virtual Reality Technology, we hope this book series can provide the researchers with a scientific platform to exchange and share the latest findings, ideas, and advances, and to chart the frontiers of intelligent manufacturing.

The series' scope includes monographs, professional books and graduate textbooks, edited volumes, and reference works intended to support education in related areas at the graduate and post-graduate levels.

Peihua Gu · Deyi Xue · Qingjin Peng · Jian Zhang

Adaptable Design

Methods and Applications

華中科技大學出版社
Huazhong University of Science and Technology Press

Springer

Peihua Gu
Tianjin University
Tianjin, China

Deyi Xue
University of Calgary
Calgary, AB, Canada

Qingjin Peng
University of Manitoba
Winnipeg, MB, Canada

Jian Zhang
Shantou University
Shantou, China

ISSN 2523-3386 ISSN 2523-3394 (electronic)
Research on Intelligent Manufacturing
ISBN 978-981-99-5868-9 ISBN 978-981-99-5869-6 (eBook)
https://doi.org/10.1007/978-981-99-5869-6

Jointly published with Huazhong University of Science and Technology Press
The print edition is not for sale in China (Mainland). Customers from China (Mainland) please order the print book from: Huazhong University of Science and Technology Press.

This Springer imprint is published by the registered company Springer Nature Singapore Pte Ltd.
The registered company address is: 152 Beach Road, #21-01/04 Gateway East, Singapore 189721, Singapore

Preface

The concept of adaptable design (AD) was first conceived between 1995 and 1997 during which I held the NSERC/AECL Industrial Chair in Advanced Design (NSERC stands for Natural Science and Engineering Research Council of Canada; AECL is Atomic Energy of Canada Limited). AECL designs and builds nuclear power plants and research reactors in Canada. For a typical nuclear power plant, its safe operational life is about 40 years, some are longer and others shorter. For any power plant or other major engineering infrastructure, its design and building reflect the current technologies. As technologies continues to advance, it is understandable that in the not-too-distant future after completion of the power plants, some parts of the nuclear power plant would need upgrading for enhancing performances and for extending operational life. To accommodate the anticipated upgrading, repair, and replacement operations for the major engineering systems, special design properties may be embedded in the design, which the special property is considered as design adaptability.

For design and building of major engineering systems, engineering, procurement, and construction (EPC) companies are normally invited to bid the projects. The proposal development and bid writing usually take significant time and resources to complete. The result of the tendering process is a formal contract to deliver completed turnkey projects with agreed costs and time frame. In order to develop successful design of the proposed major project, it is desirable that EPC companies can develop new design based on the existing design of similar projects. By utilizing the existing similar design, it not only develops the design quickly, but also estimates processes and tools as well as costs to complete the project with some confidence to avoid costs and schedules associated with the projects. A group of EPC companies sponsored my second NSERC Chair in design to develop concepts and methods to rapidly developing new designs based on the existing designs—the design reuse which is a part of adaptable design.

The ongoing discussion in manufacturing research communities such as CIRP advances manufacturing technologies and methods to continuously increase production productivity and product quality and reduce total costs and environmental impacts. From a particular product, customers would like the product with excellent functionalities, high quality, and low prices; product manufacturers would hope to increase profit margin or bigger market shares or both; the society or public wants healthy products with the minimal environmental impact and maximal resources utilization. Manufacturing research may not be able to resolve all of the above issues as some of the challenges are resulted from design. For a typical manufacturing company, the enterprise simply produces products according to their designs. Based on my manufacturing research experience and industrial practices, a near optimal design can partially address the above challenges.

The concepts, methods, and applications of AD have been developing for the last 25 years, and most of the research activities and applications have been carried out by my colleagues and graduate students in Canada and China. This book provides summaries of these efforts and aims to continue the development of AD, especially in the digital, carbon neutrality, and industrial metaverse era with rapid advancements of artificial intelligences, big data, Internet of Things, cloud computing, and VR/AR technologies.

Finally, I wish to take this opportunity to thank the coauthors who complete the manuscript and to all our colleagues and students to carry out the research and application work. I hope the publication of this manuscript will promote the future research and application of AD.

Tianjin, China Peihua Gu

Contents

Chapter 1
Introduction

Overview

The market competition requires manufacturing enterprises to be responsive and responsible in developing products for better product functionality, quality, features, environmental friendliness, and low cost. The modern manufacturing enterprises have to be flexible and agile, and well connected to networks and supply chains. In the last several decades, major progresses of product manufacturing have mainly benefited from the development and application of advanced manufacturing technologies such as computer-numerically controlled (CNC) machines [5], robotics, computer-aided design and manufacturing (CAD/CAM) systems, flexible manufacturing systems (FMS) [15], computer-integrated manufacturing (CIM), reconfigurable manufacturing systems (RMS) [51], material requirement planning systems (MRP), enterprise resource planning (ERP), new manufacturing technologies such as laser machining [65], free-form fabrications [54], advanced machining [49], precision engineering, micromachining, and nano-manufacturing [18, 62], and other advanced manufacturing technologies and tools [75, 106], as well as progress in sciences and technologies in general. However, advanced manufacturing technologies alone would not be sufficient to deal with all the challenges in product development, as some challenges are originated from product design. Therefore, advanced design technologies and tools in the early product design process are needed where important decisions are made with respect to the product functionality, quality, manufacturability, cost, and environmental performance for the optimum design and development of products.

In the last three decades, significant research efforts and progresses have been made toward the development of fundamental understanding and effective tools in various aspects of early design processes [96]. Examples of engineering design research include axiomatic design [39, 42, 88, 89], function-based design [21, 40, 48, 70, 81, 102], product family/portfolio architecture, modular, and platform design [14, 23, 25, 33, 34, 38, 63, 66–68, 71, 77, 82, 83, 99–101], design-theoretic approaches [59, 88, 89, 109, 111, 112], applications of TRIZ methodologies for engineering

P. Gu et al., *Adaptable Design*, Research on Intelligent Manufacturing,
https://doi.org/10.1007/978-981-99-5869-6_1

design and innovation [28], and design for manufacturing, assembly, service, and environment [11, 50, 61, 84, 93, 97, 115]. Manufacturing enterprises have adopted different design technologies such as design for manufacturing and assembly to improve their product design. Life cycle engineering and design have also generated substantial research, development and industrial interests [3, 4, 7, 9, 12, 16, 24, 26, 41, 45–47, 52, 53, 58, 60, 72, 78, 79, 85, 86, 94, 95, 103–105]. The key concepts of the life cycle engineering and design take a holistic approach dealing with the entire life cycle of products such as examination of various stages of the product. However, these advancements, focused on specific product design aspects, should be utilized to address optimization issues considering the entire product life cycle. A flexible and comprehensive product design approach is yet to be established for these design methodologies and technologies to provide effective design tools for design of new products that are able to meet changing design requirements.

This book reports on a new design approach, namely adaptable design (AD), for efficient and effective products design for functionality, manufacturing, customization, environment, and life cycle performance including the low carbon requirement [13]. AD was proposed to improve the competitiveness of products in the marketplace while reducing the environmental impact during their lifespans. Adaptable design can extend and change functions of a product through adaptation activities such as upgrading functional modules and adding new functions. In adaptable design, adaptability is defined as the utility extension of a system or product [7]. The measure of product adaptability is critical in the adaptable product design. Both specific adaptabilities based on the cost saving [35] and general adaptability by comparing the actual structure of the product with its ideal structure [29] are required in predicting the future adaptation of a current product in its life cycle. The adaptable design method has been used to design the product considering changes of requirements, configurations, and parameters in the product lifecycle [108]. A robust design approach can identify the optimal design configuration and parameters of an adaptable product using functional performance measures that are the least sensitive to uncertainties [114]. An integrated model was developed to obtain the optimal configurations and parameters of adaptable products with the robust performance using the AND/OR graph and two-level optimization method [114]. To meet the personalized demand in a cost-effective way, the Open Architecture Product (OAP) was proposed with three types of functional modules: platform/standard modules, customized modules, and personalized modules [43]. The customized modules are selected by users. The personalized module is a specially developed module for the individual user's need. To enhance the adaptability of interfaces effectively, some design rules and evaluation methods were discussed for the initial design of open interfaces [34, 37]. Various efforts have also been devoted to the implementation of adaptable design. This book aims to report the progress of the adaptable design theory and its applications. Limitations and further trends of adaptable design will also be discussed.

1.1 Product Life Cycle and Engineering Design

1.1.1 Product Life Cycle

Development of a product undergoes a sequence of activities from product requirements to the end life of the product as shown in Fig. 1.1. We classify these activities into five major phases:

- *Requirements*: identification of the design needs,
- *Design*: creation of the design solution,
- *Manufacturing*: implementation of the design solution,
- *Operation*: utilization of the product functions, and,
- *Retirement*: treatment of the product at its end of life.

Phase I—Requirements

- *Customer needs*: A design is usually initiated from customer needs. These customer requirements can be achieved from the market analysis through the collection of customer feedbacks to the existing products and expectations to the

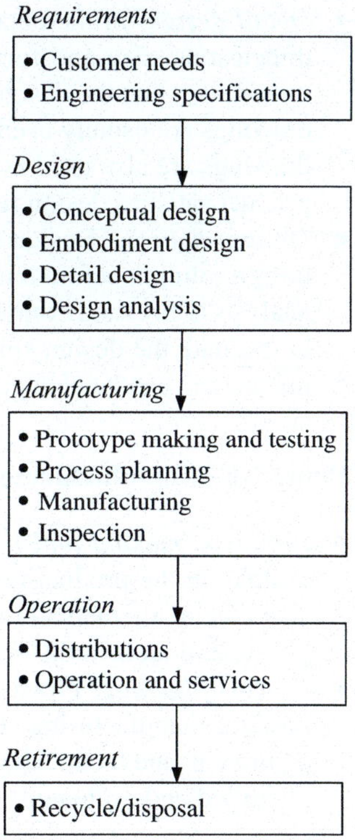

Fig. 1.1 Product development activities in various life cycle phases

new products. The customer requirements are usually qualitative and subjective representing various preferences of the customers. For example, the light weight is a typical customer requirement for a cell phone.

- *Engineering specifications*: Engineering specifications are the technical descriptions of customer requirements. Engineering specifications are usually quantitative and used as criteria to evaluate whether customer needs are satisfied. For example, the battery life of video playback over 10 h for a cell phone is a typical engineering specification.

Phase II—Design

- *Conceptual design*: Conceptual design is a process to identify a concept solution to satisfy design requirements. In conceptual design, scientific and engineering principles are selected based on design requirements. Multiple conceptual design candidates can be created and compared to identify the best one for further development. A design candidate can be completely new or modification to an existing design. Textbooks, patents, and design repositories are resources for creation of conceptual design candidates.
- *Embodiment design*: Product architecture, configuration, and major parameter values are determined in embodiment design. Optimization is often used to achieve the best values of key parameters.
- *Detail design*: In the detail design, all the descriptions of the design, including dimensions, size and form tolerances, material, heat treatment, etc., are determined such that the design can be produced from these descriptions. Engineering drawings are usually used for documentation of the detail design solution. These drawings are also used as legal documents to evaluate whether the manufactured product satisfies design requirements.
- *Design analysis*: Design analysis is conducted to evaluate performances of the design solution before it is produced. Design analysis tools, such as finite element analysis (FEA) and computational fluid dynamics (CFD) systems, are often used to evaluate the design solution modeled by geometry. Design configuration and parameters can be changed based on results of the design evaluation.

Phase III—Manufacturing

- *Prototype making and testing*: Physical prototyping and testing aim at further evaluating the product design in addition to virtual prototyping and evaluation, since the mathematical models in the virtual system may not be the same as the physical models in the real-life world.
- *Process planning:* Process planning serves as a link between product design and product manufacturing. In this stage, raw material, manufacturing operations, equipment and tools, labor resources, production lines and material flows, etc. are identified and evaluated.

- *Manufacturing:* Manufacturing is the process to produce products with resources including materials, equipment, labors, and material handling systems. Manufacturing is usually accomplished through the mass production.
- *Inspection:* Inspection is conducted for the quality control of manufacturing processes and manufactured products. Either some samples or all the products are inspected such that the wrongly manufactured products are corrected or rejected, and the process parameters are adjusted to improve quality of the products.

Phase IV—Operation

- *Distribution*: Distribution serves as a link between manufacturers and end-user customers of the product. Distribution is often conducted through sales activities.
- *Operation:* The product eventually comes to the operation stage to deliver the intended functions for the customer. This phase is the longest among all phases of the product life cycle. Maintenance and repair services are conducted to retain and restore the intended functions. The operation data can be collected through sensor technologies such as those used in the Internet of Things (IoT) or cyber-physical systems to evaluate operation performance for future improvement of the product.

Phase V—Retirement

- *Recycle/disposal*: When the product is not functioning or its functional performance no longer satisfies customer needs, the product reaches to its end-of-life stage. Components of the product can be disassembled for reuse, remanufacturing and/or recycling. The components or the whole product can be disposed into landfill when the cost for recycle is too high. Recyclability and disposability are two important evaluation measures in environment conscious design.

1.1.2 Engineering Design

In the traditional product development process and life cycle, various activities in the different life cycle phases are carried out sequentially as shown in Fig. 1.1. The information acquired in an earlier design stage is used for decisions made in later stages. In this sequential engineering process, since conditions and constraints in the downstream phases, including manufacturing, operation, and recycle/disposal, are not well considered in the design phase, the design may require significant efforts in manufacturing, operation, and recycle/disposal phases. Design modifications have to be conducted iteratively when problems in the downstream phases have to be solved, leading to the long product development time and high product development cost.

It was estimated that 80% of the product life cycle costs are determined during the early design stage, while the design only takes about 20% of the whole product development efforts [69]. Any lifecycle characteristics such as green and low carbon

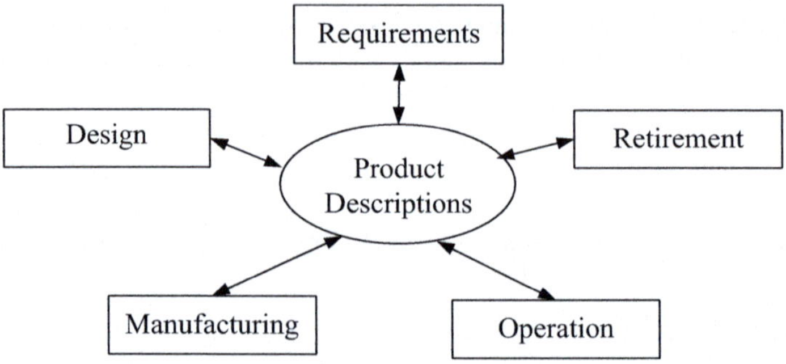

Fig. 1.2 Product development in concurrent engineering

aspects have also been defined at design stages. Therefore, the improvement in design can reduce efforts and costs significantly in downstream product life cycle phases.

The design with low costs in downstream product life cycle phases is usually achieved through design-for-X and concurrent engineering methods [55, 73]. In concurrent engineering as shown in Fig. 1.2, there are various considerations in different life cycle phases from customers, manufacturing engineers, and service persons in the design process. Constraints and feedbacks from these life cycle phases are considered concurrently in design decision making. Since concurrent engineering is primarily employed to improve quality of design, concurrent engineering is often called concurrent engineering design.

Typical downstream life cycle phases considered in concurrent engineering design include:

- *Manufacturing:* casting, forging, machining (e.g., milling and turning), welding, additive manufacturing, and assembly,
- *Operation:* maintenance, repair and upgrading, and,
- *Retirement:* reuse, remanufacturing, recycle and disposal.

Many methods and tools have been developed in the research of design-for-X including design for manufacturing [21, 22], design for assembly [8], design for maintenance [31], and design for recycling/disposal [3, 113].

Despite the progress in design-for-X and concurrent engineering design considering the different downstream product life cycle phases, functions of the presently designed products usually cannot be changed much in the product operation phases. Maintenance and repair services are carried out only to bring the degraded functions back to the expected levels. A product reaches its end of life when the degraded function no longer satisfies user needs, costs to bring the product functions back to the desired level are too high or new products with better functions are available in the market. To extend the lifespan of a product, a new design approach is required such that functions of the existing product can be upgraded with improved or new functions added. The adaptable design [35] discussed in this book is such an approach to design products that can be relatively easily changed in operation stages with the improved and new functions.

1.1.3 Design Methodologies and Tools

The research on design was initiated in 1970s [70] in the early work through systematic studies on various design activities in design such that design knowledge and design process could be modeled in the scientific way. Since then, many design methods and tools have been developed to improve design quality and efficiency [20, 98] Tomiyama et al. [96] reported a comprehensive review of design methodologies in educational and industrial applications. In this section, some important and common methods and tools used in various steps of design are summarized.

Identification of Requirements

- *Quality function deployment (QFD)*: QFD [1] is a method to convert customer needs to engineering specifications for design of new products or improvement of existing products. In this method, based on customer requirements, the existing products are evaluated against the customer requirements for identification of the key requirements to be satisfied or improved. The engineering specifications are then selected, and relations between customer requirements and engineering specifications are defined by a matrix. The tradeoff among the engineering specifications is also considered for identifying the final results of these engineering specifications.

Conceptual Design

- *Theory of inventive problem solving (TRIZ)*: TRIZ [27] is a tool developed based on studies of over 1.5 million patents for systematic problem solving in innovative design. In TRIZ, problem-solving methods are summarized as 40 inventive principles. Contradictions between engineering parameters, called elements, are defined by a contradiction matrix. TRIZ provides an effective way to obtain inventive solutions by identifying and resolving contradictions.
- *Morphological chart method*: In this method [116], design requirements are organized as functions and sub-functions. For each function or sub-function, multiple design concepts are defined. Complete design solution candidates are created by combinations of design concepts for these functions/sub-functions.
- *Axiomatic design*: In axiomatic design [88, 89], functional requirements (FRs) in the functional domain and design parameters (DPs) in the physical domain are associated by a matrix:

$$\boldsymbol{FR}_{n\times 1} = [A]_{n\times m}\boldsymbol{DP}_{m\times 1} \tag{1.1}$$

where $\boldsymbol{FR}_{n\times 1}$ is a vector with n FRs, $\boldsymbol{DP}_{m\times 1}$ is a vector with m DPs, and $[A]_{n\times m}$ is a design matrix to define the relations between FRs and DPs. In axiomatic design, the independence axiom demands that the functional requirements should be maintained to be independent. Therefore, m must be equal to or greater than n for $[A]_{n\times m}$. The ideal design is achieved with $n = m$, when the design matrix is either a diagonal one (called uncoupled design) or a triangular one (called decoupled design). A decoupled design satisfies the independence axiom when the design parameters are selected in a certain sequence. When the design matrix is a full matrix, the design is called a coupled design. A coupled design doesn't satisfy independence axiom.

Embodiment Design

- *Modular design, product platform design, and product family design*: These methods are primarily used for selection of product configurations [2, 44]. A product configuration is modeled by its major components and their relations. Modular design [2] is a common method in configuration design to decompose the whole product into functionally and physically independent modules. A module can be used in multiple products such that the design and manufacturing efforts for various products are reduced. When a common major module is used for a series of products, this module is called a platform of these products [44]. A series of products with a common platform is also called a product family.
- *Optimization*: Numerical optimization is often used to identify optimal values of selected design variables such that a certain goal (e.g., to maximize a performance measure or minimize a cost measure) can be achieved [6]. An optimization problem is defined by:

$$\begin{aligned} &\min_{\text{w.r.t.}\boldsymbol{X}} f(\boldsymbol{X}) \\ &\text{Subject to: } H_i(\boldsymbol{X}) = 0, i = 1, 2, \ldots \\ &\qquad\qquad\quad G_i(\boldsymbol{X}) \le 0, i = 1, 2, \ldots \end{aligned} \tag{1.2}$$

  where $\boldsymbol{X}$ is the vector with design variables, $f(\boldsymbol{X})$ is the optimization objective function, $H_i(\boldsymbol{X}) = 0$ is a set of equality constraints, and $G_i(\boldsymbol{X}) \le 0$ is a set of inequality constraints. A constrained optimization problem is usually converted into an unconstrained optimization problem by introducing penalty functions that are defined using constraints. Many numerical methods have been developed for solving unconstrained optimization problems. When multiple optimization objective functions (e.g., to maximize a performance and minimize a cost) have been selected, the optimization is then formulated as a multi-objective optimization problem [17]. Global optimization methods such as genetic algorithm [32] are used for preventing the optimal solution from falling into local optimum positions.
- *Robust design*: Robust design aims at identifying the design such that the influences of the noise parameters on the selected performance measure are minimized [19]. In robust design, the signal-to-noise ratio is usually selected as the measure

to maximize. Among various robust design methods, the Taguchi method [19, 91, 92] is a common method developed based on design of experiment (DOE) to identify the optimal values of design variables to maximize the signal-to-noise ratio.

- *Design for-X and concurrent engineering design*: In design-for-X and concurrent engineering design methods, the downstream product life cycle aspect *X*, such as manufacturing, maintenance, repair, recycle or disposal, is considered in the design stage. Typical design-for-X methods include design for manufacturing [21, 22], design for assembly [8], design for maintenance [31], and design for recycling/disposal [3, 113].

Detail Design

- *Computer-aided design (CAD)*: CAD systems have been widely used in the product detail design for modeling of 3D parts, assembling parts, and 2D drawings of parts and assemblies [57, 110]. In addition, parametric design, feature-based design, free-form surface modeling, and rendering are typical functions of CAD systems. The new CAD systems have more comprehensive functions including the product life cycle data management.

Design Analysis

- *Computer-aided engineering (CAE)*: CAE systems are used for various engineering analysis tasks through extensive computing [57]. Structure analysis based on the finite element analysis (FEA) method and fluid analysis based on the computational fluid dynamics (CFD) are typical CAE functions integrated with CAD systems.

Manufacturing

- *Design for manufacturing (DFM) and design for assembly (DFA)*: Design methods considering manufacturing life cycle aspect are primarily in the categories of design for manufacturing [21, 22] and design for assembly [8]. Various guidelines and evaluation measures, such as to reduce the total number of parts, use standard components, and design parts with multiple use, have been developed for DFM and DFA methods. Many methods considering specific manufacturing processes, such as casting, forging, machining (e.g., milling and turning), and welding, have been introduced and used.

Operation

- *Design for service*: In design for service methods [31], reliability of a product or a system is usually predicted such that the proper maintenance is conducted to prevent the product or system from costly failures. The development of product service systems (PSS) [64] provides a new design direction such that the real-time operating data of the products can be collected and analyzed for the product improvement.

Retirement

- *Design for recycle/disposal*: Design for recycle/disposal methods [3, 113] focus on design of the product that can be disassembled easily when the product reaches its end of life. The design for recycle/disposal methods can reduce the impact of product on environment. This approach is also called green product design or environment conscious design.

The adaptable design approach is a kind of design-for-X approach by incorporating considerations of product adaptability into the design stage. By designing a product such that the existing functions can be upgraded or new functions can be added when the product reaches its end of life, the lifespan of this product can be extended and the environment impact of the product can be reduced.

1.1.4 Requirements for Adaptable Design

Adaptable design (AD) [35] is a new design paradigm that aims at creating designs and products that can be easily adapted for different requirements in design and product life cycle stages. Both economic and environmental benefits can be obtained by using AD. The underlying philosophy of AD is the ability to adapt to new requirements and reuse a product and design when circumstances change such as replacing multiple products with one adaptable product with a set of add-on accessories or attachments. Adaptation prolongs the service of a product in time, or extends it into other applications. Such extension of life and applicability of an adaptable product usually requires additional work, called the process of "adaptation". Adaptation, wherever justified, could be more superior to recycling because it uses most components as they are, while recycling reclaims materials and reinserts it at the early stages of the product supply chain [37].

Willems et al. [107] reported that the adaptability of a product at the end of its life cycle is significantly influenced by its design. They proposed a dedicated Methodology for Assessing the Adaptability of Products (MAAP) which addresses the whole list of product demands. MAAP evaluates a product from four main categories of adaptation toward the end of a product life cycle: maintenance, repair, remanufacturing, and up and downgrading, and provides potential improvement areas in the product design where such adaptability can be considered at the product design stage. Several levels of metrics were used to provide numeric analysis of the level of adaptation. Although somewhat subjective in nature, this approach provides an interesting and treatise of the adaptability of products at the end of their life cycle.

A key element in AD is the reuse of the "utility" or "service" of a product. In the context of engineering systems and products, the utility of a product is determined by its physical functions described by its effects on material, energy, and signal. Therefore, in adaptable design, functions (both realized and potential functions) of a product comprise the most important information about the product. The functions are realized by physical structures of the product.

The second key element in the proposed approach to AD is a quantifiable definition of adaptability so that it can be considered in comparable measures along with other design criteria during decision making processes. It is proposed based on the definition of adaptability on "savings" achieved by adapting an existing product instead of producing a new one.

The third key element of AD is a model of the design process that is applicable to AD. We introduce a function-based design process model for the design stemming similarly to Suh's axiomatic design process model. In the proposed model, however, we distinguish between functions and other requirements for the reason explained in the first key element.

1.2 Principles of Adaptable Design

1.2.1 Adaptable Design

A design process usually results in the description or blueprint (or CAD model) of an entity that, when materialized (realized) according to its description (blueprint or CAD model), can fulfill a set of required objectives. This description is the "design", and the realized design is the "product". Therefore, the design process generally results in the creation of two entities, the design and the product. The adaptabilities are therefore defined with products and their designs as shown in Fig. 1.3.

Design Adaptability refers to the adaptability in the design (blueprint) of a product, so that this design can be modified (adapted) to produce another product. Products created via the adaptation of the same design are usually similar.

Adaptability in the design of a product is generally important to the producer. The producer can adapt the same design to different requirements and produce different products.

Design adaptability aims at reusing the same "design" for the creation of different product designs. Therefore, its applicability and importance in the design process directly depend on the possibility of creating new products with similar designs. For example, if the design of a special product is not similar to any other product and that design cannot be adapted to create other products, the design adaptability is irrelevant. Adaptability in design is useful only if a "population" of designs and their realizations (products) exist.

Product Adaptability refers to the ability of a product to be adapted to various usages or capabilities. The adaptation is performed by the user (or contractors

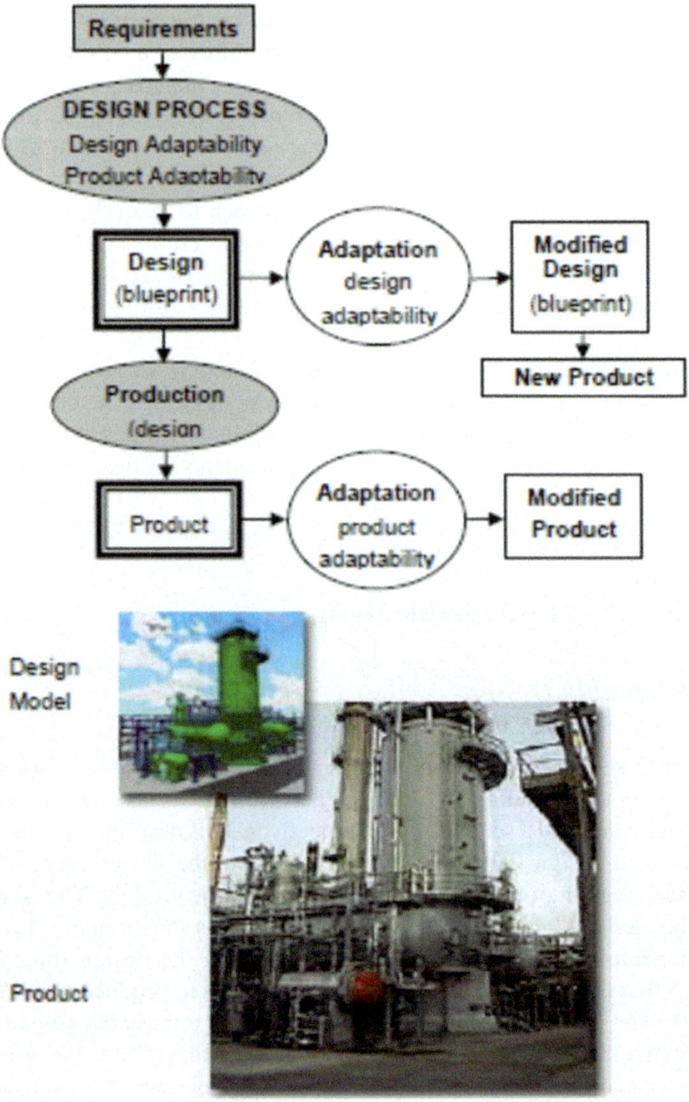

Fig. 1.3 Design adaptability and product adaptability (Courtesy of VECO Canada)

requested by the user). In case of mechanical engineering design, product adaptability refers to the ability of the user (or contractors) to modify a machine to perform various functions or to enhance its performance by the inclusion of new features.

The importance and applicability of product adaptability directly depend on the possibility of using the same product for proposed other usages or extension of its service life as required. Product adaptability is useful even if there is only one design and one product made from this single design.

Obviously, a product would not be adaptable if its design is not made adaptable. However, this distinction would not only facilitate the discussions but also have practical implications.

Adaptable design can therefore be defined as a design paradigm that extends the intended utility of products and design through design and product adaptation.

1.2.2 Extension of Utility in Adaptable Design

The very purpose of any product is to perform the intended functions, which can be called normal operational mode (Fig. 1.4). We use the word "utility" to refer to the usefulness or service of a product. Thus, the utility of a product is to perform its expected functions at its normal operational mode.

When the circumstances change, or new functionalities are needed, or more efficient technologies are to be utilized, the current operational mode of a product is no longer satisfactory. A new product can be obtained for the new operational mode. Adaptable design suggests an alternative solution, which is to adapt the existing product to the new operational mode, thus extending its utility into new operational modes.

The "extension of utility" is the underlying principle of AD. It means that adaptable design aims at extending the utility of the product into operational modes for which the product, if not made adaptable, would not be useful. AD offers greater savings by eliminating the need to redo most of the production activities, and by directly extending the utility of products. Thus, it is potentially superior to design for recycling methods or design for reuse of components. While such methods mainly aim at reusing the material in a new production process, adaptable design aims at reusing the product in its current state, with the required "adaptations", for new operational modes.

Figure 1.5 shows an example of a bicycle U-lock that is also a carrier rack. This is simultaneous extension of utility and enables the user to replace two products with

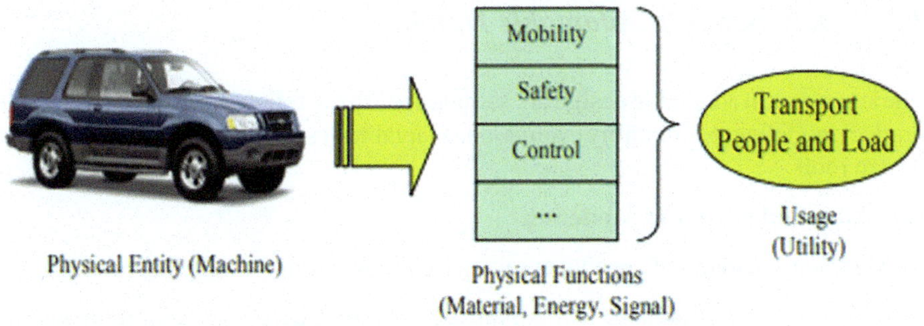

Fig. 1.4 Utility of mechanical products for performing physical functions

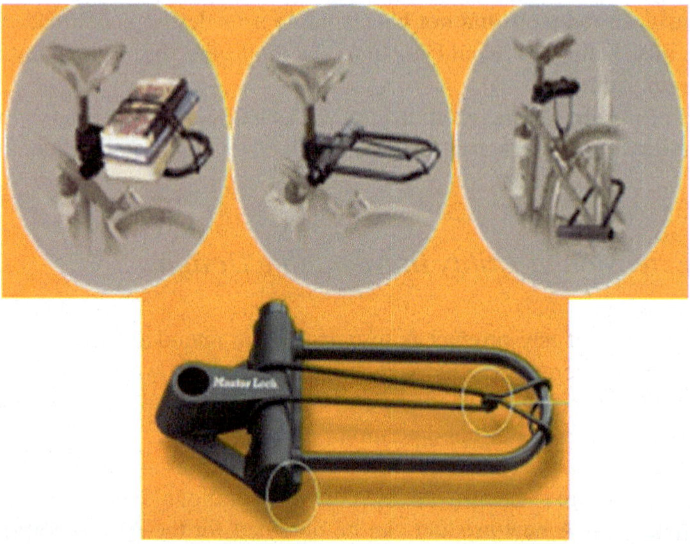

Fig. 1.5 Product adaptability performed by the user results in multi-purpose versatile machines (Courtesy of Master Lock)

one. This adaptation is reversible, and the alternation between two usages happens very frequently; therefore, the manufacturer's design has made this adaptation very easy to be performed by the user.

The above discussion established the objective of AD as the extension of utility, also established the performance of physical functions (material, energy, signal) as the utility of engineering products. Adaptable design is defined as the extension of utility into other operational modes. This extension may happen in various operations using an adaptation process during the current usage of the product, or it can happen in the course of time as circumstances change.

1.2.3 Key Issues in Adaptable Design

The key issues in adaptable design are summarized in the following aspects: adaptable product modeling, adaptability evaluation, adaptable design processes, and adaptable design tools.

(1) Adaptable Product Modeling

Modeling of adaptable products serves as a foundation for adaptable design.

- *Functions and structures of adaptable products*: The design requirements are usually modeled by functions and sub-functions. For adaptable design, function changes should also be modeled. A design solution is modeled by structure of the product. A structure is composed of components, assemblies, and parameters of these components and assemblies. For adaptable design, changes of the product in operation stage should also be modeled. Since multiple design solutions are usually created from design requirements, modeling of the generic product considering alternative design solutions should also be conducted. In addition, relations between functional components and structural components of the adaptable product should also be defined.

- *Product architecture*: Since modules of an adaptable product need to be detached, attached, modified, relocated, and replaced in the operation stage, modular design, platform design, and product family design methods are effective in adaptable design to select architecture of the adaptable product. Therefore, modular design, platform design, and product family design in the context of adaptable design need to be investigated. In addition, open architecture, which allows modules developed by third-party vendors to be connected with the platform, should also be studied.
- *Interfaces:* Since modules in adaptable product need to be detached, attached, and upgraded, interfaces between modules have to be considered to ensure interaction among these modules and ease of their disassembly/assembly operations. In addition, the interface design also plays an important role in modular design, platform design, and product family design.

(2) Adaptability Evaluation

Adaptability and other measures considering various product life cycle aspects are used to evaluate whether design requirements are satisfied. When multiple design solutions are created, these evaluation measures are used for comparing these design solutions for identification of the optimal design.

- *Adaptability*: As introduced in Sect. 1.1.1, two types of adaptation tasks are considered in adaptable design: product adaptation to change the existing products with improved or new functions, and design adaptation to change the existing designs with improved or new functions. Therefore, adaptability measures are classified into two categories: product adaptability and design adaptability. In addition, according to whether the future required functions can be predicted at the design stage, adaptability measures are classified into two categories: specific adaptability when the future functions are specifically provided, and generic adaptability when future functions are not provided. Therefore, four types of adaptability measures need to be formulated: specific product adaptability, generic product adaptability, specific design adaptability, and generic design adaptability.
- *Adaptabilities in design evaluation*: Adaptabilities are used to evaluate capabilities that products and designs can be changed with improved or new functions. In engineering design, various evaluation measures in different life cycle phases also need to be considered. Typical evaluation measures include functional performances (e.g., efficiency and power output) in design phase, manufacturability and assemblability in manufacturing phase, serviceability in operation phase, and recyclability and disposability in retirement phase. Since these evaluation

measures are in different units, integration of these evaluation measures has to be conducted such that various design solutions can be evaluated and compared for identification of the optimal design.

Adaptability as a property of any product design can be defined and measured with a more scientific approach. The efforts have been made using entropy to define adaptability [90].

(3) Adaptable Design Processes

Since design research aims at formulating design activities, particularly design processes, in scientific and systematic ways, studies on adaptable design processes have to be conducted.

- *Adaptable design processes*: Since both product adaptation and design adaptation are considered in adaptable design, the adaptable design processes are also classified into two categories: processes for adaptable design considering product adaptability and processes for adaptable design considering design adaptability.
- *Elements in adaptable design processes*: The key elements in adaptable design processes should be investigated. These elements include design requirements, design candidates, design evaluation, and design optimization. Differences of these elements between the adaptable design and traditional design should be studied.
- *Adaptable design considering specific design aspects*: Adaptable design considering specific design aspects also needs to be studied. Typical specific design aspects include robustness of adaptable design to reduce influences of noise parameters on selected performances, efforts for change of configurations through disassembly and assembly operations to improve product adaptability, and mass personalization of adaptable products to produce personalized products based on requirements of individual customers.

(4) Adaptable Design Tools

Adaptable design tools are used for improving efficiency and quality of adaptable design. Typical tools for adaptable design include:

- *Modular design tools*: These tools are used to cluster components into modules based on similarities in functions and structures of adaptable products.
- *Optimization tools*: These tools are used to achieve the optimal designs of adaptable products. Typical tools include constrained optimization tools based on numerical search, multi-objective optimization tools considering evaluation measures in different life cycle aspects, multi-level optimization tools for identification of optimal configurations and parameters, and global optimization tools to improve optimization quality by preventing the optimal solution from falling into local optimal locations.
- *Web-based tools*: These tools allow engineers from different disciplines and different departments working together through the collaborative development of adaptable products.

- *Virtual reality tools:* These tools allow customers to use virtual reality systems to evaluate the designed adaptable products for further improvement of the design.

1.3 Benefits of Adaptable Design

1.3.1 Users: Extended Product Utility

In the product adaptability for users, the users adapt the product so that it can be used satisfactorily under different or new circumstances. Generally, users adapt products that already exist, thus are concerned with the product adaptability. A well-designed adaptable product must be relatively easy to be adapted by users (or contractors requested by the users). It should be noted that the user adaptation process should also be cost-effective. The user-related benefits of adaptability result from the fact that the adaptable product provides greater utility, extends its service life, and replaces several products thus reducing costs.

Figure 1.5 in Sect. 1.2.2 shows an example of a bicycle U-lock that is also a carrier rack. This is simultaneous extension of utility and enables the user to replace two products with one. This adaptation is reversible, and the alternation between two usages happens very frequently, and therefore the manufacturer's design has made this adaptation very easy to be performed by the user.

Another example of user benefits can be seen from the development of adaptable homes shown in Fig. 1.6, where adaptations happen continually through time as the needs or the lifestyle of the residents change. Most adaptations of houses are reversible. The structure of an adaptable house is, in such way, to have the ability to be altered or extended to conform to the changing needs (or lifestyle) of the residents. This structure allows for future additions, extra space for possible future use of wheelchairs, and the change in features and functions of various parts in and around the house. The adaptable design housing etiquette requires that modifications be simple and cost-effective to carry out when they are planned into the initial design of the house.

1.3.2 Producers: Extended Design Utility

In the adaptability for producers, the producers use the design, its associated process plans, manufacturing set-ups, and even existing parts and assemblies to produce different products. Thus, producers are generally concerned with design adaptability. In some cases, the producers may create new product functionalities or extend their lifespans by adapting existing products requested by users.

The producer-related benefits of adaptability result from the fact that one adaptable design can be used to create new products for different customers, eliminating the redundancy in the complete recreation of similar new designs. Thus the same basic

1. Covered front entrance
2. Private north facing courtyard
3. Security sensor light
4. Carport to side of house with a covered side or rear entrance
5. Straight driveway with area adjacent for width enlargement

Fig. 1.6 An example of adaptable homes

design, existing common parts and assemblies, process plans, production set-ups, inventories and supply networks, and expertise can be used in several production scenarios; thus reducing costs and development time, and increasing efficiency.

Comparing with the traditional design approach for bringing the functionality of products to a predefined level through maintenance and repair, as illustrated in Fig. 1.7a, adaptable design can increase the functionality through modification of the product, such as upgrading, as shown in Fig. 1.7b. When adaptable design is used to create new designs, existing designs are reused to save time and effort in the new product development process.

The producer's benefits can be classified as:

- Reuse of design, engineering, and other knowledge,
- Reduction in manufacturing cost and time (via standardization),
- Reduction in the cost of post-sale services,
- Marketing advantage via user benefits and environmental friendliness.

Design adaptability is most applicable to manufacturing companies that produce a family of products with similar functions, such as auto makers. Various models can be different adaptations of a basic design where every model includes a different set of features and functionalities. Figure 1.8 shows three Ford vehicles that share not

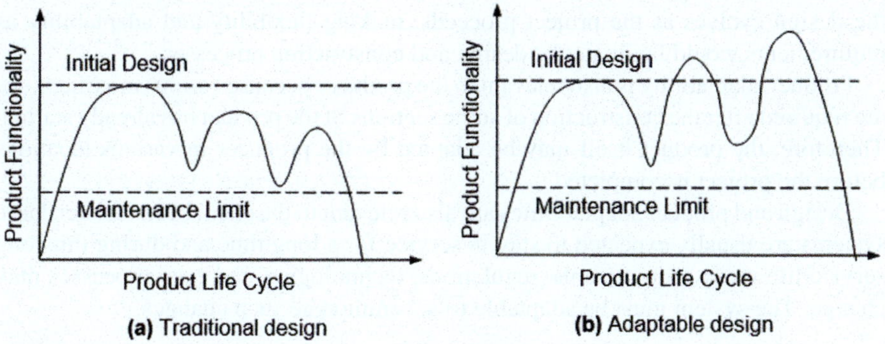

Fig. 1.7 Traditional design and adaptable design considering product functionality

Fig. 1.8 Design adaptability for product variety and customization

only the same general design, but also various parts (chassis, headlights, engines, accessories, etc.).

1.3.3 User and Producer Benefits of Large Systems

The design and production of some products can be a major undertaking such as nuclear and other power plants, large production and assembly lines, factories, mineral and food processing plants, steel mills, shipyards, and massive civil projects. The common features among these projects are substantial cost, time, and complexity in both design, manufacturing, and construction of these large facilities. The design and construction of such large systems belong to an exceptional case in adaptable design because both design and product adaptabilities are relevant to both the user and the producer.

Design adaptability is important to the producer for two reasons. First, the producer is usually involved in similar projects. Substantial capital, time, and expertise are invested in these projects. It is essential to reuse the designs and engineering knowledge as much as possible.

Second, due to the complexity of these projects, it is often impossible to predict all design requirements and finalize the design before the construction begins. Therefore, the design evolves as the project proceeds, making flexibility and adaptability as requirements would facilitate the design and construction processes.

Product adaptability is also relevant to the producer because various modifications are required after the construction of some sections of the product has already started. Therefore, the product itself may be adapted by the producer in various locations before the project is completed.

Design and product adaptabilities are also relevant to the client because such large systems are usually expected to stay in service for a long time and during this long service life many requirements, regulations, technologies, and even processes may change. The system must be adaptable to accommodate such changes.

1.3.4 Environmental Benefits

The engineering production cycle can be described as the flow of natural resources such as materials and energy into a production supply chain, and the flow of waste and pollution from this chain back to nature. Adaptable design aims at redirecting the flow of used resources back into the production supply chain. This is illustrated in Fig. 1.9.

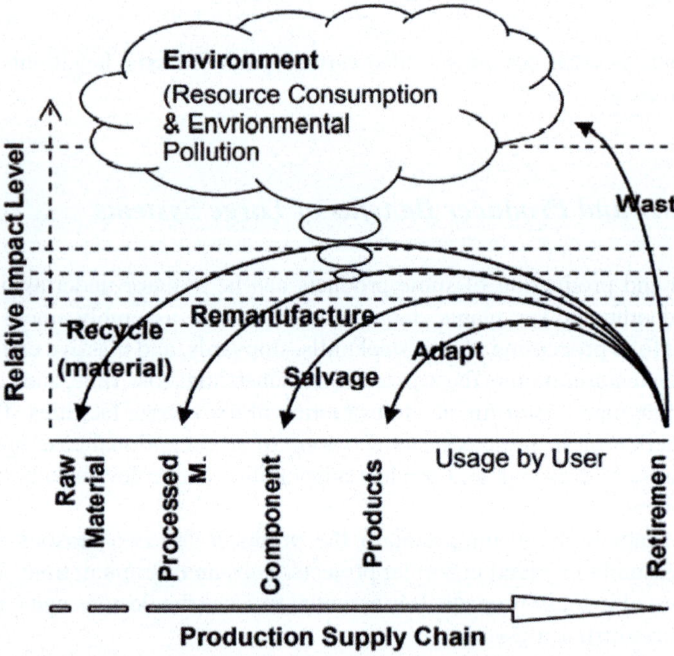

Fig. 1.9 Environmental impact of adaptable design after retirement

The most common method in this chain is "recycling", which typically has the sole objective of reclaiming material. A more effective approach, wherever applicable, is the remanufacture of used parts and components. This process redirects the flow of waste into later stages of the production chain and is potentially superior to recycling. An even better approach, whenever possible, is to salvage parts and assemblies of a product and use them in other copies of the same product. A common example is the recovery and reuse of parts from automobiles.

AD reuses the "utility" of products, not focusing on their parts or materials. Therefore, an intelligent adaptable design for a product can have potentially superior environmental impacts compared to recycling or reuse of parts because it redirects the product to a later stage of the supply chain. Figure 1.9 indicates that AD encourages the faster return of products to service than other methods, and thus yields greater environmental benefits.

1.4 Applicability of Adaptable Design

Products in manufacturing engineering are referred as discrete products such as machine tools and automobiles. The discrete products can be classified into various categories according to specific characteristics. These characteristics can be production volumes, product varieties, product architectures, and so on. For example, considering production volume, products can be classified into: one-of-a-kind products such as ships and oil refineries, small volumes such as aircrafts and custom-made racing cars, medium volumes which cover a very wide range of products such as machine tools and cars, and large volumes that are in the order of millions. Examples include simple screws and nuts. Products can also be classified according to varieties such as low, medium, and large numbers of varieties. If using the product life, products can be classified into short-, medium-, and long-life products. For most of the product categories shown in Table 1.1, adaptable design can be used to reduce product design and development costs by reusing existing design and associated engineering knowledge as well as manufacturing processes and facilities. For example, if products are made in a medium number of varieties and small volumes, adaptable design can be used to create product families with modules and variants for product varieties. For one-of-a-kind products or systems, adaptable design can be used to adapt an existing design to a new application. This would substantially reduce product design and development time and costs.

In design process, adaptable design can be applicable from the early stage of the design, when adaptability can be made as a design requirement. As design proceeds to later stages, there is less freedom to make function-related changes in design and adaptable design becomes less effective as shown in Fig. 1.10.

Life cycle design, on the other hand, is more relevant at later stage of the design process because at early design stages little information is known about parts, material types, repair frequency, etc. As design proceeds to further levels of detail, more information is available and life cycle design becomes increasingly relevant.

Table 1.1 Product categorizations [35]

Product characteristics	Very low	Low (small)	Medium	High (large)
Volumes	One	L	M	H
Varieties	One	L	M	H
Life	Once	L	M	H (long life)
Features	Standard (S)	L	M	H

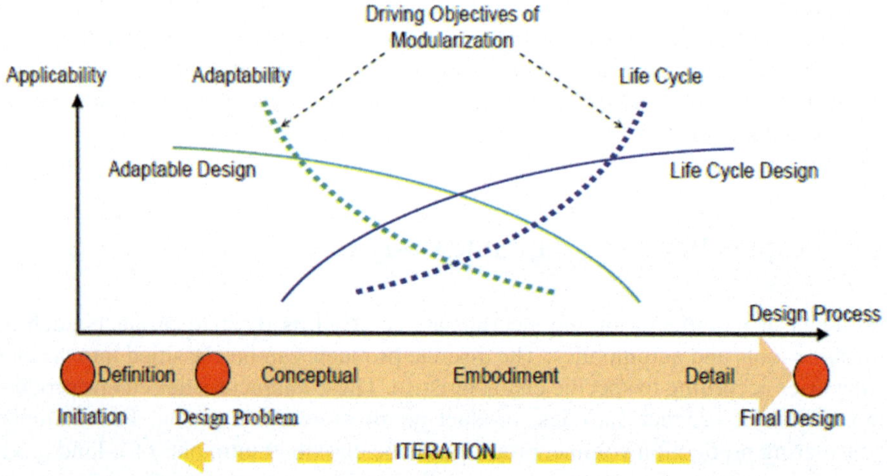

Fig. 1.10 Applicability of adaptable design [35]

Therefore, adaptable design can be performed at earlier stages of design. This suggests a sequence in procedures and eliminates the conflict, which is very fortunate. That is, adaptable design should be performed first, and then when further detail information is available, other life cycle aspects can be considered although it is more desirable that the life cycle objectives should be considered as early as possible during the design process. Iterations, as an inherent part of any design process, may be inevitable.

Although adaptable design was initiated from a mechanical engineering perspective considering upgrading, reconfiguration, customization, etc. of products and systems, this new design approach can also be applied in other engineering fields to design products, processes, and systems. Possible applications of the adaptable design approach in these engineering areas are summarized briefly as follows.

- *Electrical engineering:* When an electrical system, such as the power generation, is designed using the adaptable design approach, the effort to upgrade modules of this system can be reduced. If there is a need to add new power generation capacity, it can be done by adding more power generation modules.

- *Chemical engineering:* When a chemical process is decomposed into relatively independent sub-processes, the modification to a sub-process does not strongly impact on other sub-processes. Associated chemical plants can be designed using the adaptable approach, and facility upgrading and additional capacity can be achieved. In addition, a different process can be achieved through reconfiguring these process modules.
- *Civil engineering:* When a building is designed using the adaptable design approach, the building can be expanded with the minimal effort. In addition, adaptable design also helps to reconfigure the layout of rooms for different functions. An adaptable house is a typical adaptable design in civil engineering [30].
- *Industrial engineering*: Flexible manufacturing systems [76] are typical systems that can be created using the adaptable design approach, allowing manufacturing cells to be easily added, removed and reconfigured.
- *Software engineering*: Using object-oriented programming techniques, relatively independent objects can be reused easily in developing different application systems.
- *Computer engineering*: Components of a computer, such as CPU, memory, internal/external data storage devices and input/output devices, are designed and manufactured as relatively independent modules to allow easy configuration and upgrading.

1.5 Differences Between Adaptable Design and Other Design Methods

Many design theories and methodologies have been developed since the 1970s, and many of these approaches can be used directly or indirectly for designing adaptable products. However, adaptable design is a fairly new concept in design research [35]. This section provides a brief comparison between adaptable design and other major design approaches that are relevant to adaptable design.

- *Adaptable design versus modular design*: Although products developed using adaptable design have the modular architecture, products developed using the modular design approach are not necessarily adaptable and able to respond to changes in functional requirements. Modular design is often used to reduce the effort of design and manufacturing for producers.
- *Adaptable design versus product platform/family design*: Platform design is an extension of modular design through the sharing of a common module—the platform—in all the designs of a product family. Although product platform and family design can better satisfy customer needs with a variety of products, customer needs in the form of changes in functional requirements of the purchased products are not addressed. Adaptable design with a platform can address these new functional changes effectively.

- *Adaptable design versus mass customization design*: Mass customization design aims at developing products based on the requirements of individual customers with near mass production efficiency. Mass customization is primarily achieved by sophisticated computer-based design systems and production planning/control systems. However, products created using this approach are usually not adaptable.
- *Adaptable design versus reconfigurable design*: Reconfigurable products, such as reconfigurable machines, are considered adaptable products created to replace multiple products with single ones [56, 80]. However, other objectives of adaptable design, such as extension of additional functions and upgrade of modules, are not considered in reconfigurable design. In addition, modular design, platform design, and product family design methods may or may not be used in reconfigurable design.

1.6 Organization of This Book

This book overviews the research and applications of the adaptable design. The remaining chapters of the book are organized as follows:

- Chapter 2 introduces the key issues for adaptable design modeling. Various methods for function and structure modeling of adaptable design are also discussed.
- Chapter 3 defines measures of adaptabilities. In this chapter, classifications of adaptability measures are summarized. Evaluation methods of four types of adaptability measures are provided for evaluating an adaptable design.
- Chapter 4 discusses four major elements in adaptable design processes, including design requirements, design candidates, design evaluations, and design optimizations in detail.
- Chapter 5 describes different tools and techniques including modular design tools, optimization tools, Web-based tools, and virtual reality tools for adaptable design.
- Chapter 6 provides various case studies and industrial applications developed based on the adaptable design approach to show the effectiveness of adaptable design.
- Chapter 7 discusses the concluding remarks and future trends of adaptable design.

1.7 Summary

Manufacturing challenges resulted from increasing competition for better product functionality, quality, features, customization, environmental friendliness, lower cost, and shorter delivery time cannot be completely addressed by advanced manufacturing technologies alone as some of them are originated from product design. To address these multiple and conflict requirements and objectives, a new approach called adaptable design is proposed.

The adaptable design considers the dynamics of market demands, and introduction and innovation of new products, therefore allowing the introduction of the design innovation be easier than the conventional design. In adaptable design, the following aspects are considered as key issues: adaptable product modeling, adaptability evaluation, adaptable design processes, and adaptable design tools.

Comparing with other design methods, adaptable design can be applied in the early stage of the product development, when adaptability can be made as a design requirement. Although adaptable design was initiated from a mechanical engineering perspective, this new design approach can also be applied in other engineering fields such as electrical, chemical, civil, and software engineering areas. Differences between adaptable design and other design methods are briefly discussed in this chapter.

References

1. Y. Akao, *Quality Function Deployment: Integrating Customer Requirements into Product Design* (Productivity Press, Cambridge, MA, USA, 1990)
2. K.K. Ali, M.S. Saed, *Product Design for Modularity* (Kluwer Academic, Norwell, MA, USA, 2002)
3. L. Alting, J. Jorgensen, The life cycle concept as a basis or sustainable industrial production. Ann. CIRP **42**(1), 163–167 (1993)
4. L. Alting, Life-cycle design of products: a new opportunity for manufacturing enterprises, in *Concurrent Engineering: Automation, Tools and Techniques*, ed. by A. Kusiak (Wiley, 1992), pp. 1–17
5. Y. Altintas, *Manufacturing Automation* (Cambridge University Press, 2000)
6. J.S. Arora, *Introduction to Optimum Design* (McGraw-Hill, New York, NY, USA, 1989)
7. A. Bernard, A. Fischer, New trends in rapid product development. Ann. CIRP **51**(2), 635–652 (2002)
8. G. Boothroyd, P. Dewhurst, *Design for Assembly: A Designer's Handbook* (Boothroyd Dewhurst Inc., Wakerfield, RI, USA, 1983)
9. G. Boothroyd, P. Radovanovic, Estimating the cost of machined components during the conceptual design of a product. Ann. CIRP **38**(1), 157–160 (1989)
10. G. Boothroyd, Product design for manufacture and assembly. Comput. Aided Des. **26**(7), 505 (1994)
11. G. Boothroyd, P. Dewhurst, W. Knight, *Product Design for Manufacture and Assembly* (University of Rhode Island, Marcel Dekker Inc, New York, 1994)
12. B. Bras, J. Emblemsvag, Activity-based costing and uncertainty in design for life-cycle, in *Design for X: Concurrent Engineering Imperative*, ed. by G.Q. Huang (Chapman & Hall, London, 1995)
13. Y. Chen, Y. Han, P. Liu, J. Huang, N. Bao, P. Gu, Modeling and design method of five roller printing system without solvent. J. Eng. Des. **21**(1)
14. Q. Cheng, Y. Guo, Z. Liu, G. Zhang, P. Gu, A new modularization method of heavy-duty machine tool for green remanufacturing. Proc. Inst. Mech. Eng. C J. Mech. Eng. Sci. **232**(23), 4237–4254 (2018)
15. G. Chryssolouris, Flexibility and its measurement. Ann. CIRP **45**(2), 587–588 (1996)
16. D. Clausing, *Total Quality Development* (ASME Press, New York, USA, 1994)
17. Y. Collette, P. Siarry, *Multiobjective Optimization: Principles and Case Studies* (Springer, 2004)

18. J. Corbett, P.A. McKeown, Nanotechnology: international developments and emerging products. Ann. CIRP **49**(2), 523–546 (2000)
19. K. Dehnad, *Quality Control, Robust Design, and the Taguchi Method* (Springer, 1989)
20. G.E. Dieter, L.C. Schmidt, *Engineering Design*, 5th edn. (McGraw-Hill, New York, NY, US, 2012)
21. J. Dixon, C. Poli, Engineering design and design for manufacture. Field Stone, St Paul, MN, USA (1995)
22. Z. Dong, Design for automated manufacturing, in *Concurrent Engineering: Automation, Tools, and Techniques*, ed. by Kusiak (Wiles, New York, NY, USA, 1993), pp. 207–234
23. X. Du, J. Jiao, M. Tseng, Architecture of product family: fundamentals and methodology. Concurr Eng: Res Appl **9**(4), 309–326 (2001)
24. H.A. ElMaraghy, S. Chen, A general model for mechanical design. Ann. CIRP **39**(1), 111 (1990)
25. G. Erixon, A. Von Yxkull, A. Arnstrom, Modularity: the basis for product and factory reengineering. Ann. CIRP **45**(1), 1
26. W.J. Fabrycky, B.S. Blanchard, *Life-Cycle Cost and Economic Analysis* (Prentice Hall Inc., New Jersey, 1991)
27. V. Fey, E. Rivin, *Innovation on Demand: New Product Development Using TRIZ* (Cambridge University Press, Cambridge, UK, 2005)
28. V. Fey, E. Rivin, *The Science of Innovation: A Managerial Overview of the TRIZ Methodology* (TRIZ Group, USA, 1997)
29. D. Fletcher, R. Brennan, P. Gu, A method for quantifying adaptability in engineering design. Concur Eng-Res Appl **17**(4), 279–289 (2009)
30. A. Friedman, *The Adaptable House* (McGraw-Hill, 2002)
31. J. Gershenson, K. Ishii, Life cycle serviceability design, in *Concurrent Engineering: Automation, Tools, and Techniques*, ed. by A. Kusiak (Wiles, NY, USA, 1993), pp. 363–384
32. D.E. Goldberg, *Genetic Algorithms in Search, Optimization, and Machine Learning* (Addison-Wesley Professional, 1989)
33. J.P. Gonzalez-Zugasti, Otto, A method for architecting product platforms. Res. Eng. Des. **12**(61–72) (2000)
34. P. Gu, M. Hashemian, S. Sosale, An integrated modular design methodology for life-cycle engineering. Ann. CIRP **46**(1), 71–74 (1997)
35. P. Gu, M. Hashemian, A.Y.C. Nee, Adaptable design. Ann. CIRP **53**(2), 539–557 (2004)
36. P. Gu, A. Kusiak (eds.), *Concurrent Engineering* (Elsevier, Oxford, UK, 1993)
37. P. Gu, Adaptable design using bus system, in *Proceedings of the 10th International Manufacturing Conference in China (IMCC2002), Panel-2 158*, Xiamen, China, Oct. 11–12 (2002), p 11.
38. P. Gu, S. Sosale, Product modularization for life cycle engineering. J. Robot. CIM **15**, 387–401 (1999)
39. P. Gu, Recent development in design theory and methodology research, in *Proceedings of International Conference on Manufacturing Sciences* (Wuhan, China, 1998), pp. 21–26
40. M. Hashemian, P. Gu, A function representation scheme for conceptual mechanical design, in *Proceedings of the11th International Conference on Engineering Design, ICED97*, vol. 2, issue 1 (1996), pp. 311–314
41. Y. Hatamura, T. Nagao, M. Mitsuishi, M. Nakao, Actual conceptual design process for an intelligent machining center. Ann. CIRP **44**(1), 123 (1995)
42. F. Hillstrom, Applying Axiomatic design to interface analysis in modular product development. Adv. Des. Autom. ASME **69**(2), 363 (1994)
43. A. Jose, M. Tollenaere, Modular and platform methods for product family design: literature analysis. J. Intell. Manuf. **16**(3), 371–390 (2005)
44. Y. Koren, S. Hu, P. Gu, M. Shpitalni, Open-architecture products. Ann. CIRP **62**(2), 719–729 (2013)
45. L.K. Keys, System Life Cycle Engineering and DFX. IEEE Trans. Components Hybrids Manuf. Technol. **13**(1), 83–93 (1990)

46. F. Kimura, H. Suzuki, Representing background information for product description to support product development process. Ann. CIRP **44**(1), 113 (1995)
47. F. Kimura, H. Suzuki, A CAD system for efficient product design based on design intent. Ann. CIRP **38**(1), 149 (1989)
48. Kirschman C, Fadel GM, Almonte CC (1996). Classifying Functions for Mechanical Design. ASME Design Engineering Technical Conferences and Computers in Engineering Conference
49. F. Klocke, G. Eisenblatter, Dry cutting. Ann. CIRP **46**(2), 519–526 (1997)
50. W.A. Knight, Design for manufacture analysis: early estimates of tool costs for sintered parts. Ann. CIRP **40**, 131–134 (1991)
51. Y. Koren, U. Heisel, F. Jovane, T. Moriwaki, G. Pritschow, G. Ulsoy, H. Van Brussel, Reconfigurable manufacturing systems. Ann. CIRP **49**(2), 527–540 (2002)
52. F.L. Krause, T. Kiesewetter, S. Kramer, Distribute product design. Ann. CIRP **43**(1), 149 (1994)
53. F.L. Krause, A. Ulbrich, R. Woll, Methods for quality driven product development. Ann. CIRP **42**(1), 151 (1993)
54. J.P. Kruth, M. Leu, T. Nakagawa, Progress in additive manufacturing and rapid prototyping. Ann. CIRP **47**(2), 525–540 (1998)
55. A. Kusiak, *Concurrent Engineering: Automation, Tools, and Techniques* (Wiley, NY, USA, 1993)
56. R.G. Landers, B.K. Min, Y. Koren, Reconfigurable machine tools. Ann. CIRP **50**(1), 269–274 (2001)
57. K. Lee, *Principles of CAD/CAM/CAE* (Addison-Wesley, 1999)
58. S.C. Lu, M. Shpitalni, R. Bar-Or, R. Gadh, Virtual and augmented reality technologies for product realization. Ann. CIRP **48**(2), 471–496 (1999)
59. O. Maimon, D. Braha, On the complexity of the design synthesis problem. IEEE Trans. Syst. Man Cybern. Part A **26**(1), 0–151 (1996)
60. M.D. Marks, C.F.K. Eubanks, Life-cycle clumping of product designs for ownership and retirement (1993).
61. M.V. Martin, Design for variety: a methodology for developing product platform architectures (2000).
62. T. Masuzawa, State of the art on micromachining. Ann. CIRP **49**(2), 473–488 (2000)
63. A.J. Maupin, L.A. Stauffer, A design tool to help small manufacturers reengineer a product family, in *Proceedings of DETC, DTM-14568* (ASME, 2000)
64. N. Maussang, P. Zwolinski, D. Brissaud, Product-service system design methodology: from the PSS architecture design to the products specifications. J. Eng. Des. **20**(4), 349–366 (2009)
65. J. Meijer, K. Du, A. Gillner, D. Hoffmann, laser machining by short and ultrashort pulses. State of the art and new opportunities in the age of the photons. Ann. CIRP **51**(2), 531–550 (2002)
66. K.G. Mertens, E. Greve, D. Krause, M. Meyer, Reviewing the intellectual structure of product modularization: toward a common view and future research agenda. J. Product Innov. Manage. (2022)
67. M.H.R. Meyer, Product platforms in software development. Sloan Manage. Rev. Fall 61–74 (1998)
68. P.J. Newcomb, Implications of modularity on product design for the life cycle, in *Proceedings of the 1996 ASME Design* (1996)
69. P. O'Grady, R.E. Young, Issues in concurrent engineering systems. J. Des. Manuf. **1**, 27–34 (1991)
70. G. Pahl, W. Beitz, *Engineering Design: A Systematic Approach* (Springer, Berlin, Germany, 1988)
71. Q. Peng, Y. Liu, J. Zhang, P. Gu, Personalization for massive product innovation using open architecture. Chin. J. Mech. Eng. (2018). https://doi.org/10.1186/s10033-018-0239-0
72. J. Peters Poli, D. Van Campenhout, Manufacturing oriented and functional design. Ann. CIRP **37**(1), 153 (1988)

73. B. Prasad, *Concurrent Engineering Fundamentals* (Prentice-Hall, Englewood Cliffs, NJ, USA, 1996)
74. G. Pritschow, Y. Altintas, F. Jovane, Y. Koren, M. Mitsuishi, Open controller architecture—Past, present and future. Ann. CIRP **50**(2), 463 (2000)
75. K.P. Rajurkar, D. Zhu, J.A. McGeough, J. Kozak, A. De Sikva, New developments in electrochemical machining. Ann. CIRP **48**(2), 567–580 (1999)
76. A. Raouf, M. Ben-Daya, *Flexible Manufacturing Systems: Recent Developments* (Elsevier Science, 2005)
77. J. Sand, P. Gu, G. Watson, HOME: house of modular enhancement for product modularization. Concurr. Eng. Res. Appl. **10**(2), 153–164 (2002)
78. M. Santochi, G. Dini, F. Failli, Computer aided disassembly planning: state of the art and perspectives. Ann. CIRP **51**(2), 509–527 (2002)
79. G. Seliger, Product innovation—industrial approach. Ann. CIRP **50**(2), 425–443 (2001)
80. R. Setchi, N. Lagos, Adaptive, responsive and reconfigurable product support for future manufacturing. Int. J. Innov. Comput. Inf. Control **4**(3), 615–625 (2008)
81. Y. Shimomura, S. Tanigawa, H. Takeda, Y. Umeda, T. Tomiyama, Functional evaluation based on function content, in *Proceedings of ASME Design Engineering Technical Conference* (ASME, California, DETC/DTM-1532, 1996)
82. Z.D.W. Siddique, D. Rosen, Product family configuration reasoning using discrete design spaces, in *Proceedings of ASME 2000 International Design Engineering Technical Conferences and Computers and Information in Engineering Conference* (Baltimore, Maryland, 2000)
83. T.W. Simpson, J.R.A. Maier, F. Mistree, Product platform design: method and application. Res. Eng. Design **13**(1), 2–22 (2001)
84. D. Singh, P. Gu, Product life cycle serviceability analysis for supporting engineering design: Part two: serviceability evaluation for product redesign. J. Eng. Des. Autom. **3**(3), 275–298 (1997)
85. L. Sivard, E. Agerman, Customer based design with constraint reasoning. Ann. CIRP **42**(1), 139 (1993)
86. G. Sohlenius, Concurrent engineering. Ann. CIRP **41**(2), 645 (1992)
87. R. Spicer, Y. Koren, M. Shpitalni, D. Yip-Hoi, Design principles for machining system configurations. Ann. CIRP **51**(1), 275–280 (2002)
88. N.P. Suh, *The Principles of Design* (Oxford University Press, Oxford, UK, 1990)
89. N.P. Suh, *Axiomatic Design: Advances and Applications* (Oxford University Press, Oxford, UK, 2001)
90. Z. Sun, K. Wang, Y. Chen, D. Xue, P. Gu, Information entropy method for adaptable design evaluation. Chin. J. Eng. Des. **28**(1), 1–13 (2021)
91. G. Taguchi, *On-line Quality Control During Production* (Japanese Standards Association, Tokyo, Japan, 1981)
92. G. Taguchi, A.E. ElSayed, T. Hsiang, *Quality Engineering in Production Systems* (McGraw-Hill, New York, NY, USA, 1988)
93. M.M. Tseng, X. Du, Design by customers for mass customization products. Ann. CIRP **47**(1), 103 (1998)
94. S. Tichkiewitch, M. Véron, Integration of manufacturing processes in design. Ann. CIRP **47**(1), 99 (1998)
95. T. Tomiyama, Y. Umeda, A CAD for functional design. Ann. CIRP **42**(1), 143 (1993)
96. T. Tomiyama, P. Gu, Y. Jin, D. Lutters, C. Kind, F. Kimura, Design methodologies: educational and industrial applications. Ann. CIRP **58**, 543–565 (2009)
97. M. Tseng, J. Jiao, Design for mass customization by developing product family architecture, in *Proceedings of ASME Design Engineering Technical Conference*, Atlanta, GA (1998)
98. D.G. Ullman, *The Mechanical Design Process* (McGraw-Hill, New York, NY, USA, 1997)
99. K. Ulrich, D. Robertson, Planning for product platforms. Sloan Manage. Rev. Summer(19–31) (1998)

100. K. Ulrich, The role of product architecture in the manufacturing firm. Res. Policy **24**(1), 419–440 (1993)
101. K.T. Ulrich, K. Tung, Fundamentals of product modularity, in *ASME Winter Annual Meeting*, vol. DE, issue 39 (1991), pp. 73–80
102. Y. Umeda, H. Takeda, T. Tomiyama, H. Yoshiokawa, Function, behavior, and structure, in *AIENG'90 Applications of AI in Engineering* (Springer, 1990), pp. 177–193
103. E. Westkamper, L. Alting, G. Amdi, Life cycle management and assessment approaches and visions towards sustainable development. Ann. CIRP **49**(2), 501–522 (2000)
104. E. Westkamper, K. Feldmann, G. Reinhar, G. Seliger, Integrated development of assembly and disassembly. Ann. CIRP **48**(2), 557–566 (1999)
105. H. Weule, Life-cycle-analysis: a strategic element for future products and manufacturing technologies. Ann. CIRP **42**(1), 181 (1993)
106. H.P. Wiendahl, S. Lutz, Production in networks. Ann. CIRP **51**(2), 573–586 (2002)
107. B. Willems, G. Seliger, J. Duflou, B. Basdere, Contribution to design for adaptation: method to assess the adaptability of products (MAAP), in *Proceedings of EcoDesign: Third International Symposium on Environmentally Conscious Design and Inverse Manufacturing*, Tokyo, Japan, December, vol. 8, issue 11 (2003), pp. 589–596
108. D. Xue, G. Hua, V. Mehrad, P. Gu, Optimal adaptable design for creating the changeable product based on changeable requirements considering the whole product life-cycle. J. Manuf. Syst. **31**(1), 59–68 (2012)
109. H. Yoshikawa, General design theory and a CAD system. Man Machine Communication in CAD/CAM, in *Proceedings of the IFIP WG5.2–5.3 Working Conference*, Tokyo, Japan (1980)
110. I. Zeid, *CAD/CAM: Theory and Practice*, 2nd edn. (McGraw-Hill, 2009)
111. Y. Zeng, P. Gu, A science-based approach to product design theory Part I: formulation and formalization of design process. J. Robot. Comput. Integr. Manuf. **15**(4), 331–339 (1999)
112. Y. Zeng, P. Gu, A science-based approach to product design theory Part II: Formulation of design requirements and products. J. Robot. Comput. Integr. Manuf. **15**(4), 341–352 (1999)
113. H.C. Zhang, T.C. Kuo, H. Lu, S.H. Huang, Environmentally conscious design and manufacturing: a state-of-the-art survey. J. Manuf. Syst. **16**(5), 352–371 (1997)
114. J. Zhang, D. Xue, P. Gu, Adaptable design of open architecture products with robust performance. J. Eng. Des. **26**(1–3), 1–23 (2015)
115. H.C. Zhang, T.C. Kuo, Disassembly model for recycling—personal computer, in *Technical Paper of the NAMRI/SME*, Michigan (1996), pp. 139–144
116. F. Zwicky, *Discovery, Invention, Research, Through the Morphological Approach* (The Macmillan Company, Toronto, Canada, 1969)

Chapter 2
Modeling for Adaptable Design

As discussed in Chap. 1.2.3, modeling of adaptable products serves as a foundation for adaptable design. Key issues for adaptable design modeling include:

- Modeling of functions and structures for adaptable design.
- Modeling of product architectures for adaptable design.
- Modeling of adaptable interfaces.

Details of these key issues are discussed in the following sections.

2.1 Rationalized Functional Structures for Adaptable Design

Design is usually considered as a mapping process from design requirements in functional space to design solutions in structural space as shown in Fig. 2.1.

Modeling of design requirements in the functional space, modeling of design solutions in the structural space, and mapping from the functional space to the structural space have been discussed in many design theories. In the axiomatic design developed by Suh [31, 32], the functional space was defined by functional requirements (FRs), the structural space was defined by design parameters (DPs), and the mapping from the functional space to the structural space was defined by the design matrix. In the general design theory (GDT) developed by Yoshikawa [37], functional space was modeled by functional sets, structural space (called attribute space in GDT) was modeled by attribute sets, and the mapping was accomplished through computation with set operations.

In this chapter, modeling of design requirements in the functional space, modeling of design solutions in the structural space, and relations between functional space and structural space are discussed. Mapping from the functional space to the structural space through adaptable design processes will be discussed in Chap. 4.

P. Gu et al., *Adaptable Design*, Research on Intelligent Manufacturing,
https://doi.org/10.1007/978-981-99-5869-6_2

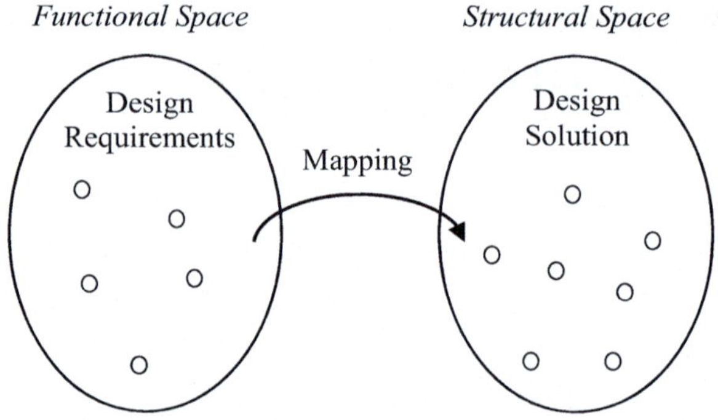

Fig. 2.1 Design through mapping from functional space to structural space

2.1.1 *Function Modeling for Adaptable Design*

Modeling of design functions has been extensively studied in the last three decades [7, 26]. The major requirements for function modeling in adaptable design are as follows:

- When an adaptable design is used to design a product for delivering multiple functions that are usually provided by multiple products, the modeling of multiple and/or alternative functions is required.
- Due to the complexity and scale of adaptable design, modeling of functions at multiple levels is needed. In addition, these functions may be associated by their relations. These relationships should also be modeled.
- Since functions can be added, removed, and modified in adaptable design, modularity and independence of functions have to be considered.
- In engineering design, qualitative customer needs are usually converted into quantitative engineering metrics. Therefore, both qualitative and quantitative requirements should be described in function modeling.

The major methods that can be used for adaptable design are tree-based, network-based, and AND-OR tree-based function modeling techniques as shown in Fig. 2.2.

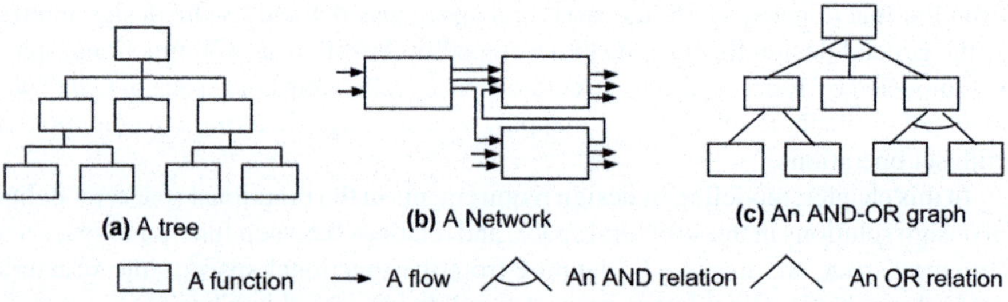

Fig. 2.2 Some functional modeling methods [13]

(1) Tree-Based Function Modeling

In this method, a complex function is decomposed into sub-functions, as shown in Fig. 2.2a. The overall function is satisfied by identifying solutions to decomposed simpler sub-functions. This scheme allows functions to be modeled at multiple levels.

In Fig. 2.3 [4], functions of a motherboard are defined. A motherboard of the personal computer is mainly composed of control chips, a basic input/output system (BIOS) chip, a central processing unit (CPU) slot, a random access memory (RAM) slot, interfaces for peripheral devices, dual in-line package (DIP), expansion slots, etc. The central task of the motherboard is to maintain the CPU and peripheral devices in a collaborative work pattern. The CPU receives various external data and orders and transmits operation results to designated peripheral devices through peripheral component interconnect (PCI), accelerated graphics port (AGP), and other bus interfaces.

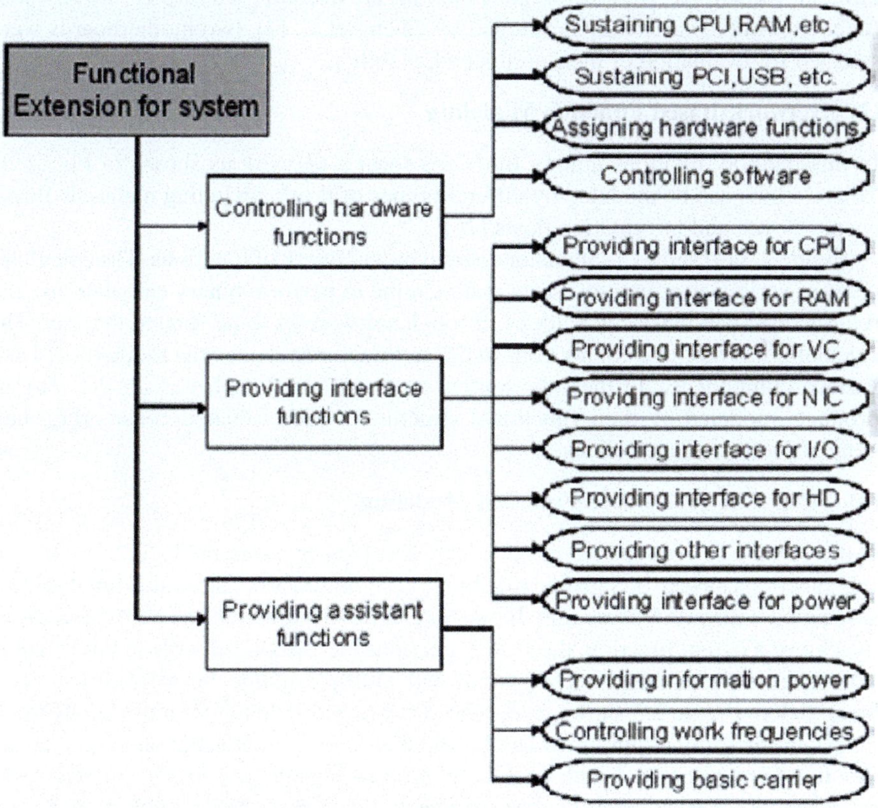

Fig. 2.3 Functions of a motherboard defined by a tree [4]

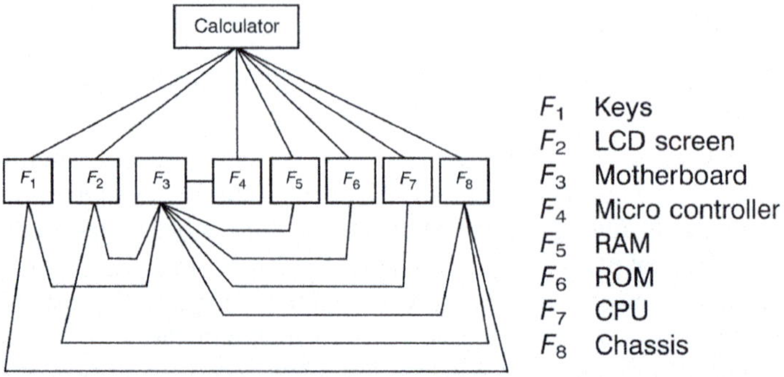

Fig. 2.4 Functions of a calculator defined by a network [9]

In this example, the overall function of the motherboard is decomposed into three sub-functions. These three sub-functions are decomposed further into sub-sub-functions. In the case study developed by Cheng et al. [4], two motherboards were selected for evaluation of their product adaptability.

(2) Network-Based Function Modeling

In this method, relations among functions form a network as shown in Fig. 2.2b. The relations can be modeled by different types of flows, including materials flows, energy flows, and information flows [26].

Figure 2.4 [9] shows a calculator defined by a network of functions. The calculator takes in various user inputs, uses a math engine to perform binary calculations, and returns results of these calculations (through software of some sort) to the user. The calculator has multiple connections in the architecture between the motherboard and various components. In the case studies developed by Fletcher et al. [9], various products modeled by their functional structures were evaluated considering their general product adaptabilities.

(3) AND-OR Tree-Based Function Modeling

In this AND-OR tree scheme introduced by Xue [36] as shown in Fig. 2.2c, if all of its sub-functions have to be satisfied to achieve a parent function, these sub-functions are associated with an AND relation. If only one of its sub-functions needs to be satisfied to achieve a parent function, these sub-functions are associated with an OR relation. This scheme allows design requirements with multiple and/or alternative functions to be modeled. Figure 2.5 shows examples of function modeling with AND-OR trees.

When the same requirement can be satisfied by different functions, the relevant functions are associated with an OR relation in design (OR-D). In this case, the requirement for one operation phase can be modeled by multiple candidates that are achieved by different design solutions. Among all these alternative design solution candidates, only one needs to be selected as the final design solution. Figure 2.5a shows a generic AND-OR tree with functions for a cell phone system composed of a cell phone and its data plan. Either low-end cell phone or high-end cell phone is

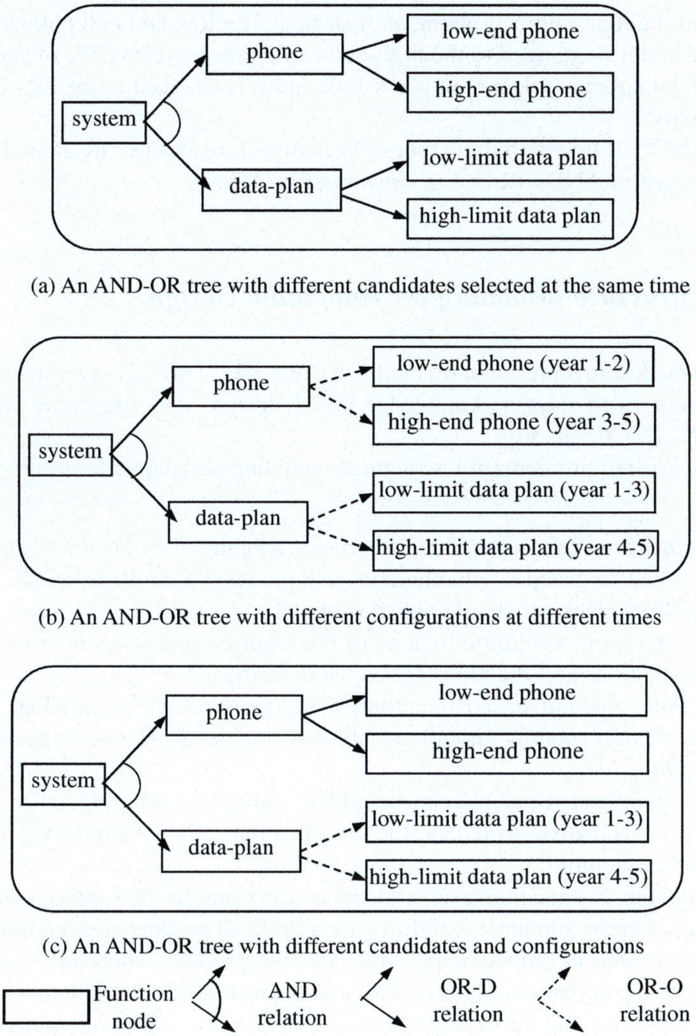

Fig. 2.5 AND-OR trees for modeling of functions

selected in decision-making, and the cell phone cannot be changed once it is selected. In addition, either low-limit data plan or high-limit data plan must be selected in decision-making.

When different functions of the adaptable product need to satisfy different requirements in different operation phases, the relevant function nodes are associated with an OR relation in operations (OR-O). In this case, design solutions to achieve these functions in different operation phases are modeled by different operation configurations. Figure 2.5b shows a generic AND-OR tree with functions of an adaptable cell phone system considering multiple operation phases. In this case, both the low-end cell phone and the high-end cell phone are selected. The low-end cell phone is used in years 1–2 and it is changed to the high-end cell phone in years 3–5. In addition, the low-limit data plan is selected in years 1–3, and it is changed to the high-limit data plan in years 4–5.

The AND-OR tree with three types of relations (i.e., AND, OR-D, and OR-O) is called the hybrid AND-OR tree as shown in Fig. 2.5c.

2.1.2 Structure Modeling for Adaptable Design

Research on design representation has been extensively carried out with the advances of computer-aided design, knowledge-based design, and life-cycle engineering design [21, 25, 26, 28, 30].

The major requirements for structural modeling in adaptable design are listed below:

- Due to the complexity of product structure in adaptable design, the design descriptions should be grouped into elements and modeled at multiple levels. Relations among these elements should also be defined.
- Since a design is a solution to achieve the required functions, relations between design solutions and functions should be described.
- Since both qualitative and quantitative descriptions are required in adaptable design, design models should be able to handle these two types of design descriptions.
- Since components/modules can be added, removed, and replaced in adaptable design, a mechanism to reduce the modification effort of product components/modules is required.
- In adaptable design, multiple products can potentially be replaced by a single product. Some components/modules are used in all product configurations, while others are used in only specific models of the product. Therefore, modeling of design using a common platform and add-on accessories may be more suitable.
- Since the same functional requirements can be achieved by alternative designs, modeling of the alternative design solutions and identification of the optimal one under constraints need to be considered.

The major methods that can be used for adaptable design are tree-based, network-based, and AND-OR tree-based structural modeling techniques as shown in Fig. 2.6. The structural modeling methods shown in Fig. 2.6 are similar to the function modeling methods given in Fig. 2.2. The generic functions in the function modeling methods are replaced by specific solutions in the structural modeling methods.

(1) Tree-Based Structural Modeling

In this method, a complex product is decomposed into sub-systems. Figure 2.7 shows the motherboard product in the tree structure corresponding to the function model of

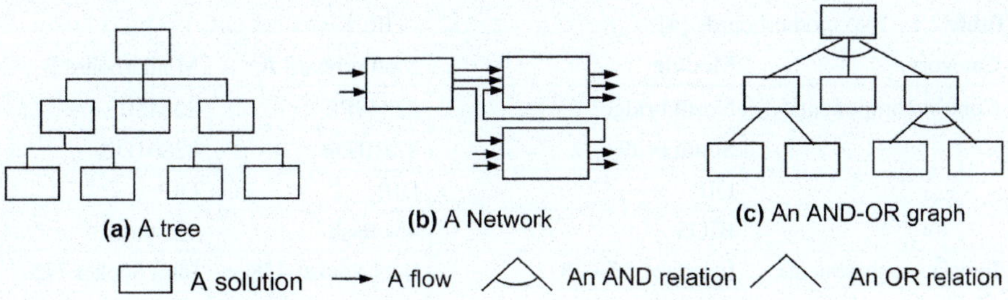

Fig. 2.6 Some structural modeling methods

this motherboard given in Fig. 2.3. From the same requirements, multiple design solutions can be selected and compared through evaluation. Two specific motherboards are shown in Table 2.1.

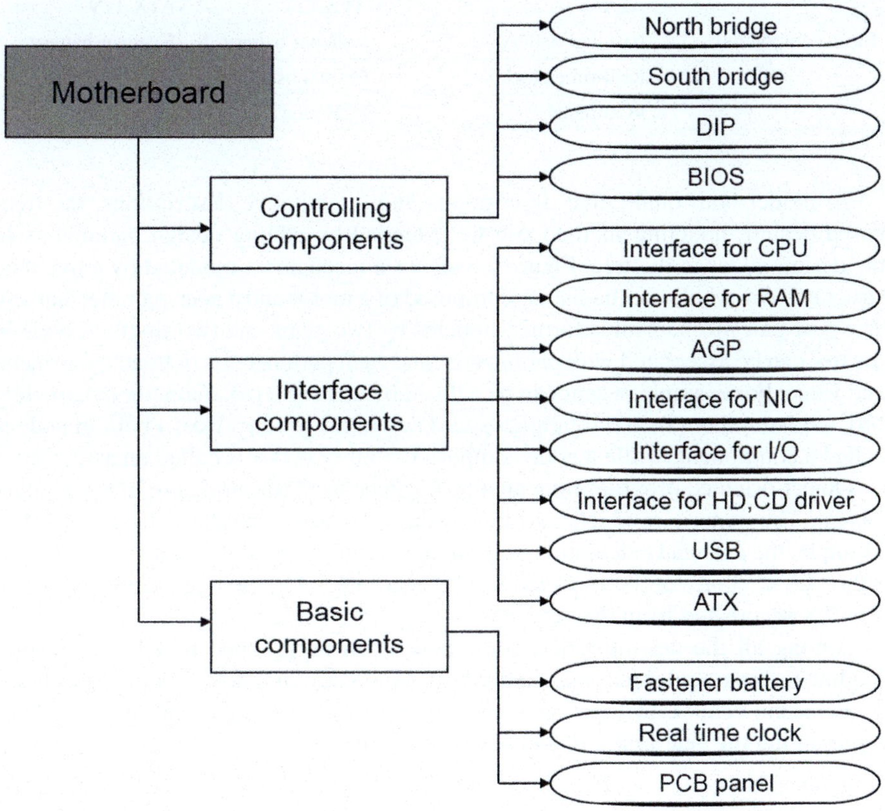

Fig. 2.7 Structures of a motherboard defined by a tree [4]

Table 2.1 Two motherboards [4]

Category	Module	Motherboard A	Motherboard B
Control components	North bridge	82845PE	82845PE
	South bridge	82801DB	82801DB
	DIP	DIP	DIP
	BIOS	4M flash	4M flash
Interface components	Interface for CPU	Intel Socket 478	Intel Socket 775
	Interface for RAM	2 DDR2 DIMM	2 DDR3 DIMM
	AGP	AGP 4X	AGP 4X
	Interface for NIC	PCI	PCI
	Interface for I/O	PS/2	PS/2
	Interface for HD CD drive	ATA33/66/100 +2xSATA	ATA33/66/100 +2xSATA
	USB	Four	Six
	ATX	ATX 12V	ATX 12V
Basic components	Fastener battery	Fastener battery	Fastener battery
	Real-time clock	Crystal oscillator	Crystal oscillator
	PCB panel	PCB panel	PCB panel

To model both qualitative descriptions and quantitative descriptions, the tree-based design modeling method can be extended by adding design parameters to the design nodes in the tree. Figure 2.8 shows a mechanism modeled by a tree data structure. The whole mechanism is composed of a motor and a gear-pair mechanism. The gear-pair mechanism is further modeled by two gears and two shafts. A node in the tree can be associated with parameters, and each parameter is defined by its name and value. For example, a gear can be defined by its height (h), diameter (d), module (m), tooth number (z), and rotational speed (n). In the product tree, a bottom node is called a component, while a node composed of other nodes is called an assembly.

Object-oriented programming approach can be used for modeling of the product design. The generic products and their behaviors can be modeled by classes. For example, the gear and calculation of its diameter d using module m and tooth number z through $d = mz$ can be modeled by the gear class. The two gears, $g1$ and $g2$, in Fig. 2.8 are created from the gear class.

Among all the design nodes, some nodes can be grouped together to form a module for a particular purpose such as to deliver a design function or to be produced by the same vendor. In Fig. 2.9, some components are grouped into two modules representing the gear box unit and bevel gear-pair unit.

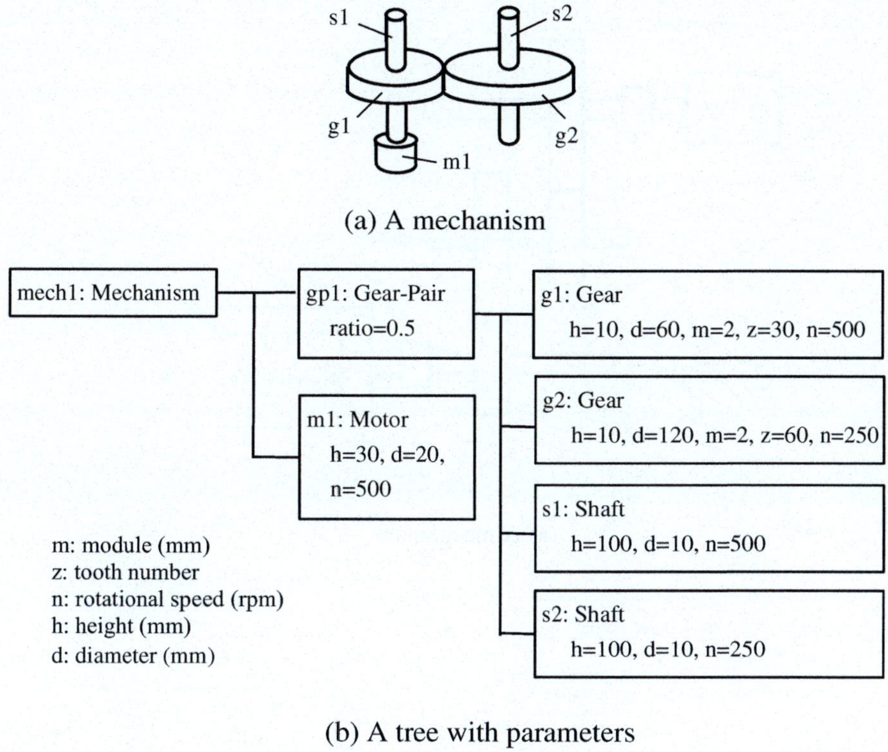

Fig. 2.8 Modeling of a mechanism in a tree and parameters of nodes in the tree

(2) Network-Based Structural Modeling

In this method, a product is modeled by nodes associated into a network as shown in Fig. 2.6b. The network-based structural model is usually derived from the functional diagram, since a functional diagram is often modeled by functions and relations between functions called flows including materials flows, energy flows, and information flows [26].

A product defined by a network-based structural model is difficult for product adaptation. Therefore, change of the design from the network-based structure to tree-based structure, called segregated structure, plays an important role in adaptable design [9].

Fletcher et al. [9] studied general product adaptability considering different structures of the considered product. Figure 2.10 shows two different product designs to satisfy the same requirements. It was found that the design modeled by a segregated structure (i.e., tree structure) provided better product adaptability than the design modeled by a full structure (i.e., network structure).

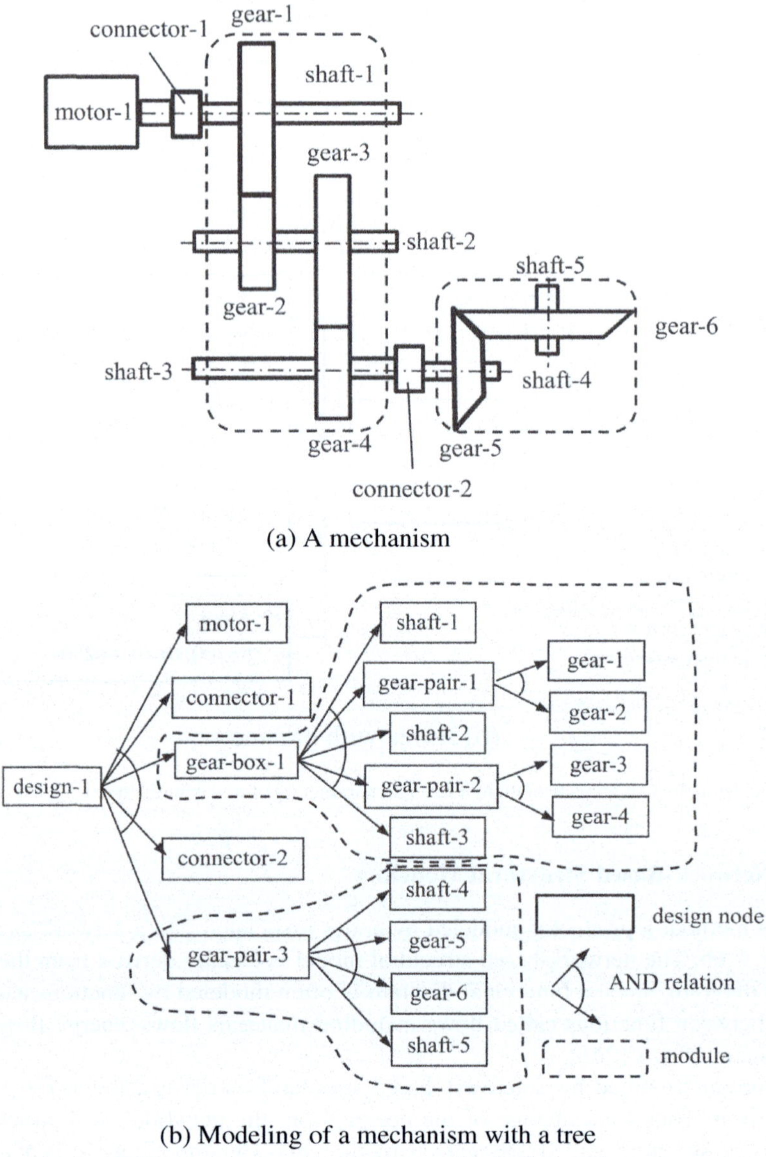

Fig. 2.9 Modeling of a mechanism by a tree with modules [23]

(3) AND-OR Tree-Based Structural Modeling

AND-OR tree-based design modeling, as shown in Fig. 2.6c, was developed from tree data structure to model multiple design candidates for the same design functions [36]. When a product structure can be decomposed into a number of sub-structures, these sub-structures are associated with an AND relation. When a product structure can be satisfied by alternative sub-structures, these sub-structures are associated with

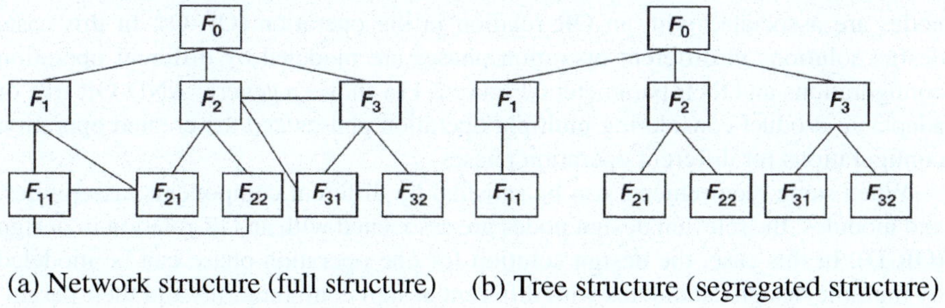

Fig. 2.10 Two different designs for the same requirements [9]

an OR relation. A design candidate is created from the AND-OR tree through the state space search [27].

For adaptable product, change of a component or a module during the product operation stage should also be considered. Figure 2.11 shows adaptable products modeled by AND-OR trees.

When different components, assemblies, and modules of the adaptable product need to satisfy different requirements in different operation phases, relevant design

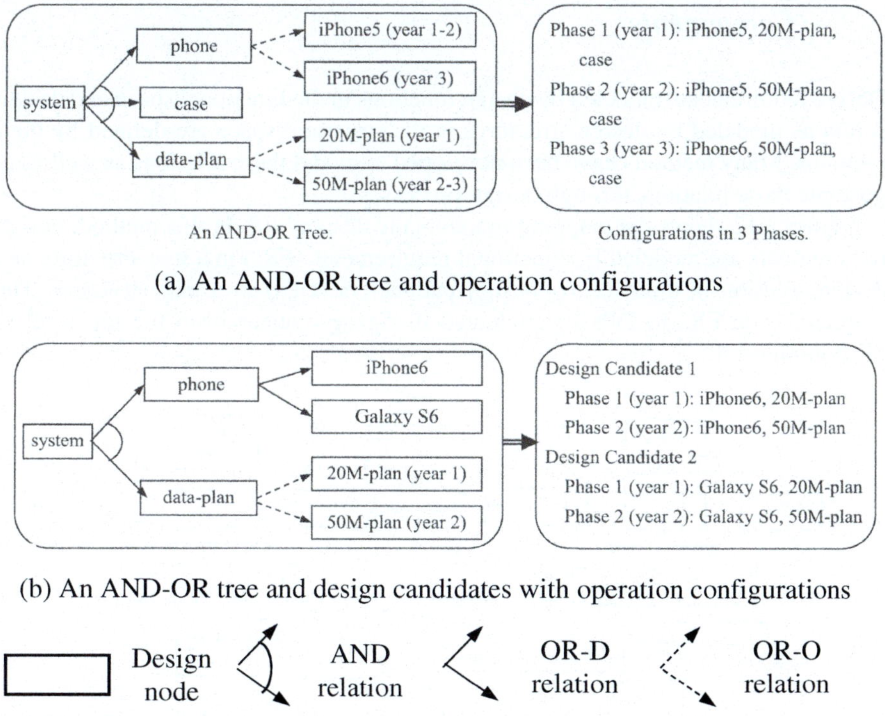

Fig. 2.11 AND-OR tree for modeling adaptable products [23]

nodes are associated with an OR relation in the operation (OR-O). In this case, design solutions in different operation phases are modeled by different operation configurations and their parameters. Figure 2.11a shows a generic AND-OR tree of adaptable product considering multiple operation phases and the created operation configurations for different operation phases.

When same requirements can be satisfied by different components, assemblies, and modules, the relevant design nodes are associated with an OR relation in design (OR-D). In this case, the design solution for one operation phase can be modeled by multiple design candidates with different design configurations and their parameters. Among all these alternative design solution candidates, only one needs to be selected as the final design solution. Figure 2.11b shows a generic AND-OR tree of an adaptable product considering alternative design candidates and the created design candidates. The AND-OR tree with three types of relations (i.e., AND, OR-D, and OR-O) is called the hybrid AND-OR tree.

Creation of design candidates and operation configurations is conducted through the solution search [27]. Design process through the search will be discussed in Sect. 4.2.

2.1.3 Relations Between Functional and Structural Components

Design requirements modeled by design functions in the functional space and design solutions modeled by design structures in the structural space are defined by their relations. Many methods have been developed to model these relations as well as to generate these relations through design processes.

Figure 2.12 shows the mapping in axiomatic design [40]. In this method, design requirements are modeled by functional requirements (FRs) in a tree structure, and design solution is modeled by design parameters (DPs) in a tree structure. The mapping from FRs to DPs is conducted in zigzag manner from the top level to the bottom level.

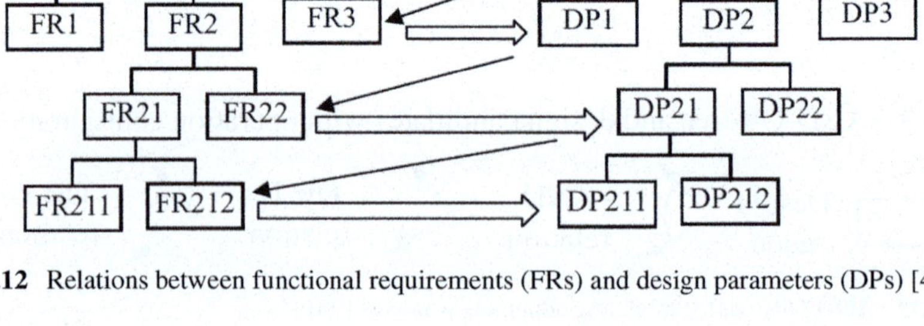

Fig. 2.12 Relations between functional requirements (FRs) and design parameters (DPs) [40]

Relations between functional requirements (FRs) and design parameters (DPs) can be modeled by a matrix based on axiomatic design [31, 32]. Equation (2.1) shows a matrix for modeling the relations between function requirements (FRs) shown in Fig. 2.3 and design parameters (DPs) shown in Fig. 2.7 for evaluation of product adaptability of a computer.

$$\begin{bmatrix} FR_{11} \\ FR_{12} \\ FR_{13} \\ FR_{14} \\ FR_{21} \\ FR_{22} \\ FR_{23} \\ FR_{24} \\ FR_{25} \\ FR_{26} \\ FR_{27} \\ FR_{28} \\ FR_{31} \\ FR_{32} \\ FR_{33} \end{bmatrix} = \begin{bmatrix} X & 0 & 0 & 0 & 0 & 0 & 0 & 0 & 0 & 0 & 0 & 0 & 0 & 0 & 0 \\ 0 & X & 0 & 0 & 0 & 0 & 0 & 0 & 0 & 0 & 0 & 0 & 0 & 0 & 0 \\ 0 & 0 & X & 0 & 0 & 0 & 0 & 0 & 0 & 0 & 0 & 0 & 0 & 0 & 0 \\ X & X & 0 & X & 0 & 0 & 0 & 0 & 0 & 0 & 0 & 0 & 0 & 0 & 0 \\ X & 0 & X & X & X & 0 & 0 & 0 & 0 & 0 & 0 & 0 & 0 & 0 & 0 \\ X & 0 & 0 & X & 0 & X & 0 & 0 & 0 & 0 & 0 & 0 & 0 & 0 & 0 \\ X & 0 & 0 & X & 0 & 0 & X & 0 & 0 & 0 & 0 & 0 & 0 & 0 & 0 \\ 0 & X & 0 & X & 0 & 0 & 0 & X & 0 & 0 & 0 & 0 & 0 & 0 & 0 \\ 0 & X & 0 & X & 0 & 0 & 0 & 0 & X & 0 & 0 & 0 & 0 & 0 & 0 \\ 0 & X & 0 & X & 0 & 0 & 0 & 0 & 0 & X & 0 & 0 & 0 & 0 & 0 \\ 0 & X & 0 & X & 0 & 0 & 0 & 0 & 0 & 0 & X & 0 & 0 & 0 & 0 \\ 0 & X & 0 & X & 0 & 0 & 0 & 0 & 0 & 0 & 0 & X & 0 & 0 & 0 \\ 0 & 0 & 0 & 0 & 0 & 0 & 0 & 0 & 0 & 0 & 0 & 0 & X & 0 & 0 \\ 0 & X & 0 & 0 & 0 & 0 & 0 & 0 & 0 & 0 & 0 & 0 & 0 & X & 0 \\ 0 & 0 & 0 & 0 & 0 & 0 & 0 & 0 & 0 & 0 & 0 & 0 & 0 & 0 & X \end{bmatrix} \begin{bmatrix} DP_{11} \\ DP_{12} \\ DP_{13} \\ DP_{14} \\ DP_{21} \\ DP_{22} \\ DP_{23} \\ DP_{24} \\ DP_{25} \\ DP_{26} \\ DP_{27} \\ DP_{28} \\ DP_{31} \\ DP_{32} \\ DP_{33} \end{bmatrix} \tag{2.1}$$

2.2 Product Architecture for Adaptable Design

2.2.1 Modular Products

Modular design aims at developing product architecture by using relatively independent modules [1, 10, 11, 15]. A module is defined as a component or a group of components that can be disassembled non-destructively from the product as a unit. Modular design is evaluated by modularity. Since modules in a modular product are relatively independent, these modules can be designed and manufactured separately. Each module can be attached, detached, modified, relocated, and replaced easily for upgrading, repair, recycling, or reuse. Modular design serves as basis for adaptable design.

In modular design, similar components are grouped into modules according to their functions, technologies, or physical structures [14]. Function-based modular design in adaptable design is effective for satisfying requirements of different customer groups, each demanding a certain function. When different or new functions are expected, modules with the required functions are then used to modify the product. The function-based modular design allows one product to deliver multiple/alternative functions to satisfy needs of different groups of customers. For example, the CD-RW drive of a computer is a module with the function of external data access/storage. When larger capacity of data access/storage is needed, this CD-RW drive can be upgraded with a DVD-RW drive or a cloud storage.

Modules of a product can also be identified based on the similarity of technologies with similar lifespans. For example, since the CPU and data communication bus in a computer are designed based on VLSI technology, these two components are grouped in the motherboard module of this computer. When different components need to be grouped together due to their physical structures, these components can be designed and manufactured as a module. For example, since a computer's graphics process unit (GPU) and its cooling fan have to be located close to each other, these two units can be grouped as a module.

Many methods have been developed to organize components with similar functions, technologies, and structures into modules. Gershenson et al. [10] classified the modular design methods into four main categories: checklist methods, design rules, matrix manipulations, and step-by-step measure and redesign methods. Modular designs have also been achieved using advanced computing tools, including fuzzy mathematics [33], optimization [12], genetic algorithms [16], and simulated annealing [6].

2.2.2 *Product Platforms and Families*

In the platform design, common components of different products are grouped as the platform to be shared by these products [14, 15, 30]. The products sharing the same platform usually form a family of products. Platform design is considered as the extension of modular design by using the platform—the main module—in all the products of this family. For example, Honda Civic models DX, EX, and LX use the same platform.

In adaptable design, functions of different products can be achieved using the platform design approach. When certain functions are required, modules with these functions are attached to the platform. For example, a hand vacuum cleaner with various accessory attachments for dry dust and wet spills is a simple adaptable product developed using the platform design approach.

Considerable effort has been devoted to platform design and product family design [3, 14, 15, 30]. Product platforms can be classified into two categories: modular platforms and scalable platforms [30]. A modular platform is a collection of components that are shared by all the products in a product family. Different functions in this family are achieved by reconfiguring different attachment modules to the platform [24]. A scalable platform has scalable variables to "stretch" or "shrink" the platform to satisfy different requirements [29]. A scalable platform can be achieved through parametric design.

A methodology of proposing a platform differentiation plan (PDP) to provide a platform leverage strategy across market niches was reported [3]. PDP is designed to provide a consistent balance between mutual differentiation and components sharing for the leveraged platforms at the early product planning stage. Two indices—differentiation index (DI) and commonality opportunity index (COI)—have been presented to represent dimensions of the product platform differentiation and commonality, respectively. Through a case study of electronic test equipment, the usefulness of PDP was demonstrated for the platform leverage strategy across market niches.

2.2.3 Open Architecture Products

Product architecture decides the integration of functional elements and physical components of a product. Generally, product architecture can be defined as "the scheme by which the function of a product is allocated to physical components" [34]. There are two types of products, closed architecture products and open architecture products, based on the owner of the architecture [2, 18, 38].

For a closed architecture product, different product modules/components are specified by the original equipment manufacturer (OEM). Most of traditional products including reconfigurable products belong to this category. The open architecture product can be considered as the one with a platform such that add-on modules developed by different vendors can be connected to the platform through the interfaces [17, 39]. For example, a personal computer can be considered as an open architecture product, because various devices developed by different vendors can be connected to this personal computer through USB interfaces. Excavator is another example of open architecture product because different front execution devices such as hammer, digger, and bucket provided from various producers can be connected to the excavator through the interface.

Characteristics of open architecture products include:

- An open architecture product (OAP) is composed of a platform, add-on modules, and open interfaces to connect the platform and add-on modules.
- Specifications and constraints of open interface parameters are open to the public.
- The platform and add-on modules are connected by open interfaces through relationships defined by input and output parameters.
- Add-on modules can be specific ones that need to be designed during product development stage and unknown ones that can be designed and added in the future.
- Add-on modules can be provided by both the original equipment manufacturer (OEM) and the third-party vendors.

For an open architecture product, since different manufacturers/customers can participate in the design of add-on modules, the product variety can be achieved. Product sustainability, adaptability, upgradeability, and extendibility can also be achieved from an open architecture product [8, 18]. At the product purchasing stage, customers can select the required add-on modules from all available ones based on their needs. At the product operation stage, customers can change, upgrade, and extend functions by using different add-on modules.

Based on the analysis of characteristics of open architecture products, a comparative study of the open architecture product with mass-produced product, mass-customized product, upgradable product, and reconfigurable product was conducted by Zhang et al. [39] as shown in Table 2.2.

For the mass-produced products, functions are determined by the manufacturer and cannot be changed. For the mass-customized products, different options are available for customers to select at the product purchasing stage. The mass-customized products, however, usually cannot be changed in the product operation stage. Both reconfigurable products and upgradeable products enable customers to change product configurations during the product operation stage to satisfy different needs of customers. A reconfigurable product is usually used to replace multiple products with a single one. New components/modules are not added to the reconfigurable product during product operation stage. For an upgradable product, new components/modules are usually used to replace the old ones for improving functions of the product.

In general, an open architecture product as shown in Fig. 2.13 can be modeled by a platform and open interfaces where different add-on modules from different sources can be connected to satisfy requirements of customers.

The platform (M^{P}) of an open architecture product is designed and manufactured by the original equipment manufacturer (OEM). The platform of an OAP can have several open interfaces to connect different add-on modules. In an open architecture product, both the specific add-on modules designed at product development stage and unknown add-on modules that could be added later on during the product operation stage should be considered.

The specific add-on modules, $M^{\mathrm{S}}_{im_i}$ $(i = 1, 2, \ldots, l)$, are the ones whose configurations and parameters need to be determined at product design stage. For each interface of the platform in an OAP, usually several add-on modules need to be designed to

Table 2.2 Comparison among different types of products (Modified from Zhang et al. [39])

Type of product	Achievement of different needs		New components and/or modules added at operation stage	Open interfaces provided for connections of personalized add-on modules
	Different options at product purchasing stage	Different options at product operation stage		
Mass-produced product	No	No	No	No
Mass-customized product	Yes	No	No	No
Reconfigurable product	No	Yes	No	No
Upgradeable product	No	Yes	Yes	No
Open architecture product	Yes	Yes	Yes	Yes

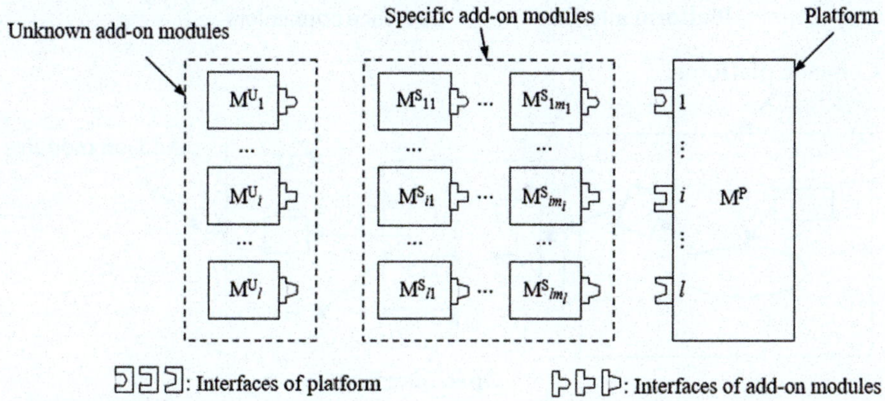

Fig. 2.13 Platform and add-on modules of an open architecture product

achieve different functions. Since all the design details of the specific add-on modules are determined at the design stage, performance measures and their variations of an OAP with specific add-on modules can be calculated for evaluation of the product. The unknown add-on modules, M_i^{U} ($i = 1, 2, \ldots, l$), are those that could be added later on in product operation stage. Since configurations and parameters of these unknown add-on modules cannot be decided at the design stage, performance measures of the OAP with unknown add-on modules are not described by specific values and large variations of these performance measures are also expected. Usually, an unknown add-on module is defined by constraints on open interface parameters such as the available value sets for discrete interface parameters and upper/lower boundaries for continuous interface parameters.

Interactions between platform and add-on modules in an OAP are defined by input and output parameters of the open interfaces. Since both the specific add-on modules and unknown add-on modules are considered, input/output parameters of interfaces for interactions between the platform and add-on modules should be defined differently as follows:

- Input/output parameters of interfaces for specific add-on modules: Since design configurations and parameters of the add-on modules are determined at the product development stage, values of the input/output interface parameters can be calculated from parameters of the platform and add-on modules.
- Input/output parameters of interfaces for unknown add-on modules: Since design configurations and parameters of the add-on modules are not determined at the product development stage, values of the input/output interface parameters have to be defined by constraints. Constraints for discrete and continuous parameters are considered.

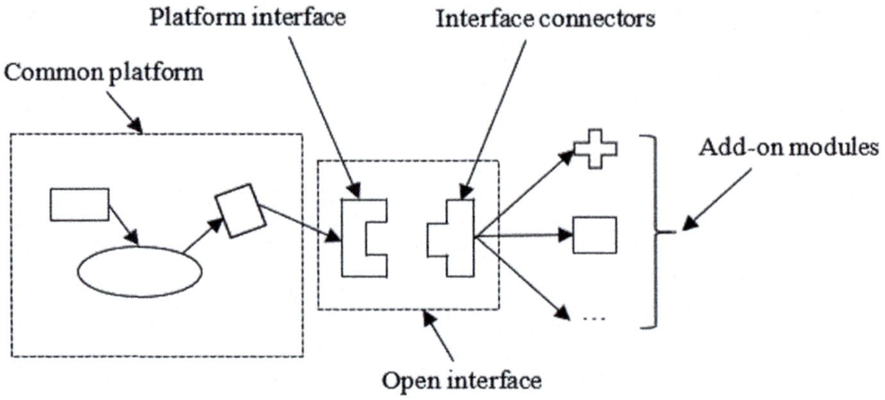

Fig. 2.14 Interface of an open architecture product

The open interface is critical to the adaptable design of an open architecture product. As shown in Fig. 2.14, an open interface is modeled by a platform interface and an interface connector. Furthermore, both platform interface and interface connector are modeled by parts and assembly relations among the parts. The following characteristics of an open interface have been observed:

- *Openness*: The detailed information of the open interface can be accessed by the public for the development of individualized personal add-on modules.
- *Adaptability*: Various add-on modules can be connected and disconnected to a common platform through an open interface to satisfy changeable customer requirements.

Therefore, the interface has to be open for adaptability in an open architecture product. An excavator interface to connect front execution devices is an open interface, because devices including hammer, digger, bucket, etc., from third-party vendors can be connected to the excavator through the interface to satisfy various functional requirements of customers during the excavator operation. Detailed information of the excavator interface can be accessed by the public.

2.3 Adaptable Interface

Since components and modules in adaptable design need to be detached, attached, and upgraded, interfaces have to be considered to ensure interactions among these components/modules and ease of their disassembly/assembly. In addition, the interface design also plays an important role in modular and platform designs.

Adaptable interfaces would facilitate the reuse of modules, platforms, and variants. If adaptable interfaces are provided within modules, the reuse of sub-modules and critical components within the modules can also been facilitated. For example, a base platform is designed and used for a family of products, and the add-on modules are assembled to the base platform using the adaptable interfaces. Such interfaces can also be used for upgrading of products.

Different categories of interfaces, including open interfaces, adaptable interfaces, and standardized interfaces, were introduced to benefit the product personalization, sustainability, adaptability, and product variety [13, 18, 38]. A comparative study of different types of interfaces was conducted as shown in Table 2.3.

Although similarities among the adaptable interface, open interface, and standardized interface can be observed, according to Table 2.3, an adaptable interface is different from the open interface and standardized interface.

An adaptable interface may not be an open interface, because modules connected by the adaptable interface may not be allowed to be changed to add-on modules provided by third-party vendors. For example, a lathe interface to connect with its tailstock is an adaptable interface rather than an open interface because the tailstock provided by a third-party vendor is not allowed to connect with the lathe. On the other hand, an open interface is an adaptable interface. Add-on modules are allowed to be connected to the open interface in both product manufacturing stage and operation stage. For an adaptable interface, only adaptations during the product operation stage are considered. Therefore, open interfaces (e.g., the USB interface and excavator interface) are adaptable interfaces.

Interface standardization supports adaptability of the open and adaptable interfaces. However, open and adaptable interfaces may not be necessary to be standardized interfaces. For example, the excavator interface is an open interface but not a standardized interface as open interfaces designed by different excavator manufacturers are different from each other. On the other hand, since add-on modules from different sources may not be allowed to be connected to the original product, a standardized interface may not be an open interface. For example, the mobile phone charger interface is considered as a standardized interface as the charger interfaces of different types of mobile phones from a specific manufacturer are the same. However, the mobile charger interface is not an open interface as some chargers provided by third-party vendors are not allowed to be connected to the mobile phone.

Table 2.3 Interface classifications

Interface	Definition	Example	Implication
Closed interface	Interface used to connect modules specified in the product design process	• An automobile interface to connect engine • A printer interface to connect cartridge	• Not for the user assembly and disassembly • Closed innovations for both product and technology • Not required for standardization • Closed business model • Lack of the product personalization
Open interface	Interface used to connect unknown modules that can be provided by third-party vendors	• An excavator interface to connect front execution devices • An USB interface to connect different functional device	• Easy for user assembly and disassembly • Potential open innovations for product and technology • Require adaptability • Standardization • Open business model • Personalized functions • For product variety
Adaptable interface	Components and their relationships of the interface can be easily adapted in the product operation stage to satisfy different requirements	• A camera interface to connect different lens • A lathe interface to connect tailstock	• Satisfy changeable requirements • Meet product variety • Good for product upgradability
Un-adaptable/fixed interface	Components and their relationships of the interface cannot be adapted in the product operation stage to satisfy different requirements	• An automobile interface to connect engine • A television interface to connect screen	• Specific product functional requirements • Optimal performance • Lower product adaptability • Compact structure and function sharing of components

(continued)

Table 2.3 (continued)

Interface	Definition	Example	Implication
Standardized interface	The interface is standardized in the industry and/or international organization	• An electricity socket interface to connect plug • An USB interface to connect different functional devices	• Reduced cost as mass-produced components • Easy for product development management • Good for components exchange • Good for product upgradability
Non-standardized interface	The interface is not standardized either in the industry or international organization	• A car interface to connect headlight • A back cover of wrist watch	• Relative high manufacturing cost • Not for component exchange • Optimal performance

2.4 Assembly and Disassembly for Adaptation Operations

2.4.1 Modeling of Assembly Operations

2.4.1.1 Definitions and Assumptions

Definitions and assumptions for modeling of assembly operations are summarized as follows [22].

Definitions

- Support platform module: A module used as the assembly base of other modules in product assembly operations is called a support platform module. Other modules are added on this module sequentially to complete the assembly of the entire product.
- Base component: A component used as the assembly base in the component assembly inside a module is called the base component. Other components are added on this base component sequentially to complete the assembly of the entire module.

Assumptions

- Modules that cannot be assembled in a sub-assembly can be divided into several sub-assemblies or components. They still belong to the same level of the bill of materials (BoM). After a module is divided, separated components are processed as modules to plan the assembly sequence with other modules.
- Module types of the assembly are known from the product design.

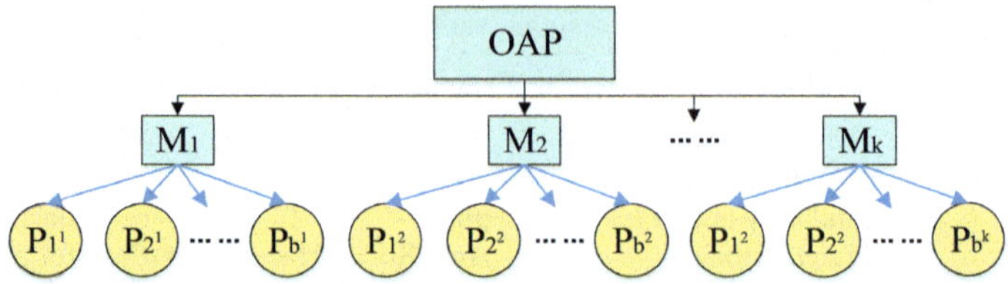

Fig. 2.15 Bill of materials

- Only six translation directions of the assembly are considered. Global and local coordinate systems are used for a product and its modules, a module and its components, respectively.

2.4.1.2 Bill of Materials

A bill of materials (BoM) is used to describe hierarchies of a product, as shown in Fig. 2.15, where M_k is a module k. P_b^k is a component b in M_k. b is the total number of components in M_k, while k is the total number of modules in the product.

2.4.1.3 Product Structure Diagram and Assembly Constraint Matrix

A structure diagram $G = \{P, E, W\}$ is proposed to describe the precedence constraint knowledge of the product, where $P = \{P_{1^1}, P_{2^1}, \ldots, P_{b^1}, P_{1^2}, P_{2^2}, \ldots, P_{b^2}, \ldots, P_{1^k}, P_{2^k}, \ldots, P_{b^k}\}$ represents components of a product, called notes. $E = \{E_1, E_2, \ldots, E_n\}$ is a set of undirected edges used to connect the nodes, and $W = \{W_1, W_2, \ldots, W_m\}$ is a set of directed edges used to connect the nodes. An assembly is defined by $\{P, \text{AC}\}$. $\text{AC} = [\text{ac}_{ij}]$ represents the relation of two components, where $i, j \in \{P_{1^k}, P_{2^k}, \ldots, P_{b^k}\}$, $\text{ac}_{ij} = 1, -1, 2, 9, -9$. $\text{ac}_{ij} = -\text{ac}_{ji}$. When $i = j$, $\text{ac}_{ij} = ac_{ji} = 0$. ac_{ij} represents the assembly constraint relationship as follows:

- $\text{ac}_{ij} = 1$: A connection relation exists between P_i and P_j, but P_i must be assembled before P_j.
- $\text{ac}_{ij} = -1$: A connection relation exists between P_i and P_j, but P_i must be assembled after P_j.
- $\text{ac}_{ij} = 2$: A connection relation exists between P_i and P_j, but P_i can be assembled before or after P_j.
- $\text{ac}_{ij} = 9$: There is no connection between P_i and P_j, but P_i must be assembled before P_j.
- $\text{ac}_{ij} = -9$: There is no connection between P_i and P_j, but P_i must be assembled after P_j.

In Eq. (2.2), AC_k is an assembly constraint matrix to represent assembly constraint relations among components in module M_k. SM_k defines assembly constraint relations of parts in M_k, and components in other modules called the simplified matrix shown in Eq. (2.3). MM_{ij} is the assembly constraint relation between parts P_{1^i}, P_{2^i}, ..., P_{b^i} in module M_i and parts P_{1^j}, P_{2^j}, ..., P_{b^j} in module M_j as described in Eq. (2.4).

$$\mathrm{AC}_k = \begin{array}{c} \\ P_{1^k} \\ P_{2^k} \\ \vdots \\ P_{b^k} \end{array} \begin{array}{c} \begin{array}{cccc} P_{1^k} & P_{2^k} & \cdots & P_{b^k} \end{array} \\ \begin{bmatrix} 0 & \mathrm{ac}_{1^k 2^k} & \cdots & \mathrm{ac}_{1^k b^k} \\ \mathrm{ac}_{2^k 1^k} & 0 & \cdots & \mathrm{ac}_{2^k b^k} \\ \vdots & \vdots & \ddots & \vdots \\ \mathrm{ac}_{b^k 1^k} & \mathrm{ac}_{b^k 2^k} & \cdots & 0 \end{bmatrix} \end{array}, \tag{2.2}$$

$$\mathrm{SM}_k = \begin{array}{c} \\ p_{1^k} \\ p_{2^k} \\ \vdots \\ p_{b^k} \end{array} \begin{array}{c} \begin{array}{cccc} p_{1^k_r} & p_{2^k_r} & \cdots & p_{b^k_r} \end{array} \\ \begin{bmatrix} 0 & \mathrm{sm}_{1^k 2^k_r} & \cdots & \mathrm{sm}_{1^k c^k_r} \\ \mathrm{sm}_{2^k 1^k_r} & 0 & \cdots & \mathrm{sm}_{2^k c^k_r} \\ \vdots & \vdots & \ddots & \vdots \\ \mathrm{sm}_{b^k 1^k_r} & \mathrm{sm}_{b^k 2^k_r} & \cdots & 0 \end{bmatrix} \end{array}, \tag{2.3}$$

$$\mathrm{MM}^{ij} = \begin{array}{c} \\ P_{1^i} \\ P_{2^i} \\ \vdots \\ P_{b^i} \end{array} \begin{array}{c} \begin{array}{cccc} P_{1^j} & P_{2^j} & \cdots & P_{b^i} \end{array} \\ \begin{bmatrix} 0 & \mathrm{ac}_{1^i 2^j} & \cdots & \mathrm{ac}_{1^i b^j} \\ \mathrm{ac}_{2^i 1^j} & 0 & \cdots & \mathrm{ac}_{2^i b^j} \\ \vdots & \vdots & \ddots & \vdots \\ \mathrm{ac}_{b^i 1^j} & \mathrm{ac}_{b^k 2} & \cdots & 0 \end{bmatrix} \end{array}, \tag{2.4}$$

where $M_k = \{P_{1^k}, P_{2^k}, \ldots, P_{b^k}\}$, $P_{1^k_r}, P_{2^k_r}, \ldots, P_{c^k_r}$ are components except those in M_k. $i = \{P_{1^k}, P_{2^k}, \ldots, P_{b^k}\}$, $j = \{P_{1^k_r}, P_{2^k_r}, \ldots, P_{c^k_r}\}$, and sm_{ij} has same values as ac_{ij}. AC_k, SM_k, and MM_{ij} are sub-matrices of AC.

For the product module and its structure diagram shown in Fig. 2.16, if parts P_{1^1} and P_{2^1} are connected, and P_{2^1} is assembled before P_{1^1}, a solid line is used in the diagram with a direction from P_{2^1} to P_{1^1}. $\mathrm{ac}_{ij} = 1$, $i = 1^1, j = 2^1$ in the matrix. If there is no connection relation between parts P_{1^1} and P_{4^1}, and P_{4^1} is assembled after P_{1^1}, a dash line is used in the diagram with a direction from P_{1^1} to P_{4^1}, and $\mathrm{ac}_{ij} = 9$, $i = 1^1, j = 4^1$ in the matrix. There is a connection without the sequence requirement in the assembly between parts P_{1^1} and P_{3^1}, $\mathrm{ac}_{ij} = 2$, $i = 1^1, j = 3^1$ in the matrix as shown in Eq. (2.5).

$$\mathrm{AC}_1 = \begin{array}{c} \\ P_{1^1} \\ P_{2^1} \\ P_{3^1} \\ P_{4^1} \end{array} \begin{array}{c} \begin{array}{cccc} P_{1^1} & P_{2^1} & P_{3^1} & P_{4^1} \end{array} \\ \begin{bmatrix} 0 & -1 & 2 & 9 \\ 1 & 0 & 1 & 0 \\ 2 & -1 & 0 & 1 \\ -9 & 0 & -1 & 0 \end{bmatrix} \end{array}. \tag{2.5}$$

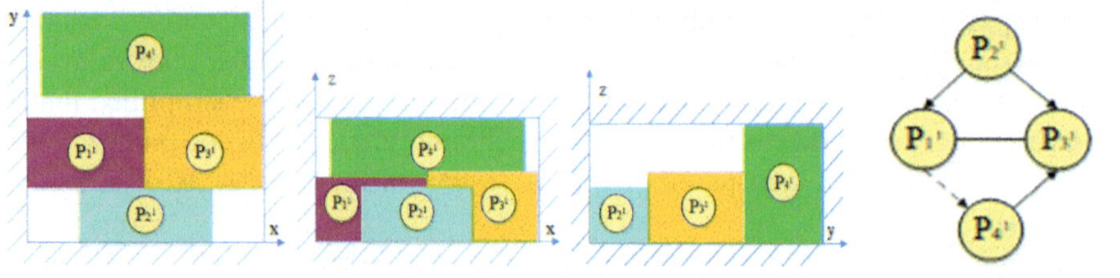

(a) A module and parts (b) Structure diagram

Fig. 2.16 Product module and its structure diagram

2.4.1.4 Direction Matrix

Direction matrix (DM) describes feasible directions of a component assembly. Six translation directions are used as $x, -x, y, -y, z, -z$ in an x-y-z coordinate system. $\text{DM} = [\text{dm}_{oi}]^{6\times i} = \{x, -x, y, -y, z, -z\}$ represents a direction set of the assembly. $\text{dm}_{oj} = 0$, or 1, where 1 represents that the part can be assembled in the direction and 0 represents that the part cannot be assembled in the direction. For example, a direction matrix of the module shown in Fig. 2.2 is as follows:

$$\text{DM} = \left[\text{dm}_{oj}\right]^{6\times b_k} = \begin{matrix} \\ x \\ -x \\ y \\ -y \\ z \\ -z \end{matrix} \begin{matrix} P_1^1 \; P_2^1 \; P_3^1 \; P_4^1 \\ \begin{bmatrix} 0\,0\,0\,0 \\ 0\,0\,0\,0 \\ 1\;1\;1\;1 \\ 0\,0\,0\,0 \\ 0\,0\,0\,0 \\ 0\,0\,0\,0 \end{bmatrix} \end{matrix}. \tag{2.6}$$

2.4.1.5 Feasible Analysis of a Module to Be Assembled in a Single Sub-Assembly

A single sub-assembly means that a set of components can be installed into the product independently, or a product module can be assembled in a single operation without the interference with other modules or components. A module can be checked for the independent operation based on Eq. (2.3) in following steps:

(1) Each component in a module is examined for its embedment by two or more other components that belong to different modules.
(2) For any column elements in SM_k, if they contain not only positive but also negative values except 2, M_k cannot be assembled in one operation.
(3) Based on the direction matrix, a module cannot be operated in one sub-assembly unless there exist one row where all elements are 1.

If a module can be assembled as a sub-assembly, the product BoM is directly used as the hierarchal structure diagram to guide the assembly sequence of modules. Otherwise, the module has to be divided into several sub-modules or components to be assembled independently. The dividing process uses following rules:

(1) If there is a component P_i in M_k surrounded by components of other modules, P_i is to be separated from the module.
(2) If an element of SM_k is negative, the corresponding component of this element must be assembled after corresponding parts in the column of this element. Therefore, corresponding parts in rows with negative elements should be separated.
(3) The separated part is added to the same structure level of the module. The process continues until all modules are checked. The product structure graph is adjusted.

For example, a gear box consists of four modules based on the product BoM of the modular design: three shafts and one housing, M_1, M_2, M_3, and M_4. Based on its structure graph shown in Fig. 2.17, the gear box is formed by four sub-assemblies. Module M_4 is a support platform module to install other modules. But, M_4 cannot be assembled as one unit in a single operation as M_1, M_2, and M_3 are embedded in M_4. Therefore, module M_4 should be decomposed. The adjusted structure graph is shown in Fig. 2.18. The separated components are located in the same level of M_4 in the assembly operation.

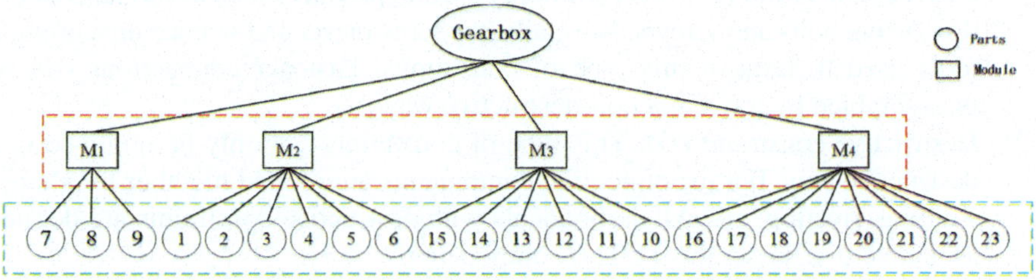

Fig. 2.17 Original structure graph of the gear box

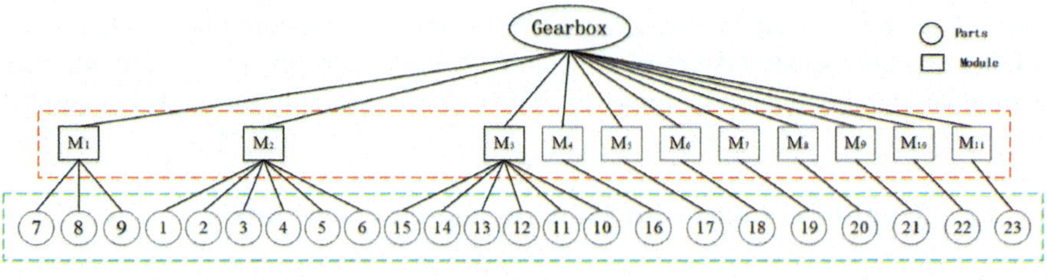

Fig. 2.18 Adjusted structure graph of the gear box

2.4.2 *Modeling of Disassembly Operations*

Disassembly is an operation in either product maintenance to replace the failed part of the product, or for recycling reusable parts of an end-of-life product to reduce the impact on environments and maximize the utilization of resources. A product must be separated into individual components in the repairing or recycling process, which is called the disassembly operation. Disassembly is a key operation to divide a product into separated parts. Disassembly planning ensures feasible operations of a disassembly based on criteria of the solution search [19].

2.4.2.1 Constraints of Product Disassembly

Constraints are identified to represent component connections of a product. Normally, multi-level constraint matrices can be constructed to represent a product [20]. However, without considering destructive disassembly methods, the disassembly sequences may have the detour problem for bypassing constraints that can only be removed by destructive ways. Therefore, constraint types should be extended to include constraints to be removed by destructive methods. This information is integrated in the product representation. Constraints therefore include two categories based on connection relations. They can be released by corresponding disassembly operations as follows [35]:

(1) Fastener constraint (*F*): It is commonly used in products such as thread connections using bolts and screws. Normally, the screwdriver and wrench are common tools used to remove this type of constraints. Fastener connections can be disassembled by tools in a non-destructive way.
(2) Destructive constraint (*D*): This type of constraints can only be removed in a destructive way. For example, two components are jointed together by gluing, welding, riveting, etc. All components with destructive constraints are defined as unable to disassembly (UTD) components.

Therefore, for constraints that can only be removed by destructive operations, they will be identified as type *D* constraints during the product representation. In the disassembly sequence searching process, they will be either destroyed by destructive operations or bypassed by non-destructive operations. Although destructive disassembly operations may take more time and effort, it can reach the target component with fewer components to be removed. The total disassembly time can be reduced.

2.4.2.2 Multi-level Constraint Matrices and Fastener Matrix

Product details can be recorded in a tree structure. Hierarchical relations of the tree structure are formed based on the bill of material (BoM) of a product [20]. Multi-level constraint matrices can then be built. A multi-layer constraint matrix represents constraints between one component and other components along $\pm X$, $\pm Y$, $\pm Z$ directions in Cartesian coordinates of the product model [35].

For an assembly $A = \{A_1, A_2, \ldots, A_n\}$ with n components, the corresponding constraint matrix is an $n \times n$ matrix and each element in the constraint matrix is a 6-digital array. Element a_{ijd} represents constraint relations between component i and component j along direction d. $i, j = 1, 2, \ldots n$, $d \in (\pm X, \pm Y, \pm Z)$, where $a_{ijd} = 1$ means that component j stops component i moving out of the product along direction d. $a_{ijd} = 0$ means that component j does not stop component i moving out of the product along direction d. When $i = j$, $a_{ijd} = 0$ in any direction. Therefore, all elements along diagonal of the constraint matrix are 000000.

In the first step of selective disassembly operations, the lowest level sub-assembly including the target component will be removed. In order to remove a sub-assembly or component out of a product, at least one feasible disassembly direction is required. In other words, the sub-assembly or component must be disassembled along at least one direction without being blocked by other components. The equation to evaluate component i along $d \in (\pm X, \pm Y, \pm Z)$ direction is as follows, where n is the number of components in the sub-assembly.

$$a_{id} = a_{i1d} + a_{i2d} + \cdots + a_{ind}. \tag{2.7}$$

Only when $a_{i1d} = a_{i2d} = \cdots = a_{ind} = 0$, the value of a_{id} will be 0, and component i can be removed along direction d. Otherwise, component i is temporarily blocked in direction d.

When all elements in the first column are 000000, the column will be deleted. The first row is also deleted after first sub-assembly is removed. Therefore, the size of the constraint matrix is reduced. Using a smaller matrix, the calculation complexity is reduced and the disassembly sequence planning efficiency can be improved.

A fastener-component matrix is built to represent constraints to components from fasteners. In a fastener-component matrix, intersection elements of rows and columns represent components and fasteners, respectively. "1" means that the component is constrained by a fastener that can be removed by non-destructive disassembly operations, while "0" represents that there is no constraint between the component and fastener. "2" shows that the constraint can only be released by destructive operations. Fasteners should be removed before disassembling components.

2.5 Summary

Adaptable product modeling is one of the key issues in adaptable design. Generally, adaptable design modeling includes three aspects: modeling of rationalized functions and structures, modeling of adaptable product architecture, and modeling of adaptable interfaces.

In this chapter, requirements and various methods for the rationalized function structures modeling are first introduced. The methods to model the adaptable product architectures are then explained considering different types of modules and interfaces between these modules. The interface modeling method is subsequently provided for defining the connection and interaction relations between different types of modules. Methods to model product assembly and disassembly for adaptation operations are also discussed in this chapter.

In an adaptable design process, various design candidates could be developed for design engineers to select. Therefore, evaluations to different adaptable design candidates are needed for the selection of the best design one. The evaluation methods for adaptable design will be discussed in the next chapter.

References

1. K.K. Ali, M.S. Saed, *Product Design for Modularity* (Kluwer Academic, Norwell, 2002)
2. Y. Altintas, W. Munasinghe, A hierarchical open-architecture CNC system for machine tools. Ann. CIRP **43**(1), 349–354 (1994)
3. Y.L. Cai, A.Y.C. Nee, W.F. Lu, Platform differentiation plan for platform leverage across market niches. Ann. CIRP **57**(1), 141–144 (2008)
4. Q. Cheng, G.J. Zhang, Z.F. Liu, P. Gu, L.G. Cai, A structure-based approach to evaluation product adaptability in adaptable design. J. Mech. Sci. Technol. **25**(5), 1081–1094 (2011)
5. C. Chung, Q. Peng, A novel approach to the geometric feasibility analysis for fast assembly tool reasoning. Int. J. Adv. Manuf. Technol. **31**(1–2), 125–134 (2006)
6. C. da Cunha, B. Agard, A. Kusiak, Design for cost: module-based mass customization. IEEE Trans. Autom. Sci. Eng. **4**(3), 350–359 (2007)
7. M.S. Erden, H. Komoto, T.J. Van Beek, V. D'Amelio, E. Echavarria, T. Tomiyama, A review of function modeling: approaches and applications. Artif. Intell. Eng. Des. Anal. Manuf. **22**(2), 147–169 (2008)
8. G. Ferrer, Open architecture, inventory pooling and maintenance modules. Int. J. Prod. Econ. **128**(1), 393–403 (2010)
9. D. Fletcher, R. Brennan, P. Gu, A method for quantifying adaptability in engineering design. Concurr. Eng. Res. Appl. **17**(4), 279–289 (2009)
10. J.K. Gershenson, G.J. Prasad, Y. Zhang, Product modularity: measures and design methods. J. Eng. Des. **15**(1), 33–51 (2004)
11. J.K. Gershenson, G.J. Prasad, Y. Zhang, Product modularity: definitions and benefits. J. Eng. Des. **14**(3), 295–313 (2003)
12. P. Gu, S. Sosale, Product modularization for life cycle engineering. Robot. Comput.-Integr. Manuf. **15**(5), 387–401 (1999)
13. P. Gu, D. Xue, A.Y.C. Nee, Adaptable design: concepts, methods, and applications. Proc. Inst. Mech. Eng. Part B J. Eng. Manuf. **223**(B11), 1367–1387 (2009)
14. J. Jiao, T.W. Simpson, Z. Siddique, Product family design and platform-based product development: a state-of-the-art review. J. Intell. Manuf. **18**(1), 5–29 (2007)
15. A. Jose, M. Tollenaere, Modular and platform methods for product family design: literature analysis. J. Intell. Manuf. **16**(3), 371–390 (2005)
16. A.K. Kamrani, R.A. Gonzalez, Genetic algorithm-based solution methodology for modular design. J. Intell. Manuf. **14**(6), 599–616 (2003)
17. Y. Koren, *The Global Manufacturing Revolution: Product-Process-Business Integration and Reconfigurable Systems* (Wiley, New York, 2010)
18. Y. Koren, S.J. Hu, P. Gu, M. Shpitalni, Open-architecture products. Ann. CIRP **62**(2), 719–729 (2013)

19. Y. Luo, Q. Peng, Disassembly sequence planning for product maintenance, in *Proceedings of ASME International Design Engineering Technical Conferences & Computers and Information in Engineering Conference*. DETC2012-70430 (2012)
20. Y. Luo, Q. Peng, P. Gu, Integrated multi-layer representation and ant colony search for product selective disassembly planning. Comput. Industr. **75** (2016). https://doi.org/10.1016/j.compind.2015.10.011
21. G. Lyu, X. Chu, D. Xue, Product modeling from knowledge, distributed computing and lifecycle perspectives: a literature review. Comput. Ind. **84**, 1–13 (2017)
22. H. Ma, Q. Peng, J. Zhang, P. Gu, Assembly sequence planning for open-architecture products. Int. J. Adv. Manuf. Technol. **94**, 1551–1564 (2018)
23. M. Martinez, D. Xue, A modular design approach for modeling and optimization of adaptable products considering the whole product life-cycle spans. J. Mech. Eng. Sci. **232**(7), 1146–1164 (2018)
24. M. Meyer, A.P. Lehnerd, *The Power of Product Platform—Building Value and Cost Leadership* (Free Press, New York, 1997)
25. K. Otto, K. Wood, *Product Design* (Prentice Hall, 2000)
26. G. Pahl, W. Beitz, *Engineering Design: A Systematic Approach* (Springer-Verlag, Berlin, Germany, 1988)
27. S. Russell, P. Norvig, *Artificial Intelligence: A Modern Approach* (Prentice Hall, Upper Saddle River, NJ, USA, 2002)
28. J.J. Shah, M. Mantyla, *Parametric and Feature-Based CAD/CAM: Concepts, Techniques, and Applications* (Wiley, 1995)
29. T.W. Simpson, J.R.A. Maier, F. Mistree, Product platform design: method and application. Res. Eng. Des. **13**(1), 2–22 (2001)
30. T.W. Simpson, Z. Siddique, J. Jiao (eds.), *Product Platform and Product Family Design. Methods and Applications* (Springer, 2007)
31. N.P. Suh, *The Principles of Design* (Oxford University Press, Oxford, UK, 1990)
32. N.P. Suh, *Axiomatic Design: Advances and Applications* (Oxford University Press, Oxford, UK, 2001)
33. Y.T. Tsai, K.S. Wang, The development of modular-based design in considering technology complexity. Euro. J. Oper. Res. **119**(3), 692–703 (1999)
34. K.T. Ulrich, The role of product architecture in the manufacturing firm. Res. Policy **24**(3), 419–440 (1995)
35. H. Wang, Q. Peng, J. Zhang, P. Gu, Selective disassembly planning for the end-of-life product. Procedia CIRP, 512–517 (2017)
36. D. Xue, A multi-level optimization approach considering product realization process alternatives and parameters for improving manufacturability. J. Manuf. Syst. **16**(5), 337–351 (1997)
37. H. Yoshikawa, General design theory and a CAD system, in Sata, T., Warman, E.A. (eds.) *Man-machine Communication in CAD-CAM, Proceedings of the IFIP WG5.2/5.3 Working Conference*, Tokyo, Japan, 35–58 (1980)
38. J. Zhang, Y. Chen, D. Xue, P. Gu, Robust design of configurations and parameters of adaptable products. Front. Mech. Eng. **9**(1), 1–14 (2014)
39. J. Zhang, D. Xue, P. Gu, Adaptable design of open architecture products with robust performance. J. Eng. Des. **26**(1–3), 1–23 (2015)
40. C. Zhao, Q. Peng, P. Gu, Development of a paper-bag-folding machine using open architecture for adaptability. J. Eng. Manuf. **229**(1), 155–169 (2015)

Chapter 3
Evaluation for Adaptable Design

3.1 Evaluation of Engineering Design

In engineering design, life-cycle aspects of product ranging from customer requirements to design, manufacturing and operation should be evaluated in the design processes. The evaluation measure plays an important role for identification of the optimal design candidate generated in the design processes.

Any design and manufacturing activities have two common characteristics. First, an activity is performed to achieve a set of goals. Second, an activity involves costs (Fig. 3.1). The design goals can be represented by design requirements and constraints. Obviously, cost must be minimized. Typically, the cost is often used as one of the main evaluation criteria. Referring to Suh's design evaluation using information axiom [9, 10], we call the "total cost" of an activity as "information content".

In adaptable design, adaptability is defined as the utility extension of a system or product. In the adaptable design context, "total cost" represents the consumption of resources not only for design, but for any engineering activity including the task of adapting an existing product.

In Chap. 2, different methods for modeling of product functions and structures in adaptable design were discussed. In this chapter, evaluation measures for adaptable design will be provided for the selection of the optimal adaptable design candidate.

3.2 Measures of Adaptabilities

Adapting a product usually implies modifications in the structure of the product so that it can be used under different circumstances. Adapting a design implies modifications in the design or in the realization process of a design so that the new or different products can be developed from the same design. The key element in

P. Gu et al., *Adaptable Design*, Research on Intelligent Manufacturing,
https://doi.org/10.1007/978-981-99-5869-6_3

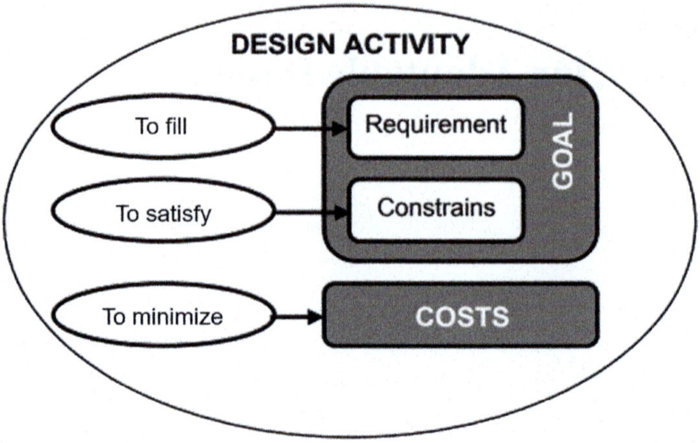

Fig. 3.1 Common characteristics of engineering activities

both *product adaptability* and *design adaptability* is "modification". Thus, in order to make a product more adaptable, these modifications must be facilitated, i.e., the design should be made adaptable.

We distinguish between two different scenarios. First, when particular adaptabilities are expected, the product can be designed to accommodate them. This is an adaptable design process in which *specific adaptabilities* are treated as design requirements. Second, when no specific adaptability is required in advance, the design can be generally made adaptable in its architecture. Adaptability without considering specific requirements is called *general adaptability*.

The classification of adaptability measures is summarized in Table 3.1. Design adaptability and product adaptability are used together with other evaluation measures such as performance and cost for identification of the best design solution from alternative design candidates.

(1) Specific Adaptabilities

Specific adaptability in adaptable design describes the ability of a product or its design to be adapted for potential applications that can be foreseen at the time the product is initially designed. Specific adaptability is applicable to both product adaptability and design adaptability.

Table 3.1 Classification of adaptabilities

		Product life-cycle phases	
		Design adaptation	Product adaptation
New requirements	Specific requirements provided	Specific design adaptability	Specific product adaptability
	General adaptability expected	General design adaptability	General product adaptability

Specific product adaptability is a measure to evaluate ability of a product to be adapted in its operation stage. Adaptable design of a product can extend its utility into different operational modes. We refer to this characteristic as "extra functionalities". Extra functions can be obtained by adding features, parts, or assemblies. Extra functions can also be obtained by utilizing the potential functions that existing parts or features can offer but these functionalities are not the reason for the existence of parts. Such utilization may require the intelligent modification of some forms or features in the construction of products. If an adaptable product replaces two or more products, the adaptation process may require module replacement, product reconfiguration, and the like. In extension of utility, functions of the product can be modified to respond to changing needs or working conditions, or enhanced by taking advantage of new technologies. Thus an adaptable product is upgradeable. There is forecast information on what functionalities may be needed in the future. Examples include the adaptable homes and upgradeable computer systems.

Specific design adaptability may be applicable only when a variety of designs are created based on a common adaptable design and requirements of these designs are known beforehand. Thus, an adaptable design may replace several designs. As an added benefit, various production processes and even manufactured parts can be shared across designs within the portfolio.

The specific adaptability of products and designs requires forecast information of adaptation requirements. The forecast information can guide specific adaptable design for versatility, variety, customization, and upgrading. The adaptability for these categories is explained below. It should be noted that the categories are not mutually exclusive; and there may be overlap between them.

Versatility is the maximum adaptability in which adaptation from one function to another does not require significant alteration of a product. The adaptation task requires very little effort because a versatile product is designed for multiple functions. Versatility is typically applicable to product adaptability for customers. An example for design of a mechanical bus was given by Gu and Slevinski [4].

Variety concerns producers (manufacturers) and refers to using the same design (blueprint) to produce a variety of products.

Customization is the adaptation of a product to customers' preferences. For example, various features and functionalities of SONY video cameras, such as image stabilizer, digital zoom, and side screen, are designed in the form of optional units. Many models are easily developed by the morphological combination of various features in response to various customer preferences. This category of adaptability is applicable to both products and designs.

Upgrading is the adaptation of existing designs and products to new needs or to enhanced technological performance. If upgrading is known during the design stage, it would be similar to versatility. A typical example is the upgrading of computer systems with rapidly expiring components, such as CPUs, designed as easily replaceable units.

Figure 3.2 summarizes the above brief discussion. It can be seen that the availability of forecast information justifies a larger initial investment as with versatile

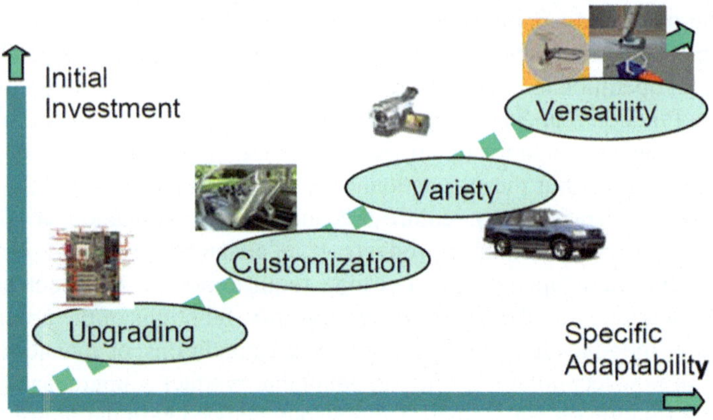

Fig. 3.2 Applications of specific adaptability [3]

products. While in the absence of such information, the initial investment is less because it is not certain that adaptations will be required after the design is finished.

(2) General Adaptabilities

The working environment of an existing product can change in an unpredictable way, or unanticipated new requirements may arise. It is desirable to adapt the product to the new operational mode to keep it in service. This becomes particularly important for more complex engineering systems such as power plants.

If a product is not designed for the possible adaptation, the necessary modifications may be very difficult or more expensive than obtaining a new product. The same statement can be made about the design of a product—sometimes it is more difficult to modify an existing design than to create a new one. The decision to adapt an existing product or develop a new one depends on the specific problem.

The cost of adaptation, as a rule, depends on the ability of the product or design to be modified relative easily. In the absence of forecast information on required adaptations, making a design easy to modify is the most effective way to increase its general adaptability.

An approach in achieving some general adaptability is based on the localization of modifications. That is, the architecture of the product is designed in such a way that prevents changes in one place from propagating into the rest of the product. Such architecture in this adaptable design is called "segregated architecture".

To make the product/design more adaptable, the interfacing between sub-assemblies is also critical to ensure the adaptability. For example, a flat surface on a part (sub-assembly or module) is more likely to facilitate product modifications than a complex surface for various reasons such as easy drilling and easy mating with other parts that may be mounted on it. Boothroyd [1] has developed various guidelines and examples for the development of physical and assembly connections for easy assembly.

3.2.1 Specific Product Adaptability

Product adaptation to change one existing product to a new product based on specific requirements is shown in Fig. 3.3.

Gu et al. [3] developed a method to measure specific product adaptability by comparing relative efforts of the product adaptation and new product creation. Suppose Tp_i is the i-th adaptation task, the effort for this task (i.e., effort for both design and manufacturing as shown in Fig. 3.3), according to the *information axiom* in axiomatic design [9, 10], can be modeled by its *information content* described by $Inf(Tp_i)$. Cost is usually used for modeling the effort. When *S1* is the current state of the existing product, *AS2* is the state after adaptation, the effort for this adaptation is then described by $\text{Inf}_{(S1\rightarrow AS2)}$. In the same way, the effort to develop a new product from scratch is described by $\text{Inf}_{(\text{ZERO}\rightarrow IS2)}$, where *ZERO* is the state to design a new product from scratch, and *IS2* is the state with only the new requirements. Since less effort is usually required to adapt a product with adaptable design than to develop a new one, the relative saving of effort is modeled as adaptable factor, $\text{AF}(Tp_i)$.

$$AF(Tp_i) = \frac{\text{Inf}_{(\text{ZERO}\rightarrow IS2)} - \text{Inf}_{(S1\rightarrow AS2)}}{\text{Inf}_{(\text{ZERO}\rightarrow IS2)}} = 1 - \frac{\text{Inf}_{(S1\rightarrow AS2)}}{\text{Inf}_{(\text{ZERO}\rightarrow IS2)}}, \quad 0 \leq AF(Tp_i) \leq 1 \tag{3.1}$$

When it takes more effort to adapt a product than to develop a new product (i.e., $\text{Inf}_{(S1\rightarrow AS2)} > \text{Inf}_{(\text{ZERO}\rightarrow IS2)}$), product adaptation should not be considered ($AF(Tp_i) = 0$). When no additional effort is required for product adaptation (i.e., $\text{Inf}_{(S1\rightarrow AS2)} = 0$), the product is a perfect adaptable product ($AF(Tp_i) = 1$).

When n product adaptation tasks, Tp_i ($i = 1, 2, \ldots, n$), and their probabilities, $\Pr(Tp_i)$, are considered, the specific product adaptability is then modeled by:

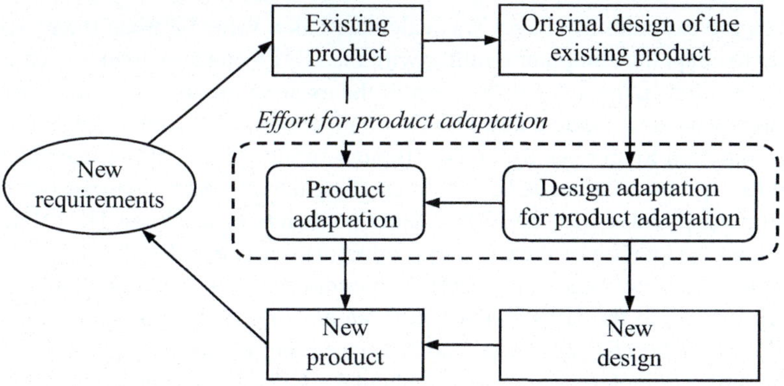

Fig. 3.3 Product adaptation considering specific requirements

$$AP(P) = \sum_{i=1}^{n} [\Pr(Tp_i)AF(Tp_i)] \tag{3.2}$$

Li et al. [6] extended this specific product adaptability evaluation method by considering three types of product adaptation tasks: *extendibility of functions, upgradeability of modules,* and *customizability of components.*

Extendibility of functions is achieved by designing a product with the potential for extension of functions. The extendibility factor, $EF(Tp_i)$, and extendibility of functions for the product, $E(P)$, are calculated by:

$$EF(Tp_i) = 1 - \frac{\text{Inf}_{(S1 \to AS2)}}{\text{Inf}_{(\text{ZERO} \to IS2)}}, \quad 0 \le EF(Tp_i) \le 1 \tag{3.3}$$

$$E(P) = \sum_{i=1}^{n} [\Pr(Tp_i)EF(Tp_i)] \tag{3.4}$$

where

S1	the current state of the existing product.
AS2	the modified state with the additional adaptable function.
ZERO	the state to design a new product from scratch.
IS2	the state with only the required adaptable function.

Li et al. [6] provided an example of an LCD computer monitor with adaptable function of HDTV monitor, Tp_1. When the cost to create a new HDTV monitor was $400, and the cost to add the HDTV function to an existing LCD computer monitor was $250, the extendibility factor, $EF(Tp_1)$, was calculated as

$$EF(Tp_i) = 1 - \frac{\text{Inf}_{(S1 \to AS2)}}{\text{Inf}_{(\text{ZERO} \to IS2)}} = 1 - \frac{\$250}{\$400} = 0.375 \tag{3.5}$$

Since the extendibility factor was calculated as a value between 0 and 1, it was cheaper to adapt the computer monitor with the HDTV function than to create a new HDTV monitor. In this case, adaptation of the existing product was considered.

When the cost to create a new HDTV monitor was $400, and the cost to add the HDTV function to an existing LCD computer monitor was $500, the extendibility factor, $EF(Tp_1)$, was calculated as 0, representing that it was more expensive to adapt the computer monitor with the HDTV function than to create a new HDTV monitor. In this case, adaptation of the existing product was not considered.

When the cost to create a new HDTV monitor was $400, and no additional cost was needed to add the HDTV function to an existing LCD computer monitor, the extendibility factor, $EF(Tp_1)$, was calculated as 1. In this case, design of the monitor with both the computer monitor function and the HDTV function can be considered.

Upgradeability is another kind of product adaptability that can achieve better performance or provide advanced technologies to meet new needs. The upgradeability factor, $UF(Tp_i)$, and upgradeability of modules for the product, $U(P)$, are calculated by:

$$UF(Tp_i) = 1 - \frac{\mathrm{Inf}_{(P1 \to UP1)}}{\mathrm{Inf}_{(P1)}}, \quad 0 \leq UF(Tp_i) \leq 1 \tag{3.6}$$

$$U(P) = \sum_{i=1}^{n} [\Pr(Tp_i) UF(Tp_i)] \tag{3.7}$$

where

P1 the state of the module without upgrading function
UP1 the state of the module with upgrading function

In this method, costs are usually selected to describe efforts for providing the upgrading function and for creating the part without the upgrading function. Generally, the cost for providing the upgrading function is calculated by the cost of the interface that allows for the upgrading of new modules.

Customizability of components is the ease of product adaptation based on requirements and preferences of individual customers. Product design with customizability can be easily reconfigured to different product configurations by combining standard components and modules based on the customer requirements. The customizability factor, $CF(Tp_i)$, and customizability of components for the product, $C(P)$, are calculated by:

$$CF(Tp_i) = 1 - \frac{\mathrm{Inf}_{(P1 \to CP1)}}{\mathrm{Inf}_{(P1)}}, \quad 0 \leq CF(Tp_i) \leq 1 \tag{3.8}$$

$$C(P) = \sum_{i=1}^{n} [\Pr(Tp_i) CF(Tp_i)] \tag{3.9}$$

where

P1 the state of the part without customization function
CP1 the state of the part with customization function

To define the specific product adaptability using these three different evaluation measures, they are first converted into dimensionless evaluation measures. The normalized measure for the extendibility of functions is calculated by:

$$NE(P_i) = \frac{E(P_i)}{E(P)_{max}} \tag{3.10}$$

where

$E(P_i)$ extendibility of functions of the i-th design candidate.
$E(P)_{max}$ the maximum value of extendibility of functions considering all the compared design candidates

In the same way, the normalized measures for upgradeability of modules and customizability of components can be calculated by:

$$NU(P_i) = \frac{U(P_i)}{U(P)_{max}} \tag{3.11}$$

$$NC(P_i) = \frac{C(P_i)}{C(P)_{max}} \tag{3.12}$$

These three values can then be combined into an overall specific product adaptability index by assigning weighting factors. The specific product adaptability index of the i-th design candidate is calculated by:

$$A(P_i) = I_E NE(P_i) + I_U NU(P_i) + I_C NC(P_i) \tag{3.13}$$

where I_E, I_U, and I_C are the weighting factors. The specific product adaptability index, $A(P_i)$, of each design candidate ranges from 0 to 1. When $A(P_i) = 0$, the design candidate does not provide any specific product adaptability. When $A(P_i) =$ 1, the design candidate provides complete specific product adaptability.

In the design of the adaptable food processor developed by Li et al. [6], a stand mixer was redesigned such that it was able to be adapted for other functions found in blender and meat grinder.

In this design, the function of a stand mixer was selected as the fundamental function of the product, and functions of a blender and a meat grinder were selected as adaptable (optional) functions [6]. The objective of this design was to create a reconfigurable product with functions of these three products. In this application, two designs were identified to fulfill requirements of a stand mixer with the potential to be adapted to the blender and meat grinder. Figure 3.4 shows the two designs, both of them can be reconfigured into three products.

In this application, the decision on what should be included in the product for sale and what should be included as add-on accessories to be sold separately was achieved based on evaluations of the different candidates for this product.

For each of these two designs, four candidates were considered as the commercial products: (1) a stand mixer, (2) a stand mixer and meat grinder, (3) a stand mixer and blender, and (4) a stand mixer, meat grinder, and blender. A total of eight candidates were created; and each candidate was evaluated by product adaptability and other three measures including the total part cost, total assembly cost, and operationability. These eight product candidates are summarized in Table 3.2.

In this case study, probabilities of using different food processing functions were identified in advance as follows:

$$\Pr(\text{Stand Mixer}) = 100\%$$

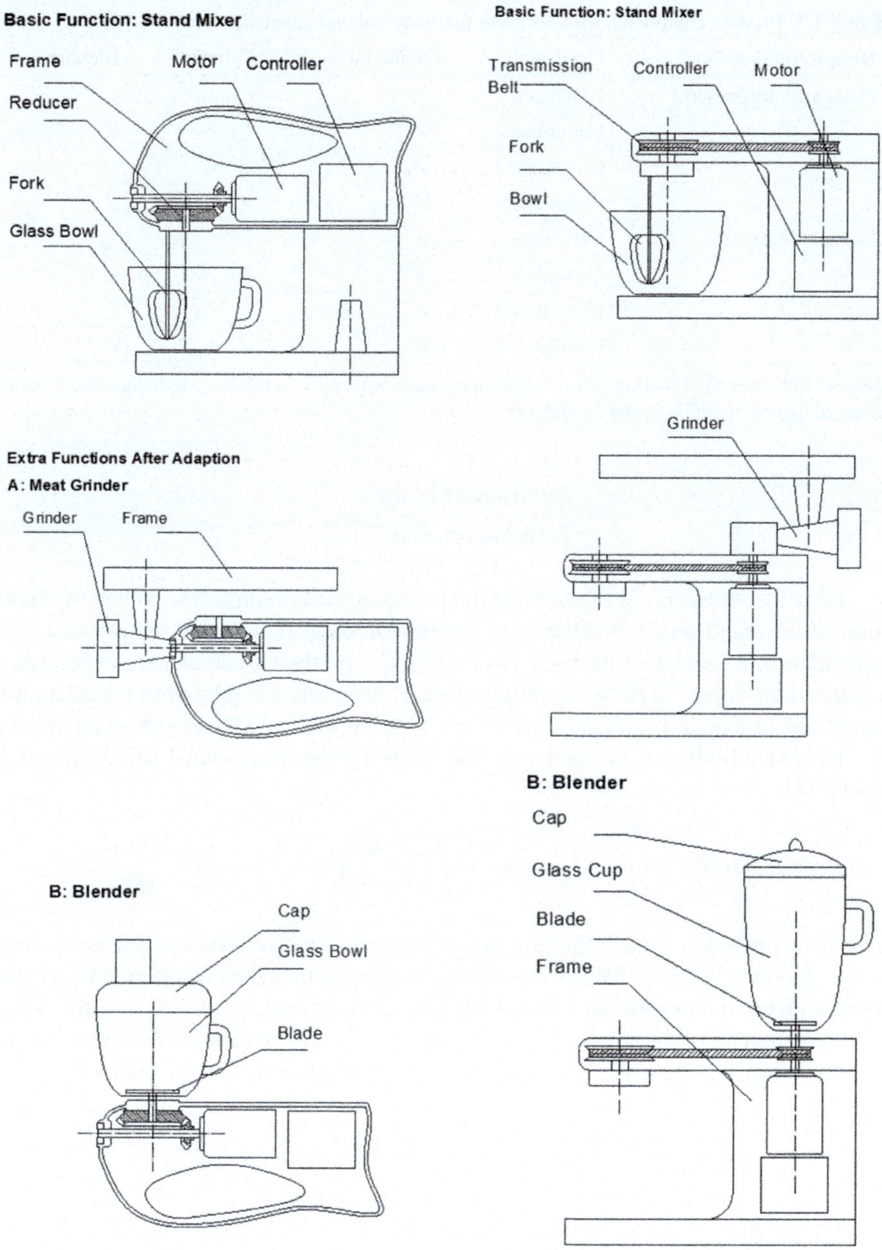

(a). Design configuration 1

(b). Design configuration 2

Fig. 3.4 Two design configurations [6]

Table 3.2 Product candidates with different fundamental and adaptable functions [6]

Design configuration	Candidate	Stand mixer	Grinder	Blender
Design configuration 1	Candidate 1	✓	○	○
	Candidate 2	✓	○	✓
	Candidate 3	✓	✓	○
	Candidate 4	✓	✓	✓
Design configuration 2	Candidate 5	✓	○	○
	Candidate 6	✓	○	✓
	Candidate 7	✓	✓	○
	Candidate 8	✓	✓	✓

Note ✓ indicates the function is a fundamental function in the candidate. ○ indicates the function is an adaptable function in the candidate

$$\Pr(\text{Grinder}) = 20\%$$
$$\Pr(\text{Blender}) = 95\%$$

Table 3.3 shows costs of parts for the two design configurations. When the function of the stand mixer is selected as the fundamental function of the product (i.e., candidate 1 in Table 3.2), the total part cost for this product is calculated as \$56. Based on the stand mixer in design configuration 1, the costs for achieving the adaptable functions of a meat grinder and a blender are identified as \$10 and \$8, respectively.

The extendibility of functions for candidate 1 is then calculated using Eqs. (3.3) and (3.4).

$$E(P_1) = 100\% \times (1 - 0) + 20\% \times \left(1 - \frac{10}{35}\right) + 95\% \times \left(1 - \frac{8}{60}\right) = 1.97$$

From Table 3.3, we can find out that some parts are selected as upgradeable parts and customizable parts. The cost to upgrade or customize the current part to another part is given in a bracket and modeled using the unit cost. In this case study, it was assumed that all of the selected parts would be upgraded or customized in the future.

The upgradeability of candidate 1 is calculated using Eqs. (3.6) and (3.7).

$$U(P_1) = 100\% \times \left(1 - \frac{3}{12}\right) + 100\% \times \left(1 - \frac{2}{8}\right) = 1.5$$

Since some parts are selected as both the upgradeable parts and customizable parts, they can use the same interfaces to be upgraded or customized. If the interface cost has been calculated considering the upgradeability, the cost for customization of the same part will be selected as 0, because both measures are achieved using the same interface. In this case study, the motor, controller, and colors of the frame were customized.

The customizability of candidate 1 was calculated using Eqs. (3.8) and (3.9).

Table 3.3 The part lists for two design configurations [6]

No.	Part	Unit cost (S)	Stand mixer	Meat grinder	Blender	Note
Panel A: The part list tor design configuration 1						
1	Motor	15(3)	✓	✓	✓	C, U
2	Controller	10(2)	✓	✓	✓	C, U
3	Reduce gear	5	✓	✓	✓	
4	Frame	10	✓	✓	✓	C
5	Bowl	10	✓			✓
6	Flat beater	2	✓			
7	Wire whip	2	✓			
8	Dough hook	2	✓			
9	Grinder	10		✓		
10	Cap	3		✓		
11	Blade	5(2)			✓	U
Panel B: The part list tor design configuration 2						
1	Motor	15(3)	✓	✓	✓	C, U
2	Controller	8	✓	✓	✓	
3	Reduce gear	5	✓	✓	✓	
4	Frame	10	✓	✓	✓	C
5	Bowl	10	✓		✓	
6	Glass cup	10			✓	
7	Flat beater	2	✓			
8	Wire whip	2	✓			
9	Dough hook	2	✓			
10	Grinder	10		✓		
11	Cap	2			✓	
12	Blade	5			✓	

Note C indicates this part is selected as customization part. U indicates this part is selected as upgradeable part. () indicates the cost to provide the upgrading or customization function

$$C(P_1) = 100\% \times \left(1 - \frac{0}{12}\right) + 100\% \times \left(1 - \frac{0}{8}\right) + 100\% \times \left(1 - \frac{0}{10}\right) = 3.0$$

Results of extendibility, upgradeability, and customizability of all the eight product candidates are summarized in Table 3.4.

To use these three different measures to evaluate the eight product candidates, these three measures should be first converted into dimensionless evaluation measures. The normalized measures for extendibility of functions, upgradeability of modules, and customizability of components are calculated using Eqs. (3.10–3.12), respectively.

Table 3.4 The results of extendibility, upgradeability, and customizability [6]

Configuration	Candidate (i)	$E(P_i)$	$U(P_i)$	$C(P_i)$	$A(P_i)$
Design configuration 1	Candidate 1	1.97	1.5	3.0	0.917
	Candidate 2	2.09	1.8	3.0	0.991
	Candidate 3	2.02	1.5	3.0	0.924
	Candidate 4	2.15	1.8	3.0	1.000
Design configuration 2	Candidate 5	1.98	0.75	2.0	0.668
	Candidate 6	2.09	0.75	2.0	0.685
	Candidate 7	2.04	0.75	2.0	0.677
	Candidate 8	2.15	0.75	2.0	0.694

$E(P_i)$ extendibility of functions, $U(P_i)$ upgradeability of modules, $C(P_i)$ customizability of components, $A(P_i)$ adaptability of product

These three normalized values can then be used to calculate the specific product adaptability index. The specific product adaptability index of the i-th design candidate is calculated by Eq. (3.13). In this case study, the weighting factor of each adaptability evaluation measure was defined equally as 33.3%, because extendibility of functions, upgradeability of modules, and customizability of components were considered the same important. The adaptabilities of all the eight product candidates are summarized in Table 3.4.

3.2.2 *Specific Design Adaptability*

Design adaptation to change one existing design to a new design based on specific requirements is shown in Fig. 3.5.

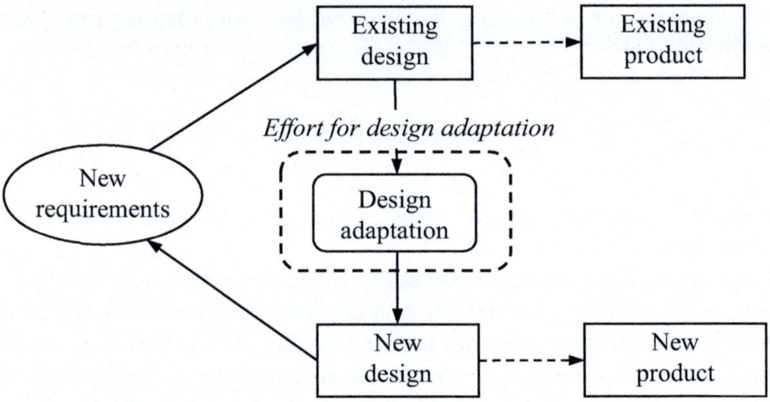

Fig. 3.5 Design adaptation considering specific requirements

Specific design adaptability *AD(D)* is defined by:

$$AD(D) = \sum_{i=1}^{n} [\Pr(Tp_i)AF(Tp_i)] \tag{3.14}$$

where Tp_i ($i = 1, 2, \ldots, n$) is a task of design adaptation, $\Pr(Tp_i)$ is the probability of this task, and $AF(Tp_i)$ is the adaptable factor representing the relative saving of effort when less effort is required to adapt an existing design than to develop a completely new one. $AF(Tp_i)$ is defined by:

$$\begin{aligned} AF(Tp_i) &= \frac{\mathrm{Inf}_{(\mathrm{ZERO}\to IS2)} - \mathrm{Inf}_{(S1\to AS2)}}{\mathrm{Inf}_{(\mathrm{ZERO}\to IS2)}} \\ &= 1 - \frac{\mathrm{Inf}_{(S1\to AS2)}}{\mathrm{Inf}_{(\mathrm{ZERO}\to IS2)}}, \quad 0 \le AF(Tp_i) \le 1 \end{aligned} \tag{3.15}$$

where *S1* is the current state of the existing design, *AS2* is the state after design adaptation, *ZERO* is the state to create a new design from scratch, and *IS2* is the state of the new design considering only the new requirements.

Comparing with product adaptation, design adaptation has been widely used in the past centuries. The evolution of phones, from rotary dial phones, to push button phones, and to the recent cell phones, is considered a typical design adaptation process. Various tools such as CAD systems have been developed for ease of design adaptation efforts.

3.2.3 General Product Adaptability

Fletcher et al. [2] developed a method to quantify general product adaptability. In this work, a product is modeled by a hierarchical data structure. When only parent nodes and their sub-nodes are associated by relations, the architecture is called *segregated product architecture*, as shown in Fig. 3.6b. When other nodes are associated by relations, the architecture is called *full product architecture*, as shown in Fig. 3.6a.

The segregated product architecture is ideal for adaptable design, since modification to a node in this architecture does not impose influence to other nodes at the same level. When an actual product is modeled by the full product architecture, its general product adaptability is measured by comparing this actual full product architecture with its ideal segregated product architecture.

Relations between two nodes are classified as interface and interaction relations. The interface describes the connection between two functional elements, while the interaction models how the two functional modules interact across the interface. Both functional relations and physical relations are considered. Therefore, the relations between two nodes are defined by four parameters, as shown in Table 3.5.

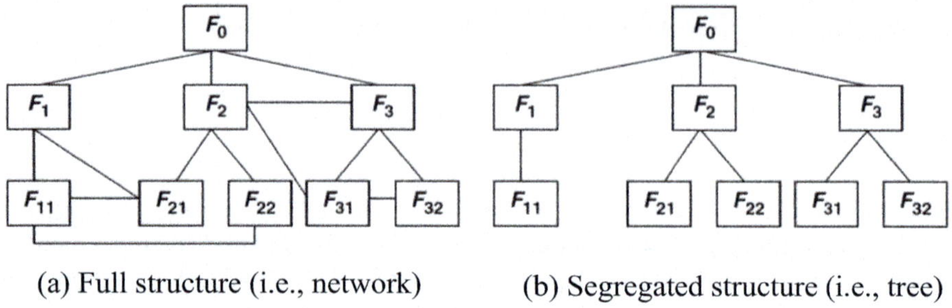

Fig. 3.6 Full structure and segregated structure [2]

Table 3.5 Interface and interaction parameters considering physical and functional relations between *i*-th node and *j*-th node [2]

Relation type	Physical relation	Functional relation
Interface parameter	A_{ij}	C_{ij}
Interaction parameter	B_{ij}	D_{ij}

A parameter is described by a value between 0 and 1, with 0 representing that a change of the *i*-th node has the minimum impact on the *j*-th node, and 1 representing that the change of the *i*-th node has the maximum impact on the *j*-th node. A parameter for modeling the relation between a parent node and its sub-node is assigned with 1. During the evaluation of impact between *i*-th and *j*-th nodes, both the relations and importance of these nodes are considered. The importance of the *i*-th node is defined by the percentage of design and manufacture effort for creating this node in the whole product. Cost is employed to measure the effort.

The impact between a parent node, *i*, and its sub-node, *j*, is modeled by:

$$R_{ij}(\min\{F_i, F_j\}) = \min\{F_i, F_j\}, \; R_{ij} \in \{A_{ij}, B_{ij}, C_{ij}, D_{ij}\} \tag{3.16}$$

where R_{ij} is an impact parameter with the value of 1, and F_i and F_j are the importance measures of the two nodes. The impact between two nodes that do not have a parent–child relation is modeled by:

$$R_{ij}(F_i + F_j), \; R_{ij} \in \{A_{ij}, B_{ij}, C_{ij}, D_{ij}\} \tag{3.17}$$

where R_{ij} is an impact parameter with a value between 0 and 1.

For a segregated product architecture, the total impact considering physical interface is achieved by:

$$k_A^{(S)} = \sum_{\substack{\text{segregated}\\\text{architecture}\\\text{connections}}} A_{ij}\left(\min\{F_i, F_j\}\right) = \sum_{\substack{\text{segregated}\\\text{architecture}\\\text{connections}}} \min\{F_i, F_j\} \tag{3.18}$$

In a full product architecture, the impact considering the extra relations is described by:

$$k_A^{(\text{Extra})} = \sum_{\substack{\text{extra}\\\text{connections}}} A_{ij}\left(F_i + F_j\right) \tag{3.19}$$

The relative adaptability by comparing the full product architecture and its segregated architecture is achieved by:

$$k_A = \frac{k_A^{(S)}}{k_A^{(S)} + k_A^{(Extra)}} \tag{3.20}$$

The total relative adaptability considering the four types of relations is calculated by:

$$k = \frac{k_A + k_B + k_C + k_D}{4} \tag{3.21}$$

Fletcher et al. [2] evaluated general product adaptabilities of a computer and a calculator by comparing the actual full product architectures and their ideal segregated product architectures. Figure 3.7 shows the full product architecture and the segregated product architecture of the computer. The cost breakdown by components is also given in this figure.

The relations of two nodes in product architecture were modeled by four parameters considering physical/functional interface/interaction. A parameter for modeling a relation between a parent node and a child node was assigned the value of 1. Other relations were assigned with values between 0 and 1, as shown in Table 3.6. The four relative adaptabilities, k_A, k_B, k_C, and k_D, were calculated as 0.67, 0.72, 0.68, and 0.65, respectively. The total relative general product adaptability was calculated as 0.68. In the same way, the full product architecture and segregated product architecture for the calculator were also developed, and relations among the nodes in the product architectures were modeled. The overall relative general product adaptability of the calculator was calculated as 0.37. Therefore, the computer is better than the calculator when considering general product adaptability based on this quantitative evaluation method.

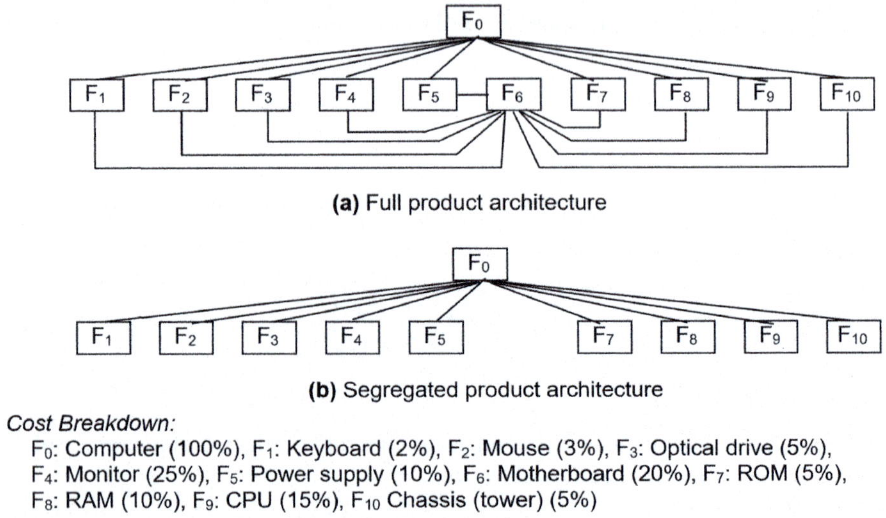

Fig. 3.7 Product architectures for the computer [2]

Table 3.6 Relation parameters for the computer [2]

Relation	Physical		Functional	
	Interface A_{ij}	Interaction B_{ij}	Interface C_{ij}	Interaction D_{ij}
F_1–F_6	0.10	0.10	0.10	0.10
F_2–F_6	0.10	0.10	0.10	0.10
F_3–F_6	0.20	0.10	0.20	0.20
F_4–F_6	0.20	0.10	0.10	0.10
F_5–F_6	0.20	0.20	0.10	0.10
F_7–F_6	0.10	0.10	0.20	0.20
F_8–F_6	0.10	0.10	0.10	0.30
F_9–F_6	0.20	0.10	0.30	0.30
F_{10}–F_6	0.10	0.20	0.10	0.10

3.2.4 General Design Adaptability

When the specific requirements for design adaptation are not provided, the capability that the existing design can be adapted to new design(s) is called general design adaptability. General design adaptability is achieved through information systems technologies (e.g., object-oriented programming, ontology modeling, design repository modeling, and CAD library modeling) such that various modules of the existing design can be adapted to other modules easily in the design process.

General design adaptability is calculated in a similar way as the general product adaptability introduced in Sect. 3.2.3. Instead of the interface and interaction parameters considering physical and functional relations between nodes as shown in Table 3.5, only one type of the dependency relations between nodes considering the influences of nodes in design changes are defined.

The impact between a parent node, *i*, and its sub-node, *j*, is modeled by:

$$R_{ij}(\min\{F_i, F_j\}) = \min\{F_i, F_j\} \tag{3.22}$$

where R_{ij} is an impact parameter with the value of 1, and F_i and F_j are the importance measures of the two nodes. The impact between two nodes that do not have a parent–child relation is modeled by:

$$R_{ij}(F_i + F_j) \tag{3.23}$$

where R_{ij} is an impact parameter with a value between 0 and 1.

For a segregated product architecture, the total impact is achieved by:

$$k^{(S)} = \sum_{\substack{\text{segregated} \\ \text{architecture} \\ \text{connections}}} R_{ij}(\min\{F_i, F_j\}) = \sum_{\substack{\text{segregated} \\ \text{architecture} \\ \text{connections}}} \min\{F_i, F_j\} \tag{3.24}$$

In a full product architecture, the impact considering the extra relations is described by:

$$k^{(\text{Extra})} = \sum_{\substack{\text{extra} \\ \text{connections}}} R_{ij}(F_i + F_j) \tag{3.25}$$

The relative general design adaptability by comparing the full product architecture and its segregated architecture is achieved by:

$$k = \frac{k^{(S)}}{k^{(S)} + k^{(\text{Extra})}} \tag{3.26}$$

3.3 Adaptabilities in Design Evaluation

Adaptabilities, together with other evaluation measures in different life-cycle phases such as functional performances in design and production costs in manufacturing, are used to evaluate a design solution. When multiple design candidates are provided, these evaluation measures are used to compare the relevant design solution candidates for selection of the best design.

The different evaluation measures, including adaptabilities, are defined by different units. One way to integrate these evaluation measures for evaluation of a design solution is to convert these evaluation measures into comparable and unit-less evaluation indices [11]. When the j-th evaluation measure of the i-th design candidate is defined by E_{ij}, its evaluation index I_{ij} is calculated by the nonlinear relation I_j between the evaluation measure and evaluation index.

$$I_{ij} = I_j\left(E_{ij}\right), i = 1, 2, \ldots, n;\ j = 1, 2, \ldots m \tag{3.27}$$

The overall evaluation index, I, is defined by:

$$I_i = \frac{\sum_{j=1}^{m}\left(W_j I_{ij}\right)}{\sum_{j=1}^{m} W_j}, i = 1, 2, \ldots, n. \tag{3.28}$$

where W_j is the weighting factor between 0 and 1 representing the importance of the j-th evaluation aspect.

Li et al. [6] employed the gray relational analysis method [8] for ranking of design candidates considering adaptability and other evaluation measures.

The gray relational analysis is a method in the gray theory to compare evaluation measures quantitatively. This method is used to establish the relations among different evaluation measures by comparing the evaluation measures of a particular design candidate with the best evaluation measures considering all the design candidates. The gray relational analysis approach needs less data and requirements on data distributions to overcome the disadvantages of traditional statistical analysis methods.

Evaluation of adaptable designs using the gray relational analysis approach is conducted through the following six steps.

(1) *Establish comparative series and standard series*

The comparative series is composed of the n evaluation measures for the m adaptable design candidates:

$$A_i = \left(x_{i1}, x_{i2}, \ldots, x_{ij}, \ldots, x_{in}\right), i = 1, 2, \ldots, m \tag{3.29}$$

These n evaluation measures for the m adaptable designs form an m by n decision matrix, **D**.

The standard series is the target series modeled by:

$$A_0 = \left(x_{01}, x_{02}, \ldots, x_{0j}, \ldots, x_{0n}\right) \tag{3.30}$$

The standard series is composed of the best values for each of all the evaluation measures.

(2) *Generate the normalized decision matrix* **K** *(dimensionless)*

To compare the different evaluation measures, these measures should be first converted into the dimensionless evaluation measures considering the following three cases.

(a) If the evaluation measure is of the larger-the-better type (e.g., the adaptability), the normalized measure is calculated by:

$$x_{ij}^* = \frac{x_{ij} - \min\limits_i x_{ij}}{\max\limits_i x_{ij} - \min\limits_i x_{ij}} \tag{3.31}$$

(b) If the evaluation measure is of the smaller-the-better type (e.g., the cost), the normalized measure is calculated by:

$$x_{ij}^* = \frac{\max\limits_i x_{ij} - x_{ij}}{\max\limits_i x_{ij} - \min\limits_i x_{ij}} \tag{3.32}$$

(c) If the evaluation measure is of the nominal-the-best type, and the target value is selected as x_{obj}, the normalized measure is calculated by:

$$x_{ij}^* = \frac{\left|x_{ij} - x_{obj}\right|}{\max\limits_i x_{ij} - x_{obj}} \tag{3.33}$$

The standard series should be converted into $A_0^* = \left(x_{01}^*, x_{02}^*, \ldots, x_{0j}^*, \ldots, x_{0n}^*\right)$ in the same manner.

(3) *Obtain differences between the comparative series and standard series*

To discover the degree of the gray relationship, differences, Δ_0, between the normalized decision matrix **K** and normalized standard series $A_0{}^*$ are achieved by:

$$\Delta_{0ij} = \left|x_{0j}^* - x_{ij}^*\right| \tag{3.34}$$

(4) *Calculate the gray relational coefficients*

The gray relational coefficients, γ_0, indicate the contiguous grades between the comparative series and standard series. A relational coefficient with higher value represents a closer relationship with the best evaluation measure considering all design candidates. The gray relational coefficient is calculated by:

$$\gamma_{0ij} = \frac{\Delta \min + \varsigma \Delta \max}{\Delta_{0ij} + \varsigma \Delta \max} \tag{3.35}$$

where Δmax and Δmin are obtained by:

$$\Delta \max = \max_i \max_j \Delta_{0ij} \tag{3.36}$$

$$\Delta \min = \min_i \min_j \Delta_{0ij} \tag{3.37}$$

ζ is called a distinguished coefficient, only affecting the relative value without changing the priority. Generally, ζ is selected as 0.5.

(5) *Determine the degree of relation with the standard series*

To achieve the degree of relation with the standard series (i.e., the ideal design), the weighting factors of the evaluation measures must first be decided. These weighting factors can be determined by expert experience or marketing strategies in the firm. The degree of relation with the standard series for a design candidate is calculated by:

$$\Gamma_{0i} = \sum_{j=1}^{n} [w_j \times \gamma_{0ij}] \tag{3.38}$$

where w_j is the weighting factor for the *j*-th evaluation measure satisfying:

$$\sum_{j=1}^{n} w_j = 1 \tag{3.39}$$

(6) *Prioritize the design candidates*

From degrees of the relation between the comparative series and standard series, design candidates can be prioritized. The design candidate with a larger Γ_{0i} can better satisfy requirements considering different product life-cycle aspects.

This gray relational analysis approach treats each design candidate as a comparative series. The degree of relations between the comparative series and standard series for each design candidate is then calculated. At the end, the best design can be selected based on the rankings of all the design candidates.

For the adaptable food processor developed by Li et al. [6] as shown in Fig. 3.4, the total part cost, total assembly cost, and operationability were also selected in addition to product adaptability. Values of the four evaluation measures for the eight design candidates given in Table 3.2 are shown in Table 3.7.

Table 3.7 Four evaluation measures for eight product candidates

Candidate	Adaptability (0–1)	Total part cost ($)	Assembly cost ($)	Operationability (0–5)
Candidate 1	0.917	56	10	4.5
Candidate 2	0.991	64	10	4.5
Candidate 3	0.924	66	12	4.5
Candidate 4	1.000	74	12	4.5
Candidate 5	0.668	54	13	5
Candidate 6	0.685	71	13	5
Candidate 7	0.677	64	15	5
Candidate 8	0.694	81	15	5
Value type	Quantitative	Quantitative	Quantitative	Qualitative
Criteria type	The larger the better	The smaller the better	The smaller the better	The larger the better

Table 3.8 Weighting factors of the four evaluation measures

Evaluation measure	Weighting factor (%)
Adaptability	45
Part cost	20
Assembly cost	20
Operationability	15

The weighting factors of the four different evaluation measures were determined from expert experience and marketing strategies. These weighting factors were identified as given in Table 3.8. Since this case study focused on adaptable design, a high weighting factor was selected for the adaptability evaluation measure.

Prioritization of the product candidates using the gray relational analysis approach was conducted as follows:

(1) Establish the decision matrix **D**

$$\boldsymbol{D} = \begin{bmatrix} 0.917\ 56\ 10\ 4.5 \\ 0.991\ 64\ 10\ 4.5 \\ 0.924\ 66\ 12\ 4.5 \\ 1.000\ 74\ 12\ 4.5 \\ 0.668\ 54\ 13\ 5.0 \\ 0.685\ 71\ 13\ 5.0 \\ 0.677\ 64\ 15\ 5.0 \\ 0.694\ 81\ 15\ 5.0 \end{bmatrix}$$

(2) Obtain the standard series

$$A_0 = \begin{bmatrix} 1 & 54 & 10 & 5 \end{bmatrix}$$

(3) Generate the normalized decision matrix **K** and the standard series $A_0{}^*$

$$\boldsymbol{K} = \begin{bmatrix} 0.750 & 0.926 & 1.000 & 0.000 \\ 0.973 & 0.630 & 1.000 & 0.000 \\ 0.771 & 0.556 & 0.600 & 0.000 \\ 1.000 & 0.259 & 0.600 & 0.000 \\ 0.000 & 1.000 & 0.400 & 1.000 \\ 0.051 & 0.370 & 0.400 & 1.000 \\ 0.027 & 0.630 & 0.000 & 1.000 \\ 0.078 & 0.000 & 0.000 & 1.000 \end{bmatrix}$$

$$A_0^* = \begin{bmatrix} 1 & 1 & 1 & 1 \end{bmatrix}$$

(4) Calculate the differences between the comparative series and the standard series

$$\Delta_0 = \begin{bmatrix} 0.250 & 0.074 & 0.000 & 1.000 \\ 0.027 & 0.370 & 0.000 & 1.000 \\ 0.229 & 0.444 & 0.400 & 1.000 \\ 0.000 & 0.741 & 0.400 & 1.000 \\ 1.000 & 0.000 & 0.600 & 0.000 \\ 0.949 & 0.630 & 0.600 & 0.000 \\ 0.973 & 0.370 & 1.000 & 0.000 \\ 0.922 & 1.000 & 1.000 & 0.000 \end{bmatrix}$$

$$\Delta \max = \begin{bmatrix} 1 & 1 & 1 & 1 \end{bmatrix}$$
$$\Delta \min = \begin{bmatrix} 0 & 0 & 0 & 0 \end{bmatrix}$$

(5) Determine the gray relational coefficients

$$\gamma_0 = \begin{bmatrix} 0.667 & 0.871 & 1.000 & 0.333 \\ 0949 & 0.574 & 1.000 & 0.333 \\ 0.686 & 0.529 & 0.556 & 0.333 \\ 1.000 & 0.403 & 0.556 & 0.333 \\ 0.333 & 1.000 & 0.455 & 1.000 \\ 0.345 & 0.443 & 0.455 & 1.000 \\ 0.339 & 0.574 & 0.333 & 1.000 \\ 0.352 & 0.333 & 0.333 & 1.000 \end{bmatrix}$$

(6) Determine the degrees of relations with the standard series

Using the weighting factors and Eq. (3.38), degrees of relations for all the eight product candidates were obtained as follows:

$$\Gamma_{01} = 0.724$$
$$\Gamma_{02} = 0.792$$
$$\Gamma_{03} = 0.576$$
$$\Gamma_{04} = 0.692$$
$$\Gamma_{05} = 0.591$$
$$\Gamma_{06} = 0.485$$
$$\Gamma_{07} = 0.484$$
$$\Gamma_{08} = 0.442$$

(7) Prioritize the eight product candidates.

From the ratings of the product candidates, candidate 2 was selected as the best one, considering all the relevant life-cycle evaluation measures. Comparing with candidate 1, candidate 2 can provide one more often-used functions (i.e., blender) with minor additional cost. Comparing with candidate 2, candidate 4 provides only 1 more function (i.e., meat grinder), which is seldom used, with the considerable additional cost.

As the results, design candidate 2 from the first design configuration was chosen after the evaluation and prioritization. The selected best candidate was designed in details.

3.4 Evaluation of Adaptabilities Based on Information Entropy

3.4.1 Functions, Structures, and Parameters of Adaptable Products

Design of a product can be considered as a mapping process from *design functions* to *design structures* as shown Fig. 3.8 [9, 10]. The mapping is usually conducted in an iterative manner. A function can be decomposed into sub-functions, and these sub-functions are mapped to sub-structures that are used to build the structure at a higher level. In addition, a structure module, such as a component or a sub-assembly, can be defined by *design parameters*.

In adaptable design, functions, structures, and parameters of the product are changed based on requirements of adaptation tasks. The changes in functions, structures, and parameters to satisfy adaptation tasks form variants in functions, structures, and parameters.

The information entropies considering changes in functions, structures, and parameters of the products are used to evaluate adaptabilities of the products.

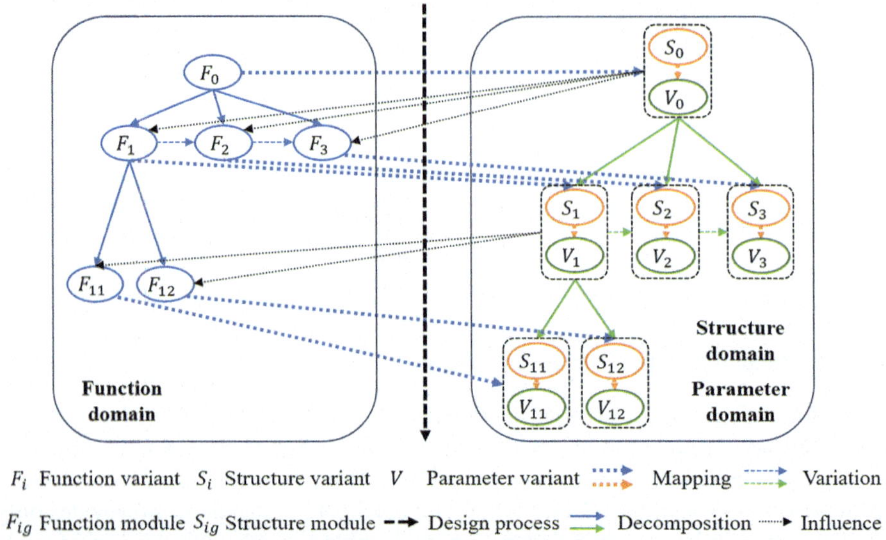

Fig. 3.8 Functions, structures, and parameters of adaptable products

3.4.2 Information Entropies for Evaluation of Adaptabilities

In information theory, for a random experiment X with N possible outcomes x_t ($t = 1, 2, \ldots, N$), its entropy $H(X)$ (i.e., the expected information content) is calculated as the sum of the information content for each possible outcome multiplied by its corresponding probability [7] as shown in Eq. (3.40).

$$H(X) = -K \sum_{t=1}^{N} (p_t \log_a p_t) \tag{3.40}$$

where p_t is the probability of the possible outcome x_t, $\sum_{t=1}^{N} p_t = 1$, and $0 \leq p_t \leq 1$. $-\log_a p_t$ is the information content for the possible result x_t. K is a positive constant, and its value is determined according to the base of the logarithm operator. When the base of the logarithm operator is selected as 2, $K = 1$.

From Eq. (3.40), for the random experiment X with large uncertainty (i.e., large standard deviation or variance of a probability distribution considering the N states), its entropy is also large. In a design process, large information entropy is estimated during the early conceptual design stage due to the large design space with large uncertainty, while small information entropy is expected in the late detail design stage due to the small design space with small uncertainty.

(1) Information Entropy Considering Product Function Changes

Suppose n_f product function variants are considered for adaptable design. The effort to extend/change the current i-th function to a new j-th function is modeled by a measure r_{fij} $(0 \leq i, j \leq n_f)$ called *degree of function change*. r_{fij} is define by a value between 0 and 1, corresponding to the most and least efforts to add the new function through adaptable design. Various degrees of function changes are illustrated in Table 3.9.

From all degrees of function changes considering any two functions in the n_f product function variants, a *function change degree matrix* R_F can be achieved as shown in Eq. (3.41).

$$R_F = \begin{bmatrix} r_{f11} & r_{f12} & \cdots & r_{f1n_f} \\ r_{f21} & r_{f22} & \cdots & r_{f2n_f} \\ \vdots & \vdots & \ddots & \vdots \\ r_{fn_f1} & r_{fn_f2} & \cdots & r_{fn_fn_f} \end{bmatrix} \quad (3.41)$$

For the i-th function, the probability to extend it to the j-th function, p_{fij}, is defined by Eq. (3.42) considering how easy it is to extend to the *j*-th function by comparing it with other functions to extend to the j-th function.

$$p_{fij} = \frac{r_{fij}}{\sum_{i=1}^{n_f} r_{fij}}, \mathrm{i}, \mathrm{j} = 1, 2, \cdots, n_f \quad (3.42)$$

$K = 1/\ln n_f$ for Eq. (3.40) when natural logarithm operator (i.e., base $e = 2.71828$) is selected. Entropy for the *i*-th function considering its extension/change to all other functions is calculated by:

$$H(R_{Fi}) = -\frac{1}{\ln n_f} \sum_{j=1}^{n_f} (p_{fij} \ln p_{fij}),\ i = 1, 2, \ldots, n_f \quad (3.43)$$

The information entropy considering all function variants is defined by:

$$H(R_F) = \sum_{i=1}^{n_f} H(R_{Fi}) \quad (3.44)$$

A product design solution with small information entropy considering function changes indicates its high product adaptability. When multiple design solutions

Table 3.9 Degrees of function changes in adaptable design

Degree of function change	Definition
1	Function can be extended directly without additional effort
0.7	Function can be extended with low effort
0.3	Function can be extended with high effort
0	Function cannot be extended

are considered, the one with the minimum information entropy should be selected considering function changes.

(2) Information Entropy Considering Product Structure Changes

Suppose n_s product structure variants are considered for adaptable design. The effort to change the current i-th structure to a new j-th structure is modeled by a measure r_{sij} $(0 \le i, j \le n_s)$ called *degree of structure change*. r_{sij} is define by a value between 0 and 1, corresponding to the most and least efforts to change to the new structure through adaptable design. Various degrees of structure changes are defined in Table 3.10.

From all degrees of structure changes considering any two structures in the n_s product structure variants, a *structure change degree matrix* R_S can be achieved as shown in Eq. (3.45).

$$R_S = \begin{bmatrix} r_{s11} & r_{s12} & \cdots & r_{s1n_s} \\ r_{s21} & r_{s22} & \cdots & r_{s2n_s} \\ \vdots & \vdots & \ddots & \vdots \\ r_{sn_s1} & r_{fn_s2} & \cdots & r_{fn_sn_s} \end{bmatrix} \tag{3.45}$$

For the *i*-th structure, the probability to change it to the *j*-th structure, p_{sij}, is defined by Eq. (3.46) considering how easy it is to change to the *j*-th structure by comparing it with other structures to change to the j-th function.

$$p_{sij} = \frac{r_{sij}}{\sum_{i=1}^{n_s} r_{sij}}, \text{i, j} = 1, 2, \cdots, n_s \tag{3.46}$$

$K = 1/\ln n_s$ for Eq. (3.40) when natural logarithm operator (i.e., base $e = 2.71828$) is selected. Entropy for the i-th structure considering its change to all other structures is calculated by:

$$H(R_{Si}) = -\frac{1}{ln\ n_s} \sum_{j=1}^{n_s} (p_{sij}\ ln\ p_{sij}),\ i = 1, 2, \cdots, n_s \tag{3.47}$$

The information entropy considering all structure variants is defined by:

Table 3.10 Degrees of structure changes in adaptable design

Degree of structure change	Definition
1	Structure can be achieved directly without additional effort
0.7	Structure can be achieved with low effort
0.3	Structure can be achieved with high effort
0	Structure cannot be achieved

$$H(R_S) = \sum_{i=1}^{n_s} H(R_{Si}) \tag{3.48}$$

A product design solution with small information entropy considering structure changes indicates its high product adaptability. When multiple design solutions are considered, the one with the minimum information entropy should be selected considering structure changes.

(3) Information Entropy Considering Product Parameter Changes

Suppose n_p adaptation tasks are considered for adaptable design, and m_p design parameters are selected to model the structures for these adaptation tasks. Product parameter value changes considering the n_p adaptation tasks and the m_p design parameters are defined by a parameter value change matrix V as shown in Eq. (3.49)

$$V = \begin{bmatrix} v_{11} & v_{12} & \cdots & v_{1m_p} \\ v_{21} & v_{22} & \cdots & v_{2m_p} \\ \vdots & \vdots & \ddots & \vdots \\ v_{n_p1} & v_{n_p2} & \cdots & v_{n_pm_p} \end{bmatrix} \tag{3.49}$$

where $v_{ij}\left(i = 1, 2, \cdots, n_p, j = 1, 2, \cdots, m_p\right)$ represents the specific value of the j-th design parameter under the i-th adaptation task.

Since these different parameters are defined at different scales with different units, normalization is first conducted to change the parameter values in Eq. (3.49) into normalized values between 0 and 1 [5].

Since parameters can be classified into the-large-the-better parameters and the smaller-the-better parameters, Eqs. (3.50) and (3.51) are used for parameter value normalization, such that all the parameter values are normalized between 0 and 1, corresponding to the worst parameter value and the best parameter value, respectively.

- For the-large-the-better parameters:

$$u_{ij} = \frac{v_{ij} - \min\left\{v_{1j}, v_{2j}, \ldots, v_{n_pj}\right\}}{\max\left\{v_{1j}, v_{2j}, \ldots, v_{n_pj}\right\} - \min\left\{v_{1j}, v_{2j}, \ldots, v_{n_pj}\right\}},$$
$$i = 1, 2, \ldots, n_p,\ j = 1, 2, \ldots, m_p \tag{3.50}$$

- For the-smaller-the-better parameters:

$$u_{ij} = \frac{\max\left\{v_{1j}, v_{2j}, \ldots, v_{n_pj}\right\} - v_{ij}}{\max\left\{v_{1j}, v_{2j}, \ldots, v_{n_pj}\right\} - \min\left\{v_{1j}, v_{2j}, \ldots, v_{n_pj}\right\}},$$
$$i = 1, 2, \ldots, n_p,\ j = 1, 2, \ldots, m_p \tag{3.51}$$

The normalized parameter values form a *parameter value change degree matrix* U:

$$U_P = \begin{bmatrix} u_{p11} & u_{p12} & \cdots & u_{p1m_p} \\ u_{p21} & u_{p22} & \cdots & u_{p2m_p} \\ \vdots & \vdots & \ddots & \vdots \\ u_{pn_p1} & u_{pn_p2} & \cdots & u_{pn_pm_p} \end{bmatrix} \tag{3.52}$$

For the j-th parameter, the probability to select the u_{pij} for the i-th adaptation task, p_{pij}, is defined by Eq. (3.53) considering how good this parameter value is considering the-large-the-better and the-smaller-the-better parameter properties by comparing it with other parameters.

$$p_{pij} = \frac{u_{pij}}{\sum_{j=1}^{m_p} u_{pij}}, i = 1, 2, \cdots, n_p, j = 1, 2, \ldots, m_p \tag{3.53}$$

$K = 1/\ln n_p$ for Eq. (3.40) when natural logarithm operator (i.e., base $e = 2.71828$) is selected. Information entropy for the j-th product parameter considering its parameter value changes is calculated by:

$$H(U_j) = -\frac{1}{\ln n_p} \sum_{i=1}^{n_p} (p_{pij} \ln p_{pij}), j = 1, 2, \ldots, \tag{3.54}$$

Due to differences in the importance of design parameters, their weights also need to be considered when calculating the information entropy considering all parameter value changes. The weight of the j-th design parameter is calculated by:

$$\lambda(U_j) = \frac{1 - H(U_j)}{\sum_{j=1}^{m_p}(1 - H(U_j))}, \mathrm{j} = 1, 2, \ldots, m_p \tag{3.55}$$

The information entropy considering all parameter value changes is achieved by:

$$H(U) = \sum_{j=1}^{m_p} (\lambda(U_j) H(U_j)) \tag{3.56}$$

A product design solution with small information entropy considering parameter value changes indicates its high product adaptability. When multiple design solutions are considered, the one with the minimum information entropy should be selected considering parameter value changes.

3.5 Summary

Development of effective evaluation methods is critical to engineering design. In order to identify the design candidate with the best adaptability, the quantitative measure for adaptable design evaluation is needed.

Presently, product adaptability is usually used to evaluate an adaptable design. The product adaptability is defined as the capability of a physical product to be adapted to satisfy the changed requirements. In adaptable design for product adaptability, a new product is achieved by modifying the existing product, such as adding new components/modules, replacing or upgrading the existing components/modules with new ones, and reconfiguring the existing components/modules.

Since the adaptation task and its probability may or may not be predictable, the adaptability of product can be classified into two categories: specific adaptability and general adaptability. When the adaptation task and its probability can be predicted, the adaptability is considered as specific adaptability. On the other hand, when the adaptation task and/or its probability cannot be predicted, the adaptability is considered as general adaptability. In this chapter, both specific adaptability based on the cost saving and general adaptability by comparing the actual structure of the product with its ideal structure are introduced for measuring product adaptability.

To provide guidelines and tools of adaptable design to improve adaptability, processes, and methods for adaptable design will be introduced in the next chapter.

References

1. G. Boothroyd, Product design for manufacture and assembly. Comput. Aided Des. **26**(7), 505–520 (1994)
2. D. Fletcher, R. Brennan, P. Gu, A method for quantifying adaptability in engineering design. Concurrent Eng. Res. Appl. **17**(4), 279–289 (2009)
3. P. Gu, M. Hashemian, A.Y.C. Nee, Adaptable design. Ann. CIRP **53**(2), 539–557 (2004)
4. P. Gu, M. Slevinsky, Mechanical bus for modular product design. Ann. CIRP **52**(1), 113–116 (2003)
5. R.A. Johnson, D.W. Wichern, *Applied Multivariate Statistical Analysis* (Prentice Hall, Upper Saddle River, NJ, USA, 2002)
6. Y. Li, D. Xue, P. Gu, Design for product adaptability. Concurrent Eng. Res. Appl. **16**(3), 221–232 (2008)
7. C.E. Shannon, A mathematical theory of communication. Bell Syst. Tech. J. **27**(3), 379–423 (1948)
8. K.C. Sih, *The Theory of Grey Information Relation* (Chun-Hwa Publishing Co, Taipei, 1997)
9. N.P. Suh, *The Principles of Design* (Oxford University Press, Oxford, UK, 1990)
10. N.P. Suh, *Axiomatic Design: Advances and Applications* (Oxford University Press, Oxford, UK, 2001)
11. H. Yang, D. Xue, Y.L. Tu, Modeling of the non-linear relations among different design and manufacturing evaluation measures for multi-objective optimal concurrent design. Concurrent Eng. Res. Appl. **14**(1), 43–53 (2006)

Chapter 4
Processes and Methods for Adaptable Design

4.1 Processes for Adaptable Design

4.1.1 Processes for Adaptable Design Considering Product Adaptability

Since the introduction of adaptable design, research has been conducted to formulate the guidelines and processes for adaptable design to improve adaptability. In this section, guidelines and processes related to product adaptability are discussed. Product adaptation is conducted by users for extension of the product utility.

Hashemian [9] introduced guidelines for adaptable design considering specific product adaptability and general product adaptability as follows:

Guidelines considering the specific product adaptability:

- Defining the primary functional requirements (FRs) that have to be satisfied and the additional FRs (AFRs) that have potential to be required for the product.
- Providing extra features and functionalities for possible future needs.
- Utilizing existing features and components to achieve extra functionalities.
- Identifying a group of products that can be developed from a shared adaptable design. Identifying common or recurring elements, either functional or structural, among products within the portfolio. These common elements are designed as a shared platform.
- Identifying differentiating features among products within a portfolio or family of products and designing them as add-on modules.
- Designing interfaces between platforms and modules for easy attachment and detachment operations.
- Facilitating the replacement of components that are likely to require upgrading.
- Identifying customizable features and designing a product for the easy alteration, replacement, or addition of these features.

P. Gu et al., *Adaptable Design*, Research on Intelligent Manufacturing,
https://doi.org/10.1007/978-981-99-5869-6_4

Guidelines considering the general product adaptability:

- An adaptable product should have a segregated architecture, so that the required modifications of an adaptation task can be localized without propagating changes throughout the product.
- Subsystems are functional modules that are designed to perform unambiguous and useful functions.
- Subsystems are autonomous and self-contained, so that they can perform their functions independently from their working environment.

Gu [8] developed a design process for product adaptability. Steps of the method are as follows (iterations can occur at any step):

- Definition of adaptable design objectives: determining primary FRs and AFRs. Also, determining the life-cycle objectives of importance for the design problem.
- Design of the product/system (conceptual and configuration): developing the product architecture (configuration) based on both the primary FRs and AFRs.
- Design of the adaptable product: developing initial configurations of the overall adaptable product that performs the FRs and AFRs. This is the most creative part of adaptable design.
- Life-cycle considerations: clustering components for life-cycle objectives.
- Evaluation: evaluating designs and adaptability.

Li et al. [17, 18] further detailed the design process of product adaptability in the following steps:

- Product planning. This step includes defining the mission statement for the adaptable design, identifying customer needs, determining primary FRs and optional FRs, and establishing the tree for modeling the FRs and tree of DPs to satisfy the FRs.
- Studying modularity. Components with dependent relations of design functions are grouped into modules.
- Establishing product architecture. Developing the product architecture based on clusters of modules. Interactions among these modules are modeled by flows of material, energy, and information. Creating the rough geometric layout.
- Adaptable interface design. Developing interfaces and connectors, so that functions can be transferred between the platform and modules and between modules.
- Conceptual design evaluation. Evaluating each design candidate with four main evaluation measures, including specific product adaptability of design, part cost, assembly cost, and operationability by customers. Prioritizing the design candidates by using the gray relational analysis approach [29].
- Detailed design. Once the best design candidate is identified, details of this design, including parts and assemblies, are then modeled using a CAD system.

4.1.2 Processes for Adaptable Design Considering Design Adaptability

In this category, the adaptable design is used in the design and production of a family of products or reused to produce new designs. For product adaptability, guidelines and processes considering design adaptability are also developed.

Gu [8] developed a design process considering design adaptability. The process steps are as follows:

1. Definition of adaptable design objectives: determining the product portfolio for products that can be developed from one adaptable design, and the life-cycle objectives.
2. Design of the product/system (conceptual and configuration): determining FRs of the target products and developing initial design for each of them based on FRs and AFRs.
3. Detection of commonalities: for a family of products, identifying the commonalities among these designs. These can be functional or structural similarities. For one-of-a-kind products, identifying reusable aspects of the design that can be shared for the new product development.
4. Common design: developing common designs that can be shared among products.
5. Life-cycle considerations: clustering components for life-cycle objectives.
6. Evaluation: design and adaptability evaluations.

4.2 Elements in Adaptable Design Processes

Among various activities in the adaptable design processes, the four major elements, including design requirements, design candidates, design evaluations, and design optimizations, are discussed in detail in this section.

4.2.1 Design Requirements

Comparing with design requirements in the traditional design processes, design requirements in adaptable design can be changed in different operation stages.

In adaptable design, requirements in different time periods of the product life cycle can be different. Table 4.1 shows the requirements for design of an adaptable device to test horizontal wind power generators [33]. In this design, the testing equipment is adapted in different phases to satisfy different requirements of the power generation, vibration simulation, and temperature control.

The life-cycle time of a product in its utilization span from birth to retirement of the product is described by time parameter T whose value is changed from T_0 to T_{max}

Table 4.1 Requirements for an adaptable power generator testing device [33]

Phase	Requirements
Phase I (years 1–3)	Power 600–1000 KW, no vibration caused by forces, no wear caused by high temperature
Phase II (years 4–6)	Power 1200–1800 KW, vibration caused by mild forces, no wear caused by high temperature
Phase III (years 7–9)	Power 2000–5000 KW, vibration caused by severe forces, wear caused by high temperature

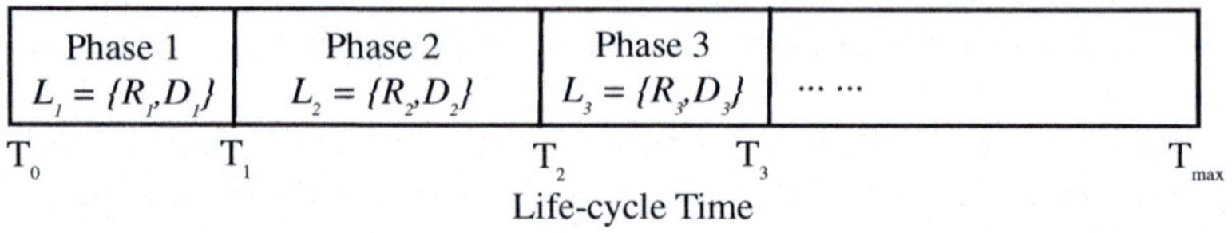

Fig. 4.1 Product operation phases [33]

representing the time at product purchase and the time at product disposal/recycle, respectively. The life-cycle time parameter T can be assigned with continuous values, discrete values, or integer values.

For an adaptable product, its whole utilization span is usually divided into operation phases as shown in Fig. 4.1. Each of these n operation phases, L_i ($i = 1, 2, \ldots n$), is modeled by its design requirements R_i and design solution D_i.

$$L_i = \{R_i, D_i\}. \tag{4.1}$$

The life-cycle time T can be assigned with different types of values.

- *Continuous T*: For example, the time that a battery used in a vehicle can be described by $T \in [0, 5]$ (year).
- *Discrete T*: For example, the time that a battery of a vehicle needs to be checked can be described by $T \in \{0, 2, 2.5, 3, 3.2, \ldots\}$ (year).
- *Integer T*: For example, the time that a vehicle needs major maintenance can be described by $T \in \{1, 2, 3, 4, \ldots\}$ (year).

Design requirements are modeled by qualitative descriptions and quantitative descriptions.

- Qualitative Design Requirements

The qualitative design requirements are described by Boolean descriptions. For example, the requirement for a Blu-ray drive in a personal computer system is described by

$$R_1(T) = \text{"Blu-ray Drive"}, 2 \le T \text{ (year)} \le 4.$$

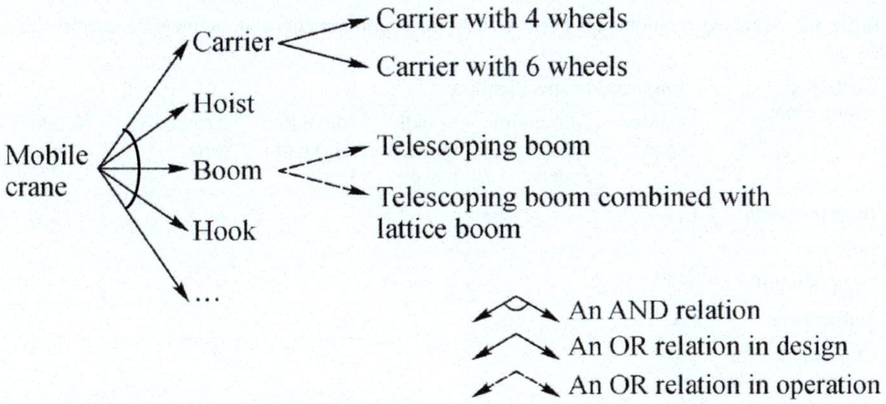

Fig. 4.2 Modeling of design requirements with AND-OR tree of functions [35]

- Quantitative Design Descriptions

The quantitative design requirements are described by numerical functions and constraints. For example, the voltage requirement for a vehicle battery is defined as follows:

$$11\text{ V} \leq \text{Voltage}(T) \leq 12.5\ V, 0 \leq T\ (\text{year}) \leq 5.$$

Function modeling methods introduced in Sect. 2.1.1 are also used to define qualitative descriptions of design requirements. Figure 4.2 shows an AND-OR tree with functions to model qualitative descriptions of design requirements in the design of an adaptable mobile crane [35].

In this AND-OR tree, both different design configuration candidates and different operation configuration states of each design configuration candidate are considered. In this tree, the requirements for partial design solutions, such as components and modules, are described by nodes. The relations among sub-nodes of a super-node are classified into three categories:

(1) *AND relation.* The AND relations are used for modeling both design configuration candidates and operation configuration states.
(2) *OR relation in design.* The OR relations in design are used for modeling different design configuration candidates.
(3) *OR relation in operation.* The OR relations in operation are used for modeling different operation configuration states.

When all the sub-nodes need to be selected to support a super-node, all of these sub-nodes have an AND relation. In the example shown in Fig. 4.2, the mobile crane is composed of a carrier, a hoist, a boom, a hook, and other components. When the super-node is supported by one of its sub-nodes in the process of design configuration candidate selection, all of these sub-nodes have an OR relation in design. For example,

Table 4.2 Mapping relationships between customer requirements and engineering specifications [5]

Customer requirement	Engineering specification					
	Gantry span	Maximum spindle speed	Spindle motor power	Maximum worktable load	Crossbeam type	Number of axes
High precision				×	×	×
Low cost			×			×
High reliability			×		×	
Stable work	×	×		×		
Convenience for human–computer interaction					×	×
Low noise		×	×	×		

only a carrier with four wheels or a carrier with six wheels needs to be selected as the carrier for a mobile crane. When the super-node is supported by one of its sub-nodes in the operation stage, all of these sub-nodes have an OR relation in operations. For example, only telescoping boom or telescoping boom combined with lattice boom is selected for a specific task in the operation stage.

Design requirements are often achieved through acquiring customer needs and converting customer needs into engineering specifications. Quality function deployment (QFD) [1] is an approach to map customer needs to engineering specifications.

In the design of an adaptable product platform for heavy-duty gantry milling machines developed by Cheng et al. [5], customer requirements were mapped into engineering specifications using relations shown in Table 4.2. Table 4.3 shows the requirements of 19 customers described by engineering specifications. In order to identify modules of the adaptable product platform, a clustering algorithm, called the K-means clustering method [21], was used to classify the requirements into common requirements and personalized requirements. Basic common modules were identified to satisfy common requirements, and add-on modules were identified to satisfy personalized requirements.

4.2.2 *Design Candidates*

The overall design solution for an adaptable product is modeled by a collection of different configurations in different operation phases as shown in Fig. 4.1, and parameters of these configurations.

A design configuration is built using components as primitives. A number of components can be grouped together as an assembly. An assembly is composed of

Table 4.3 Design requirements in engineering specifications [5]

Customer	Gantry span (mm)	Travel of gantry frame (mm)	Travel of slide carriage (mm)	Travel of ram (mm)	Spindle motor power (kw)	Maximum spindle speed (r/min)	Maximum worktable load (T/m^2)	Crossbeam type (0: fixed, 1: movable)	Number of axes (0: 3-axis, 1: 5-axis)
X1	8500	16,500	9000	3000	100	1500	25	0	1
X2	4800	13,500	6000	2000	100	1240	25	1	0
X3	10,500	17,500	11,500	3000	100	1500	25	1	0
X4	9500	16,500	10,000	2500	100	1500	25	1	1
X5	3800	11,500	5000	1500	100	2000	25	0	0
X6	8000	17,500	8000	1500	100	1500	25	0	1
X7	9500	21,500	10,000	2500	100	2000	25	1	1
X8	8000	16,500	8500	2000	100	3000	25	0	1
X9	5800	16,500	7000	1500	100	1500	25	1	0
X10	8500	17,500	10,000	2000	100	2000	25	0	1
X11	6800	17,500	8000	3000	100	1500	25	1	1
X12	10,500	16,500	11,500	2000	100	2000	25	1	1
X13	4800	11,500	6000	1500	100	1240	25	0	0
X14	6800	16,500	8000	2500	100	1240	25	0	1
X15	3800	11,500	5000	1500	100	2000	25	1	1
X16	9500	16,500	10,000	3000	100	1500	25	1	0
X17	5800	17,500	7000	2000	100	1240	25	0	1
X18	4800	17,500	6000	2000	100	1240	25	1	0
X19	7500	16,500	8000	2000	100	1240	25	1	1

components and sub-assemblies. When a group of components/sub-assemblies can be attached, detached, modified, relocated, and replaced easily in adaptable design, this group of components/sub-assemblies is also called a module. Descriptions of design configurations are qualitative in nature.

For example, configurations of the IC engine vehicle and electric vehicle are described by:

$$C(T) = \{\text{IC Engine, Gas Tank}, \ldots\}, 0 \le T(\text{year}) \le 5;$$
$$\{\text{Electric Engine, Rechargeable Batteries}, \ldots\}, 5 < T(\text{year}) \le 10.$$

Parameters are used to model attributes of components and assemblies in a configuration. A parameter can be assigned with a value. Continuous values, discrete values, integer values, and Boolean values can be used for parameters. Typical parameters are dimensions, tolerances, material properties, etc. In addition, qualitative descriptions, such as the color and texture of a component, can also be assigned to parameters.

Life-cycle time T can be used to calculate values of parameters at different life-cycle time periods. For example, the unit cost of gas can be defined by

$$\text{unit\_cost}(T) = 1 \times (1 + 2\%T)(\$/\text{liter}), 0 < T(\text{year}) \le 10.$$

A design parameter can also be defined as a numerical function of other parameters. For example, the total cost of gas for driving can be calculated by

$$\text{total\_cost}(T) = \text{unit\_cost}(T) \times d/m\ (\$),\ 0 < T(\text{year}) \le 10,$$

where d and m are distance (km) and gas mileage (km/liter) parameters, respectively. The unit_cost(T), d, and m are used as input parameters to calculate the output parameter of total_cost(T). At different life-cycle times Ts, the parameter values are different.

Figure 4.3 shows parameters modeled as numerical functions of life-cycle time parameter T.

Various design configuration candidates and operation configuration candidates can be created through meeting design requirements modeled by the hybrid AND-OR tree as shown in Fig. 4.2 through a tree-based search [28]. A design configuration candidate and an operation configuration state generated from the hybrid AND-OR tree given in Fig. 4.2 are shown in Fig. 4.4.

The following rules are used to generate different feasible design configuration candidates from the hybrid AND-OR tree:

- The root node is selected first.
- When a node is selected and all of its sub-nodes are associated with an AND relation, all these sub-nodes should be selected.
- When a node is selected and all of its sub-nodes are associated with an OR relation in design, only one of these sub-nodes should be selected.

Fig. 4.3 Modeling of design parameters as functions of time T [33]

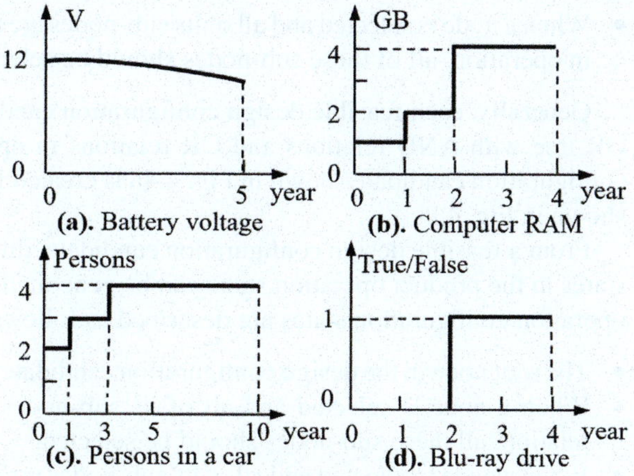

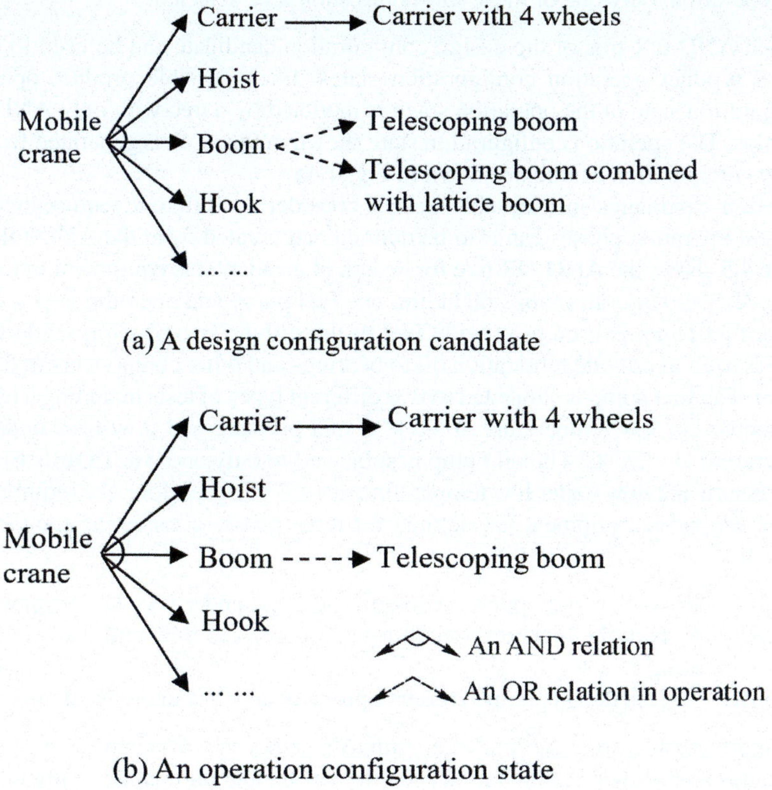

Fig. 4.4 Design configuration candidate and an operation configuration state generated from the hybrid AND-OR tree [35]

- When a node is selected and all of its sub-nodes are associated with an OR relation in operation, all of these sub-nodes should be selected.

Generally, each feasible design configuration candidate is modeled by an AND-OR tree with AND relations and OR relations in operations. The feasible design configuration candidate shown in Fig. 4.4a is created from the hybrid AND-OR tree shown in Fig. 4.2.

From a feasible design configuration candidate, different operation configuration states in the product operation stage can be generated. Rules for generating product operation configuration states are described as follows:

- The root node in the design configuration candidate AND-OR tree is selected first.
- When a node is selected and all of its sub-nodes are associated with an AND relation, all these sub-nodes should be selected.
- When a node is selected and all of its sub-nodes are associated with an OR relation in operation, only one of these sub-nodes should be selected.

Each AND-OR tree of the design configuration candidate can be used to create several product operation configuration states. Each possible product operation configuration state in the operation stage is modeled by a tree with only AND relationships. The operation configuration state shown in Fig. 4.4b is generated from the design configuration candidate shown in Fig. 4.4a.

Design candidates in adaptable design considering different requirements for different operation phases can also be defined and created from the AND-OR tree. Figure 4.5 shows the AND-OR tree for design of an adaptable equipment to test two newly designed aircraft pumps including one fuel pump and one oil pump.

The fuel pump is used to provide fuel to the turbine's combustor, while the oil pump is used to provide lubrication oil to bearings and other components in the fuel system. The fuel pump is subjected to two different types of tests including a leakage test under very low temperature at −54 °C and pressure and flow tests under low temperature at −28 °C. The oil pump is subjected to only one type of tests to check the pressure and flow under low temperature at −28 °C. Therefore, the requirements for the adaptable equipment are defined for three phases considering the different test tasks:

Phase I: Tests of the fuel pump to quantify leakage under very low temperature.
Phase II: Tests of the fuel pump to observe pressure and flow under low temperature.
Phase III: Tests of the oil pump to observe pressure and flow under low temperature.

The adaptable testing equipment is primarily composed of a chamber for housing the pumps and a refrigeration unit to keep the low temperature in the chamber. A fan is installed on the top panel of the chamber to provide air circulation. For Phase I, a window is required for the chamber. For the refrigeration unit, a cryogenic system with either liquid nitrogen (LN_2) or liquid carbon dioxide (LCO_2) is considered for Phase I, and a mechanical refrigeration system with either single stage or cascade cycles is considered for Phase II and Phase III. The hybrid AND-OR tree considering

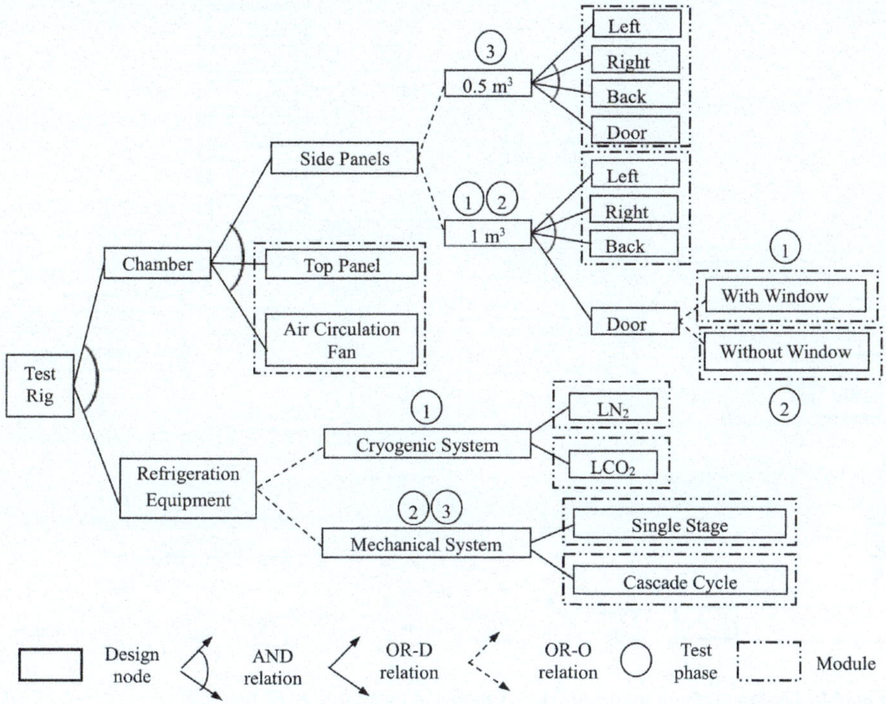

Fig. 4.5 AND-OR tree for design of an adaptable equipment to test two newly designed aircraft pumps [22]

design configuration candidates and operation configurations in different phases is shown in Fig. 4.5.

From the hybrid AND-OR tree, four design configuration candidates, as shown in Table 4.4, are generated using the OR-D relations given in Fig. 4.5. All the design nodes for the chamber are used for these candidates. Among the four design configuration candidates, the first one is selected based on evaluations to all these design configuration candidates. Figure 4.6 shows the first design configuration candidate modeled by an AND-OR tree. The three operation configurations for the first design configuration candidate are shown in Fig. 4.7.

Table 4.4 Design configuration candidates

Design candidate	Design nodes
1	Nodes for chamber, LN_2 (Phase I), single cycle (Phases II and III)
2	Nodes for chamber, LCO_2 (Phase I), single cycle (Phases II and III)
3	Nodes for chamber, LN_2 (Phase I), cascade cycles (Phases II and III)
4	Nodes for chamber, LCO_2 (Phase I), cascade cycles (Phases II and III)

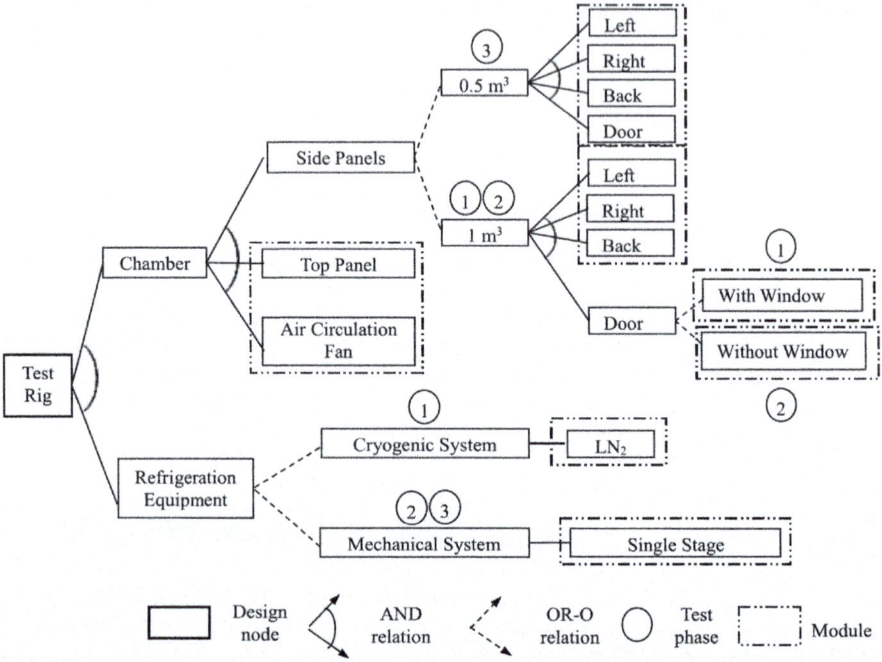

Fig. 4.6 Design configuration candidate 1 modeled by an AND-OR tree [22]

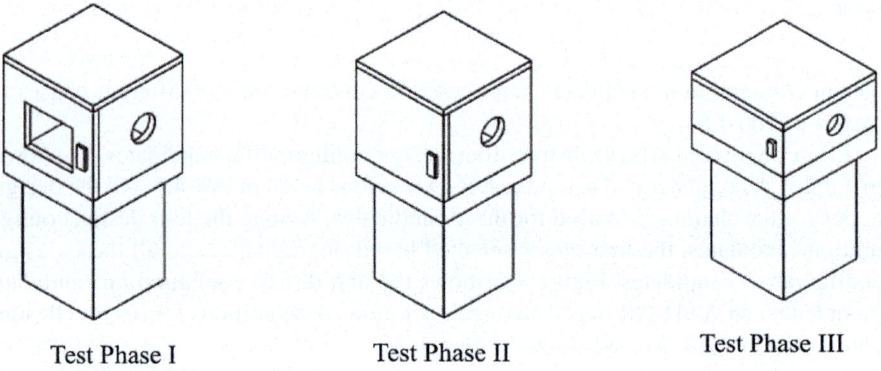

Fig. 4.7 Three operation configurations for the first design configuration candidate [22]

4.2.3 Design Evaluation

The adaptable product is evaluated by different evaluation measures considering the product utilization span. Since configurations and parameter values of an adaptable product are usually changed throughout the product utilization span, an evaluation measure in the utilization span is also usually changed. The evaluation measure at a

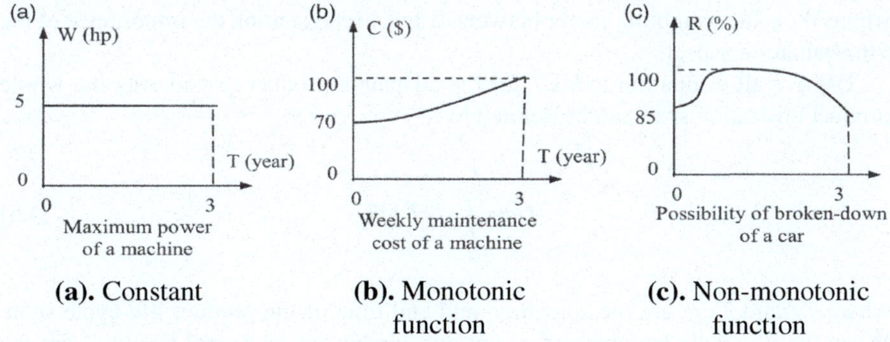

Fig. 4.8 Changes of evaluation measure in the product utilization span [22]

particular life-cycle time T in the i-th operation phase is calculated by:

$$E_j(T) = E_j(\boldsymbol{P}_i, T_i), \quad i = 1, 2, \ldots, n; \; j = 1, 2, \ldots, m, \tag{4.2}$$

where $\boldsymbol{P}_i$ is the collection of parameters for configuration in the i-th operation phase, n is the number of operation phases, m is the number of total evaluation measures, and T is the life-cycle time parameter in the i-th operation phase. An evaluation measure in the whole product utilization span can be either a constant, a monotonic function of T (i.e., increasing function or decreasing function), or a non-monotonic function of T as shown in Fig. 4.8.

To compare different evaluation measures in different units, these evaluation measures need to be converted into comparable evaluation indices between 0 and 1, representing levels of satisfaction [34].

$$I_j(T) = I_j\left[E_j(T)\right], \quad j = 1, 2, \ldots, m \tag{4.3}$$

The relation between an evaluation measure and an evaluation index can be defined by a linear relation or a nonlinear relation.

Since a higher evaluation index indicates a higher satisfaction level, the evaluation measures are classified into three categories based on relations between the evaluation measures and evaluation indices, the smaller the better, the larger the better, and the nominal the best.

When the j-th evaluation index, $I_j(T)$, is used to evaluate the adaptable product from a particular perspective at a certain life-cycle time T, the overall evaluation index, $I(T)$, considering all m evaluation aspects at the life-cycle time T can be obtained by:

$$I(T) = \frac{\sum_{j=1}^{m}\left[W_j I_j(T)\right]}{\sum_{j=1}^{m} W_j}, \tag{4.4}$$

where W_j is the weighting factor between 0 and 1 representing the importance of the j-th evaluation aspect.

The overall evaluation index I for the adaptable product considering the whole product utilization span can be defined by:

$$I = \int_{T_0}^{T_{max}} I(T)\mathrm{d}T, \tag{4.5}$$

where T_0 and $T_{\max}$ are the start time and end time of the product life-cycle span. When the life-cycle parameter T is not considered, Eq. (4.3) and Eq. (4.4) are the same as Eq. (3.27) and Eq. (3.28) given in Sect. 3.3.

In the design of the reconfigurable machine tool developed by Gadalla and Xue [7], the relation between the evaluation measure of spindle motor power P_{m} (kw) and its evaluation index $I_{P_{\mathrm{m}}}$ is defined by a cubic polynomial:

$$I_{P_{\mathrm{m}}} = -5.6 \times 10^{-6} P_{\mathrm{m}}^3 - 0.0011 P_{\mathrm{m}}^2 + 0.071 P_{\mathrm{m}} - 0.0037.$$

4.2.4 *Design Optimization*

Since a large number of design configuration candidates, operation configurations, and different parameter values for these configurations can be selected to meet the design requirements, optimization is employed in this research to identify the best design solution considering the whole product utilization span.

A multi-level optimization model [32] is employed to identify the best design configuration solution candidate and its design parameter values considering the whole product utilization span. In this optimization model, the optimal design parameter values for the i-th design configuration solution candidate are decided through the parameter optimization.

$$\begin{aligned}
&\text{Find: parameters } \boldsymbol{X}_i \\
&\text{Optimize: } I^{(i)} \\
&\text{Subject to: } \boldsymbol{X}_i^{(\mathrm{L})} \le \boldsymbol{X}_i \le \boldsymbol{X}_i^{(\mathrm{U})} \\
&\qquad h_j(\boldsymbol{X}_i) = 0,\ j = 1, 2, \ldots \\
&\qquad g_j(\boldsymbol{X}_i) \le 0,\ j = 1, 2, \ldots
\end{aligned} \tag{4.6}$$

where $I^{(i)}$ is the selected evaluation index considering the whole product utilization span, and $\boldsymbol{X}_i^{(\mathrm{L})}$ and $\boldsymbol{X}_i^{(\mathrm{U})}$ represent the lower bounds and upper bounds of $\boldsymbol{X}_i$, respectively.

Among all the p feasible product design configuration solution candidates, the optimal design solution is obtained through configuration optimization.

$$\begin{aligned}&\text{Find: the } i\text{-th design configuration candidate}\\&\text{Maximize: } I = I^{(i)}\\&\text{Subject to: } 1 \leq i \leq p\end{aligned} \tag{4.7}$$

where i represents the i-th design configuration candidate, and p is the number of all feasible design configuration candidates.

Parameter optimization can be conducted through the numerical computation and search [2]. When constraints are considered, penalty functions are used to convert a constrained optimization problem into an unconstrained optimization problem. When the number of possible design configuration candidates is not large, each of these design configuration candidates is evaluated and the best one is selected from all possible design configuration candidates. When the number of possible design configuration candidates is large, genetic programming [16] can be employed. Genetic programming is an evolutionary computing method to solve an optimization problem when solutions to the optimization problem can be modeled by tree data structures. In genetic programming, individuals (also called chromosomes) are used to describe the multiple solutions in the population of a generation. Reproduction, crossover and mutation are the three operations used to evolve individuals from a generation to the next one with a better average evaluation measure [10].

The optimization objective functions in Eq. (4.6) are defined by one of the three methods shown in Table 4.5 and Fig. 4.9.

(1) *The average-case method*

In this method, the average evaluation index considering the whole product utilization span is used as the objective function for parameter optimization considering one design configuration candidate. The average-case method is normally selected for the optimal design of adaptable product. Figure 4.9a shows an example to select a better design configuration candidate from two design configuration candidates based on the average-case method.

(2) *The best-case method*

In this method, the best evaluation index considering the whole product utilization span is used as the objective function for parameter optimization considering one

Table 4.5 Three methods to select optimization objective functions

Method	Objective function for parameter optimization
Average-case method	Maximize $I^{(i)} = \int_{T_0}^{T_{\max}} I(T)\mathrm{d}T$
Best-case method	Maximize $I^{(i)} = I(T)$
Worst-case method	Minimize $I^{(i)} = I(T)$

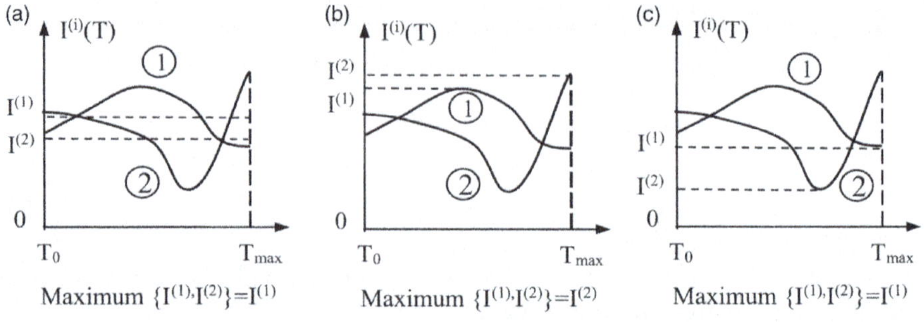

(a). Average-case method **(b).** Best-case method **(c).** Worst-case method

Fig. 4.9 Three methods for selection of optimization objective functions [22]

design configuration candidate. The best-case method is used when the maximum evaluation measure at one time point in the whole product utilization span is expected such as in the design of a racing car to achieve the maximum speed. Figure 4.9b shows an example to select a better design configuration candidate from two design configuration candidates based on the best-case method.

(3) *The worst-case method*

In this method, the worst evaluation index considering the whole product utilization span is used as the objective function for parameter optimization considering one design configuration candidate. The worst-case method is used when the minimum evaluation measure at one time point in the whole product utilization span is considered such as in the design of a satellite with the lowest risk of failure. Figure 4.9c shows an example to select a better design configuration candidate from two design configuration candidates based on the worst-case method.

In all above three methods, the objective function evaluation measure corresponding to the optimal design parameter values is selected as the evaluation measure for a design configuration candidate. Among all the design configuration candidates, the one with the best evaluation measure is selected in configuration optimization. In the design of the adaptable equipment to test two newly designed aircraft pumps developed by Martinez and Xue [22], four configurations as shown in Table 4.4 and two design parameters, vacuum pressure P (Pa) and number of radiation shields n, were considered. The optimal design is identified as follows (see Fig. 4.10):

- *Optimal configuration*: configuration 1 (i.e., nodes for chamber, LN_2 (Phase I), single cycle (Phases II and III)).
- *Optimal parameter values*: $P = 0.1$ (Pa), $n = 5$.

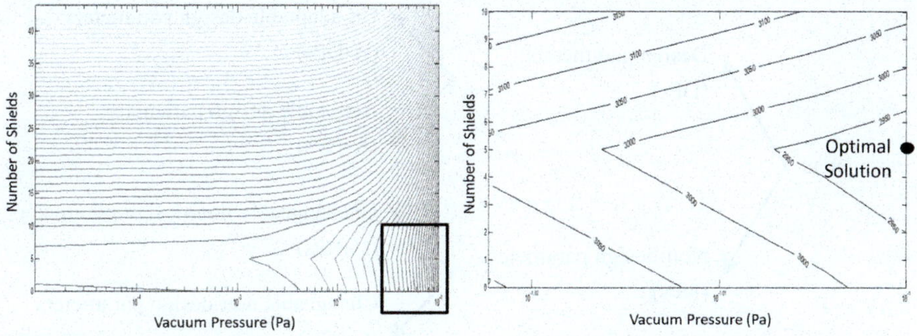

Fig. 4.10 Optimization of two design parameters [22]

4.3 Robust Adaptable Design

Adaptable design is a design paradigm that can be used for the design of adaptable products. Since quality of adaptable product can be influenced by parameter variations due to uncertainties, robustness of product design has to be considered in adaptable design. This section aims at developing an adaptable design approach such that the product is adaptable to various changes in requirements; meanwhile, the functional performance measures are the least sensitive to parameter variations. In this section, robust adaptable design at three levels in the embodiment design stage is considered: parameter, configuration, and architecture levels.

4.3.1 Robust Adaptable Design Considering Changes of Requirements and Parameters

A robust adaptable design approach is introduced for identifying the adaptable product whose functional performance is the least sensitive to variations of product/operating parameters. For an adaptable product, parameters are classified into two categories: design parameters (DPs) and non-design parameters (NPs), as shown in Fig. 4.11.

Design parameters are those to be determined by design engineers. The design parameters are further classified into two sub-categories: un-adaptable design parameters (U-DPs) and adaptable design parameters (A-DPs). While values of un-adaptable design parameters are not changed in the product operation stage, values of adaptable design parameters have to be adapted when requirements and working conditions are changed during product operations. Adaptation of adaptable design parameter values to satisfy the changed requirements and working conditions is conducted based on the following design principles.

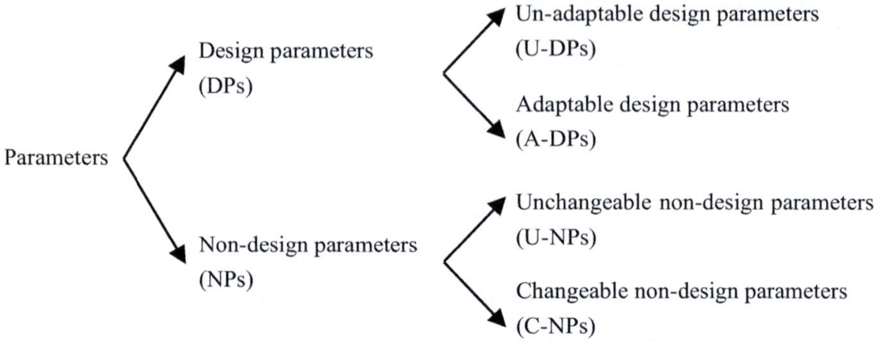

Fig. 4.11 Classification of parameters in adaptable design

The non-design parameters are those that do not need to be determined in the design process. Typical non-design parameters, such as working conditions and given product parameters, are provided before the design process. The non-design parameters are also classified into two sub-categories: unchangeable non-design parameters (U-NPs) and changeable non-design parameters (C-NPs). Unchangeable non-design parameters are the operating and/or product parameters whose values are not changed in the product operation stage. Changeable non-design parameters, on the other hand, are those whose values are changed independently during the product operation stage.

For an adaptable product, values of the parameters can be changed in the product operation stage. In adaptable design, two types of changes are considered:

- Different parameter values at same time period

In this case, these different parameter values are associated with different probabilities. For example, a laptop computer is used under power voltage of 110 V with 80% of the time and under power voltage of 220 V with 20% of the time.

- Different parameter values at different time periods

In this case, a life-cycle time parameter T ($T_{\min} \le T \le T_{\max}$) is introduced, where $T_{\min}$ is the time at product purchase, $T_{\max}$ is the time at product disposal/recycle, and these parameter values are defined as numerical functions of this time parameter. For example, the gas price is estimated at $1.2/liter in year 1-year 3 and $1.5/liter in year 4-year 6.

For an adaptable product, when values of functional requirements and changeable non-design parameters are changed at product operation stage, adaptation of adaptable design parameters has to be carried out, as shown in Fig. 4.12. The un-adaptable design parameters, adaptable design parameters, unchangeable non-design parameters, and changeable non-design parameters are defined by $\boldsymbol{P}^{(\text{U-DP})}$, $\boldsymbol{P}^{(\text{A-DP})}$, $\boldsymbol{P}^{(\text{U-NP})}$, and $\boldsymbol{P}^{(\text{C-NP})}$, respectively. Functional requirements are described by R, which is usually modeled by the target performance values. Both changeable non-design parameters, $\boldsymbol{P}^{(\text{C-NP})}$, and functional requirements, R, can be modeled by different

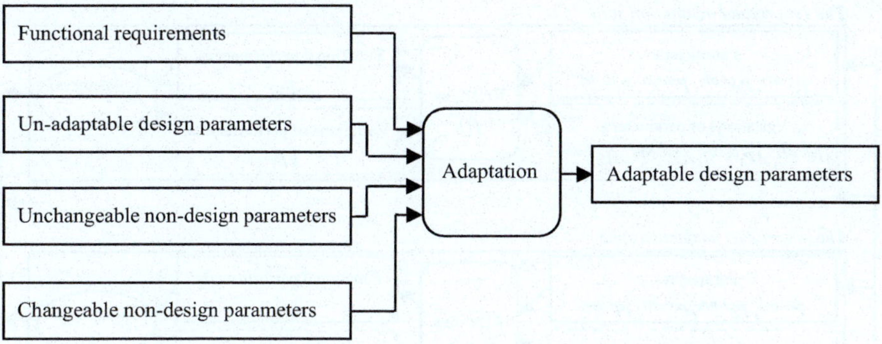

Fig. 4.12 Adaptation of adaptable design parameters in the product operation stage

values at the same time period with their probabilities or different values at the different time periods. In the product adaptation process, values of adaptable design parameters are calculated based on the newly changed functional requirements and changeable non-design parameters, as well as the un-adaptable design parameters and unchangeable non-design parameters, as shown in Fig. 4.12.

Due to uncertainties in design, manufacturing, and operation, values of product and operating parameters sometimes are deviated from their expected values, leading to deviations of functional performance measures. These deviations are also called variations. Comparing with the changes of functional requirements, changeable non-design parameters, and adaptable design parameters in the product adaptation process, variations of parameters and performance measures due to uncertainties are considerably small.

In robust adaptable design as shown in Fig. 4.13, both functional performances and variations of these functional performances are used to evaluate an identified design. Robustness is defined to evaluate the design considering both functional performances and their variations. Since adaptable design parameters are adapted with different values in the product operation stage based on different functional requirements and working conditions, robustness measures of these individual product adaptation states, S_i ($i = 1, 2, \ldots, n$), are obtained, and the overall robustness, S, is defined considering all these individual product adaptation states. The optimal un-adaptable design parameters are searched by maximizing the overall robustness through optimization. Various functional performances, $\boldsymbol{F}_i$, and their variations, $\boldsymbol{\Delta F}_i$, in the i-th product adaptation state are calculated from the four types of parameters, $\boldsymbol{P}^{(\text{U-DP})}$, $\boldsymbol{P}^{(\text{A-DP})}$, $\boldsymbol{P}^{(\text{U-NP})}$, and $\boldsymbol{P}^{(\text{C-NP})}$, their variations, $\Delta\boldsymbol{P}^{(\text{U-DP})}$, $\Delta\boldsymbol{P}^{(\text{A-DP})}$, $\Delta\boldsymbol{P}^{(\text{U-NP})}$, and $\Delta\boldsymbol{P}^{(\text{C-NP})}$, and relationships between parameters and performance measures, $\boldsymbol{f}()$. Two different types of product adaptation states are considered as follows:

- Adaptation states due to alternative functional requirements and changeable non-design parameters at the same time period associated with different probabilities.
- Adaptation states due to different functional requirements and changeable non-design parameters at the different time periods.

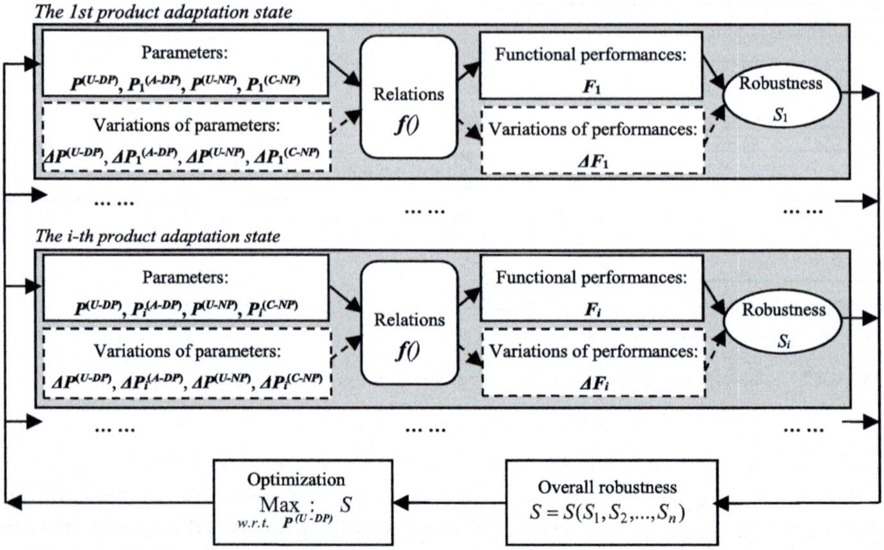

Fig. 4.13 Robust adaptable design considering robustness measures of different product adaptation states

(a) Modeling of functional requirements and product/operating parameters

As introduced above, two types of product adaptation activities are considered: adaptation activities in the same time period due to alternative requirements and adaptation activities in different time periods due to different requirements.

The requirements are modeled by functional requirements with the target performance measures and parameter requirements with the values of changeable non-design parameters. The alternative requirements in the same time period are associated with probabilities. The different requirements in different time periods are modeled as numerical functions of the life-cycle time parameter T. Values of adaptable design parameters are calculated from requirements of the target performance measures and changeable non-design parameters.

Functional requirements are defined by the target performance measures. Suppose m is the number of the functional requirements, all the functional requirements of a product, $R(T)$, at time T, can be described by a vector: $\boldsymbol{R(T)} = \{R_1(T), \ldots, R_m(T)\}$. In the operation stage of a product, the whole life-cycle time period can be divided into different phases based on different functional requirements as shown in Fig. 4.14. The whole life-cycle time period is divided into different phases for modeling different functional requirements.

The target values of functional requirements can be either discrete or continuous. Alternative functional requirements are associated with probabilities. Suppose $z_{ij}(T)$ represent the target values of the i-th functional requirement at time T in the j-th phase, functional requirements with discrete target values in the whole product life cycle can be described by:

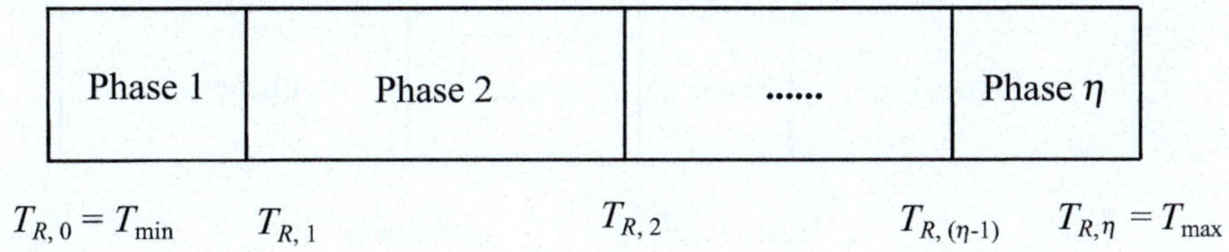

Fig. 4.14 Different phases considering different product functional requirements

$$z_{ij}(T) = \begin{cases} z_{ij1}(T), & p_{ij1}(T) \\ \vdots & \vdots \\ z_{ij\tau_{ij}}(T), & p_{ij\tau_{ij}}(T) \end{cases}, \quad i = 1, \ldots, m, \ j = 1, \ldots, \eta, \tag{4.8}$$

where τ_{ij} is an integer representing the number of possible target values of $R_{ij}(T)$, and $p_{ijk}(T)$ ($i = 1, \ldots, m, j = 1, \ldots, \eta, k = 1, \ldots, \tau_{ij}$) represents probabilities that the functional requirement $R_{ij}(T)$ should be selected as $z_{ijk}(T)$.

When the target values of the i-th functional requirement in the j-th phase can be changed continuously, and $L_{ij}^{(R)}(T)$ and $U_{ij}^{(R)}(T)$ are the lower and upper boundaries of the target values, the target values of the functional requirements, $z_{ij}(T)$, satisfy:

$$L_{ij}^{(R)}(T) \leq z_{ij}(T) \leq U_{ij}^{(R)}(T), \quad i = 1, \ldots, m, j = 1, \ldots, \eta. \tag{4.9}$$

In this case, the continuous target value $z_{ij}(T)$ is associated with a probability density function $p_{ij}[z_{ij}(T)]$.

The product and operating parameters are classified into four categories: un-adaptable design parameters (U-DPs), adaptable design parameters (A-DPs), unchangeable non-design parameters (U-NPs), and changeable non-design parameters (C-NPs).

Un-adaptable design parameters are those whose values are achieved in the robust adaptable design process through optimization. The l un-adaptable design parameters are described by a vector $\boldsymbol{P}^{(\text{U-DP})} = \{P_1^{(\text{U-DP})}, \ldots, P_l^{(\text{U-DP})}\}$. Values of un-adaptable design parameters are not changed in the product operation stage.

Changeable non-design parameters are operating and/or product parameters whose values are changed independently during the product operation stage. Suppose n is the number of the changeable non-design parameters, the vector of changeable non-design parameters, $\boldsymbol{P}^{(\text{C-NP})}(\boldsymbol{T})$, at time T, can be described by: $\boldsymbol{P}^{(\text{C-NP})}(\boldsymbol{T}) = \{P_1^{(\text{C-NP})}(T), \ldots, P_n^{(\text{C-NP})}(T)\}$. In the operation stage of a product, the changeable non-design parameters at time T can change to different values with different probabilities. The changeable non-design parameters can also change to different values at different time points. Similar to functional requirements, the whole product life cycle for changeable non-design parameters from $T_{\min}$ to $T_{\max}$ can also be divided into ψ phases as shown in Fig. 4.15.

Changes of the changeable non-design parameters can be discrete or continuous. Suppose $y_{ij}(T)$ ($i = 1, \ldots, n, j = 1, \ldots, \psi$) represents values of the i-th changeable

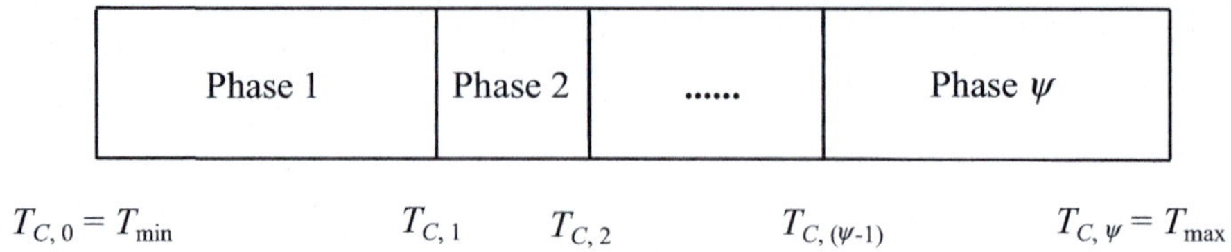

Fig. 4.15 Different phases considering different changeable non-design parameters

non-design parameter at time T in the j-th phase, when only discrete changes are considered, $y_{ij}(T)$ is described by:

$$y_{ij}(T) = \begin{cases} y_{ij1}(T), & q_{ij1}(T) \\ \vdots & \vdots \\ y_{ij\upsilon_{ij}}(T), & q_{ij\upsilon_{ij}}(T) \end{cases} \quad i = 1, \ldots, n, j = 1, \ldots \psi, \tag{4.10}$$

where υ_{ij} is an integer representing the number of possible values of $y_{ij}(T)$, and $q_{ijk}(T)$ ($i = 1, \ldots, n, j = 1, \ldots, \psi, k = 1, \ldots, \upsilon_{ij}$) represents probabilities that the $y_{ij}(T)$ should be selected as $y_{ijk}(T)$.

When values of changeable non-design parameters can be changed continuously, and $L_{ij}^{(\text{C-NP})}(T)$ and $U_{ij}^{(\text{C-NP})}(T)$ are the lower and upper boundaries of the changeable non-design parameters, the values of the i-th changeable non-design parameter at time T in the j-th phase, $y_{ij}(T)$, satisfy:

$$L_{ij}^{(\text{C-NP})}(T) \le y_{ij}(T) \le U_{ij}^{(\text{C-NP})}(T) \quad i = 1, \ldots, n, j = 1, \ldots \psi, \tag{4.11}$$

In this case, the continuous changeable non-design parameter value $y_{ij}(T)$ is associated with a probability density function $q_{ij}[y_{ij}(T)]$.

Values of adaptable design parameters are calculated based on target values of performance measures in functional requirements and assigned values of the changeable non-design parameters. When the functional requirements and changeable non-design parameters are changed at time T, values of adaptable design parameters need to be adapted. The r adaptable design parameters at time T are defined by a vector $\boldsymbol{P}^{(\text{A-DP})}(\boldsymbol{T}) = \{P_1^{(\text{A-DP})}(T), \ldots, P_r^{(\text{A-DP})}(T)\}$.

The unchangeable non-design parameters are the operating and/or product parameters whose values are not changed in the product operation stage. The t unchangeable non-design parameters are described by a vector $\boldsymbol{P}^{(\text{U-NP})} = \{P_1^{(\text{U-NP})}, \ldots, P_t^{(\text{U-NP})}\}$.

Variations of some parameters, including un-adaptable design parameters, $\boldsymbol{\Delta P}^{(\text{U-DP})}$, adaptable design parameters, $\boldsymbol{\Delta P}^{(\text{A-DP})}(\boldsymbol{T})$, unchangeable non-design parameters, $\boldsymbol{\Delta P}^{(\text{U-NP})}$, and changeable non-design parameters, $\boldsymbol{\Delta P}^{(\text{C-NP})}(\boldsymbol{T})$, can be observed in product operation stage due to uncertainties, leading to variations in functional performances. The robust adaptable product design is to identify the optimal values of the un-adaptable design parameters such that the functional performances are the least sensitive to variations of various parameters.

In the robust adaptable design, predictions of parameter changes are important. Generally, the parameter changes can be predicted according to customers' requirements, product working conditions, and/or technology advancements.

(b) Relationships between parameters and performance measures

In order to achieve the optimal values of un-adaptable design parameters considering changes of functional requirements, changeable non-design parameters, and adaptable design parameters as well as variations of various parameters, relationships between various parameters and functional performance measures need to be defined. Suppose $F_i(T)$ represents the i-th functional performance measure at time T, this performance can be calculated by:

$$F_i(T) = f_i\left[\boldsymbol{P}^{\text{(U-DP)}}, \boldsymbol{P}^{\text{(A-DP)}}(\boldsymbol{T}), \boldsymbol{P}^{\text{(U-NP)}}, \boldsymbol{P}^{\text{(C-NP)}}(\boldsymbol{T})\right], \quad i = 1, \ldots, m, \tag{4.12}$$

where $\boldsymbol{P}^{\text{(U-DP)}}$, $\boldsymbol{P}^{\text{(A-DP)}}(\boldsymbol{T})$, $\boldsymbol{P}^{\text{(U-NP)}}$, and $\boldsymbol{P}^{\text{(C-NP)}}(\boldsymbol{T})$ are vectors of un-adaptable design parameters, adaptable design parameters at time T, unchangeable non-design parameters, and changeable non-design parameters at time T, respectively, and $f_i()$ is the relationship between the various parameters and the i-th functional performance.

Under the influence of parameter variations due to uncertainties, deviation of the functional performance measure can be calculated by:

$$\begin{aligned}\Delta F_i(T) = f_i\big[&\boldsymbol{P}^{\text{(U-DP)}} + \Delta\boldsymbol{P}^{\text{(U-DP)}}, \mathbf{P}^{\text{(A-DP)}}(\mathbf{T}) + \Delta\boldsymbol{P}^{\text{(A-DP)}}(\mathbf{T}),\\ &\boldsymbol{P}^{\text{(U-NP)}} + \Delta\boldsymbol{P}^{\text{(U-NP)}}, \mathbf{P}^{\text{(C-NP)}}(\mathbf{T}) + \Delta\boldsymbol{P}^{\text{(C-NP)}}(\mathbf{T})\big] - F_i(T),\end{aligned} \tag{4.13}$$

where $\boldsymbol{P}^{\text{(U-DP)}}$, $\boldsymbol{P}^{\text{(A-DP)}}(\boldsymbol{T})$, $\boldsymbol{P}^{\text{(U-NP)}}$, and $\boldsymbol{P}^{\text{(C-NP)}}(\boldsymbol{T})$ are parameters with nominal values, $F_i(T)$ is the nominal performance calculated from $\boldsymbol{P}^{\text{(U-DP)}}$, $\boldsymbol{P}^{\text{(A-DP)}}(\boldsymbol{T})$, $\boldsymbol{P}^{\text{(U-NP)}}$, and $\boldsymbol{P}^{\text{(C-NP)}}(\boldsymbol{T})$, $\Delta\boldsymbol{P}^{\text{(U-DP)}}$, $\Delta\boldsymbol{P}^{\text{(A-DP)}}(\boldsymbol{T})$, $\Delta\boldsymbol{P}^{\text{(U-NP)}}$, and $\Delta\boldsymbol{P}^{\text{(C-NP)}}(\boldsymbol{T})$ are variations of the parameters, and $\Delta F_i(T)$ is the deviation of the i-th functional performance measure at time T due to variations of parameters.

(c) Evaluation of functional performance

The functional performance measures of a product can be classified into three categories: the larger the better, the nominal the best, and the smaller the better. In addition, multiple functional performance measures are often selected to evaluate a product [6, 19]. To solve the robust design problem considering multiple performance measures in different units, functional performance measures are converted into performance satisfaction indices between 0 and 1 using the nonlinear relations between the functional performance measures and satisfaction indices [34], as shown in Fig. 4.16.

When multiple functional performance measures of a product are considered, the value of the functional performance measure, $F_i(T)$, (i = 1, ..., m), should be converted into comparable satisfaction index, $I_i(T)$, between 0 and 1 using:

$$I_i(T) = \psi_i[F_i(T)], \quad i = 1, \ldots, m, \tag{4.14}$$

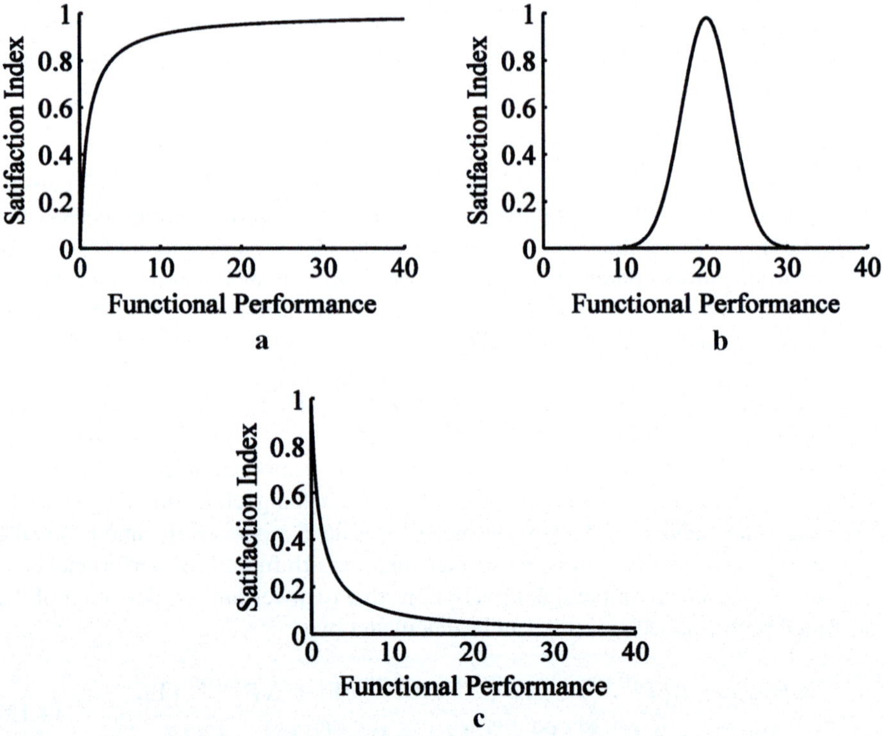

Fig. 4.16 Functional performance measures and performance satisfaction indices. **a** The larger the better. **b** The nominal the best. **c** The smaller the better

where $\psi_i()$ is the nonlinear relationship between $F_i(T)$ and $I_i(T)$.

Under the influence of uncertainties in parameters, the satisfaction indices can fluctuate in the operation stage of a product. According to Eqs. (4.13) and (4.14), the deviations of the satisfaction indices can be calculated by:

$$\Delta I_i(T) = \psi_i[F_i(T) + \Delta F_i(T)] - \psi_i[F_i(T)], \quad i = 1, \ldots, m, \tag{4.15}$$

Values of adaptable design parameters are changed when the target values of functional requirements and values of changeable non-design parameters are changed. Since the target values of functional requirements are defined in η phases as shown in Fig. 4.14, and the values of changeable non-design parameters are defined in ψ phases as shown in Fig. 4.15, the whole lifespan from $T_{\min}$ to $T_{\max}$ for the adaptable design parameters can be segmented into χ phases, as shown in Fig. 4.17, considering combinations of the η phases for functional requirements and the ψ phases for changeable non-design parameters.

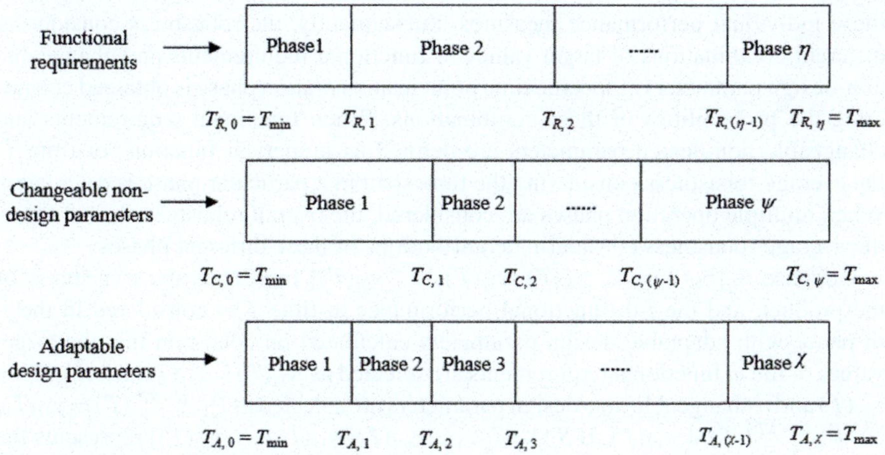

Fig. 4.17 Different phases of an adaptable product

(d) Robust design of an adaptable product

Under influence of parameter variations, the performance measures fluctuate in the product operation stage, leading to fluctuations of performance satisfaction indices. Variations of the satisfaction indices given in Eq. (4.15) are used to define the robustness of a product. When the target value of a functional requirement is achieved, the satisfaction index of this design reaches to its maximum value of 1. Therefore, the objective of robust design is to identify the values of un-adaptable design parameters to minimize the deviation of satisfaction index from 1 defined in Eq. (4.15). The robustness S is defined by:

$$S = I - 3\sigma_I = 1 - 3\sigma_I, \tag{4.16}$$

where I is the satisfaction index and σ_I is the standard deviation of this satisfaction index.

Comparing with the traditional robust design where usually only one evaluation measure is considered, the following factors are also considered in our research:

- Multiple performance measures.
- Different combinations of target values in functional requirements, changeable non-design parameters, and their probabilities at a specific time point.
- Different time points in an operation phase.
- Different operation phases.

First, the robustness considers one performance measure, and one combination of target values of functional requirements and changeable non-design parameters at a particular time point in an operation phase is calculated. Then, the robustness with multiple performance measures is achieved by considering the importance of these individual performance measures. Subsequently, the robustness considering different combinations of target values of functional requirements and changeable non-design parameters at specific time point in an operation phase is obtained considering the probabilities of these combinations. When functional requirements and changeable non-design parameters are defined as numerical functions of time T, the average robustness considering the time span in a particular phase is calculated. When multiple operation phases are considered, the overall robustness is defined as the average robustness considering contributions of these different phases.

Suppose $S_{ij}[z_{1j}(T), \ldots, z_{mj}(T), y_{1j}(T), \ldots, y_{nj}(T)]$ represents the robustness of the product, and the i-th functional performance at time T is considered in the j-th phase with adaptable design parameters calculated on condition that the target values of the m functional requirements are selected as $R_{1j}(T) = z_{1j}(T), \ldots, R_{mj}(T) = z_{mj}(T)$ and n changeable non-design parameters are selected as $P_{1j}{}^{(C-NP)}(T) = y_{1j}(T), \ldots, P_{nj}{}^{(C-NP)}(T) = y_{nj}(T)$. If $S_j[z_{1j}(T), \ldots, z_{mj}(T), y_{1j}(T), \ldots, y_{nj}(T)]$ represents the robustness, considering m functional performances at time T in the j-th phase on condition that the target values of all functional requirements are selected as $R_{1j}(T) = z_{1j}(T), \ldots, R_{mj}(T) = z_{mj}(T)$ and changeable non-design parameters are selected as $P_{1j}{}^{(C-NP)}(T) = y_{1j}(T), \ldots, P_{nj}{}^{(C-NP)}(T) = y_{nj}(T)$, $S_j[z_{1j}(T), \ldots, z_{mj}(T), y_{1j}(T), \ldots, y_{nj}(T)]$ can be calculated by:

$$S_j\left[z_{1j}(T), \ldots, z_{mj}(T), y_{1j}(T), \ldots, y_{nj}(T)\right] = \frac{\sum_{i=1}^{m}\left\{\omega_{ij} \cdot S_{ij}\left[z_{1j}(T), \ldots, z_{mj}(T), y_{1j}(T), \ldots, y_{nj}(T)\right]\right\}}{\sum_{i=1}^{m}\omega_{ij}} \quad (4.17)$$

where ω_{ij} ($i = 1, \ldots, m, j = 1, \ldots, \chi$) is the weighting factors between 0 and 1 representing importance of these functional performances.

Suppose $q[y_{1j}(T), \ldots, y_{nj}(T)]$ represents the probability when changeable non-design parameters are selected as $P_{1j}{}^{(C\text{-}NP)}(T) = y_{1j}(T), \ldots, P_{nj}{}^{(C\text{-}NP)}(T) = y_{nj}(T)$, and $q_{ij}(T)$ represents the probability when $P_{ij}{}^{(C\text{-}NP)}(T) = y_{ij}(T)$; if changes of changeable non-design parameters are discrete, the probability is calculated by:

$$q\left[y_{1j}(T), \ldots, y_{nj}(T)\right] = \prod_{i=1}^{n} q_{ij}(T), \quad j = 1, \ldots, \chi. \quad (4.18)$$

If changes of changeable non-design parameters are continuous, the probability density is calculated by:

$$q\left[y_{1j}(T), \ldots, y_{nj}(T)\right] = \prod_{i=1}^{n} q_{ij}\left[y_{ij}(T)\right], j = 1, \ldots, \chi, \quad (4.19)$$

where $q_{ij}[y_{ij}(T)]$ is the probability density of $y_{ij}(T)$.

Suppose $S_j[z_{1j}(T), \ldots, z_{mj}(T)]$ represents the robustness at time T in the j-th phase when the target values of functional requirements are selected as $R_{1j}(T) = z_{1j}(T), \ldots, R_{mj}(T) = z_{mj}(T)$, and changeable non-design parameters are selected with different possible values; if changes of changeable non-design parameters values are discrete, $S_j[z_{1j}(T), \ldots, z_{mj}(T)]$ can be calculated by:

$$S_j[z_{1j}(T), \ldots, z_{mj}(T)] = \sum_{y_{1j}=y_{1j1}}^{y_{1jv_{1j}}} \cdots \sum_{y_{nj}=y_{nj1}}^{y_{njv_{nj}}} \{q[y_{1j}(T), \ldots, y_{nj}(T)] \times S_j[z_{1j}(T), \ldots, z_{mj}(T), y_{1j}(T), \ldots, y_{nj}(T)]\} \quad j = 1, \ldots, \chi. \tag{4.20}$$

If changes of changeable non-design parameters values are continuous, $S_j[z_{1j}(T), \ldots, z_{mj}(T)]$ is calculated by:

$$S_j[z_{1j}(T), \ldots, z_{mj}(T)] = \int_{L_{1j}^{(P)}(T)}^{U_{1j}^{(P)}(T)} \cdots \int_{L_{nj}^{(P)}(T)}^{U_{nj}^{(P)}(T)} \{q[y_{1j}(T), \ldots, y_{nj}(T)] \times S_j[z_{1j}(T), \ldots, z_{mj}(T), y_{1j}(T), \ldots, y_{nj}(T)] \cdot d[y_{1j}(T)] \ldots d[y_{nj}(T)]\} \quad j = 1, \ldots, \chi. \tag{4.21}$$

Suppose $p[z_{1j}(T), \ldots, z_{mj}(T)]$ represents the probability when functional requirements are selected as $R_{1j}(T) = z_{1j}(T), \ldots, R_{mj}(T) = z_{mj}(T)$, and $p_{ij}(T)$ represents the probability when $R_{ij}(T) = z_{ij}(T)$; if the target values of functional requirements are discrete, the probability can be calculated by:

$$p\big[z_{1j}(T), \ldots, z_{mj}(T)\big] = \prod_{i=1}^{m} p_{ij}(T), \quad j = 1, \ldots, \chi. \tag{4.22}$$

If the target values of functional requirements are continuous, the probability density can be calculated by:

$$p\big[z_{1j}(T), \ldots, z_{mj}(T)\big] = \prod_{i=1}^{m} p_{ij}\big[z_{ij}(T)\big] \quad j = 1, \ldots, \chi, \tag{4.23}$$

where $p_{ij}[z_{ij}(T)]$ is the probability density of $z_{ij}(T)$.

Suppose $S_j(T)$ represents the robustness of the product at time T in the j-th phase; if changes of the target values of the functional requirements are discrete, $S_j(T)$ is calculated by:

$$S_j(T) = \sum_{z_{1j}=z_{1j1}}^{z_{1j\tau_{1j}}} \cdots \sum_{z_{mj}=z_{mj1}}^{z_{mj\tau_{nj}}} \big\{p\big[z_{1j}(T), \ldots, z_{mj}(T)\big] \cdot S_j\big[z_{1j}(T), \ldots, z_{mj}(T)\big]\big\} \tag{4.24}$$

If changes of the target values of functional requirements are continuous, $S_j(T)$ is calculated by:

$$S_j(T) = \int_{L_{1j}^{(R)}(T)}^{U_{1j}^{(R)}(T)} \cdots \int_{L_{mj}^{(R)}(T)}^{U_{mj}^{(R)}(T)} \{p[z_{1j}(T), \ldots, z_{mj}(T)] \cdot$$
$$S_j[z_{1j}(T), \ldots, z_{mj}(T)] \cdot d[z_{1j}(T)] \ldots d[z_{mj}(T)]\}. \quad (4.25)$$

Suppose S_j represents the robustness of the product in the j-th phase considering the whole time span in the j-th phase, multiple performance measures, and different values of functional requirement targets and changeable non-design parameters, and S_j can be calculated by:

$$S_j = \frac{\int_{T_{j-1}}^{T_j} S_j(T) \cdot dT}{T_j - T_{j-1}} \quad (4.26)$$

where T_{j-1} and T_j are the starting time and ending time of the j-th phase.

When all χ operation phases are considered, the overall robustness of the product, S, is calculated by:

$$S = \frac{\sum_{j=1}^{\chi} [(T_j - T_{j-1}) \cdot S_j]}{T_{\max} - T_{\min}} \quad (4.27)$$

The robust adaptable design is conducted through optimization to maximize the overall robustness S. When the product/operating parameters are not independent, relations among these parameters should be added as constraints in the optimization problem. Therefore, the optimization model for robust adaptable design can be established as follows:

Find: the un-adaptable design parameters, $\boldsymbol{P}^{(\text{U-DP})} = \{P_1^{(\text{U-DP})}, \ldots, P_l^{(\text{U-DP})}\}$.

$$\text{Maximize}: S$$
$$\text{Subject to}: h_v[\boldsymbol{P}^{(\text{U-DP})}, \boldsymbol{P}^{(\text{A-DP})}(T), \boldsymbol{P}^{(\text{U-NP})}, \boldsymbol{P}^{(\text{C-NP})}(\boldsymbol{T})] = 0, \quad v = 1, 2, \ldots$$
$$g_w[\boldsymbol{P}^{(\text{U-DP})}, \boldsymbol{P}^{(\text{A-DP})}(T), \boldsymbol{P}^{(\text{U-NP})}, \boldsymbol{P}^{(\text{C-NP})}(\boldsymbol{T})] \le 0, \quad w = 1, 2, \ldots \quad (4.28)$$

In the product operation stage, values of the changeable non-design parameters and functional requirements can be changed. The values of adaptable design parameters have to be modified to adapt to the changes in the functional requirements and/or the changeable non-design parameters. It is assumed that values of adaptable design parameters can be calculated from changeable non-design parameters and target functional requirements.

When a large number of performance measures and product/operating parameters need to be considered in a real design case, the computer program that includes different types of mathematical models can be used to define both performance measures and product/operating parameters. Designers also need to select different models from the computer program, and the calculation of performance measures can be carried out automatically according to the relations defined by Eq. (4.12).

4.3.2 Robust Adaptable Design Considering Changes of Product Configurations

In Sect. 4.3.1, the robust adaptable design method considering changes of parameter values in operation stage has been explained. Since configuration of an adaptable product may also need to be changed/adapted in the operation stage to satisfy different requirements, a robust adaptable design method considering changes of configurations in the operation stage is needed. The robust adaptable design approach considering changes of product configurations is developed. First, the method to model all configuration candidates in design is developed. Modeling of different product configurations in the operation stage for each design candidate is then discussed. Each product configuration is described by parameters. A two-level optimization model is developed to identify the optimal adaptable design with the best product configurations and parameter values.

In adaptable design, different configurations for selections in the design stage are called design configuration candidates, while the different configurations of the selected design candidate in the operation stage are called operation configuration states. For the design of an adaptable product, as shown in Fig. 4.18, different feasible design configuration candidates can be created from the same design requirements. For each design configuration candidate, different operation configuration states can be created in the product operation stage to accommodate changes in requirements. For example, the manual transmission and automatic transmission are two design configuration candidates for the transmission unit in an automobile, while the different layouts of components in the transmission unit at different transmission levels when driving the car are different operation configuration states.

In this section, a hybrid AND-OR tree is developed to model different feasible design configuration candidates and different possible operation configuration states of each design configuration candidate. Rules for creating different feasible design configuration candidates from the hybrid AND-OR tree and rules for creating different operation configuration states from each feasible design configuration candidate are also developed. Since an adaptable product can be modeled by both configurations and parameters, product configurations including the design configuration candidates and operation configuration states in the hybrid AND-OR tree are further modeled by parameters.

In order to design an adaptable product with the robust performance, the evaluation measure is then introduced by considering the robustness of an adaptable product at different operation configuration states in the operation stage. A two-level

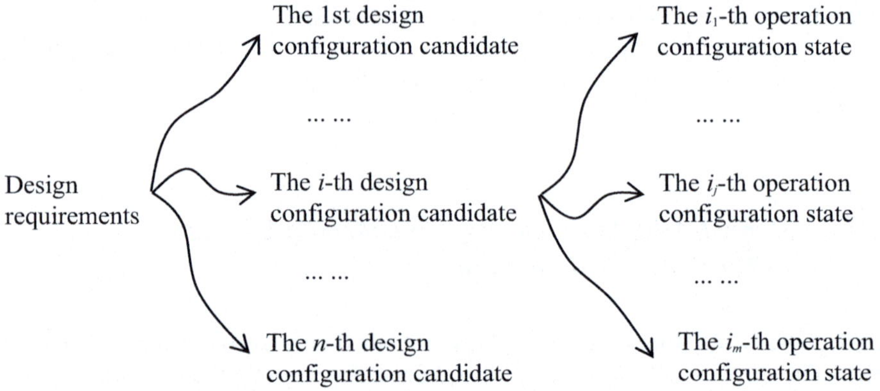

Fig. 4.18 Relations among design requirements, design configuration candidates, and operation configuration states

optimization model is subsequently developed to identify the optimal configurations and parameter values of an adaptable design. The first-level optimization is used to identify the best product configuration, and the second-level optimization is used to identify the best product/operating parameter value. In order to reduce the unnecessary complexity, it is assumed that costs of all feasible design configuration candidates are comparable, and the cost is not considered in the optimization of product configuration and its parameter values.

(1) Modeling of Product Configurations

In adaptable design, a hybrid AND-OR tree is introduced to model different design configuration candidates and different operation configuration states of each design configuration candidate. In this tree, partial design solutions, such as components and modules, are described by configuration nodes. Relations among sub-nodes of a super-node are classified into three categories:

- *AND relation.* The AND relations are used for modeling both design configuration candidates and operation configuration states.
- *OR relation in design.* The OR relations in design are used for modeling different design configuration candidates.
- *OR relation in operation.* The OR relations in operation are used for modeling different operation configuration states.

An example of the hybrid AND-OR tree is given in Fig. 4.19. When all the sub-nodes need to be selected to support a super-node, all of these sub-nodes have an AND relation. For example, since the engine, transmission unit, shaft, etc., are required for a power transmission system in an automobile, these modules are associated by an AND relation. When the super-node is supported by one of its sub-nodes in the process of design configuration candidate selection, all of these sub-nodes have an OR relation in design. For example, since only one of the two transmission units, the

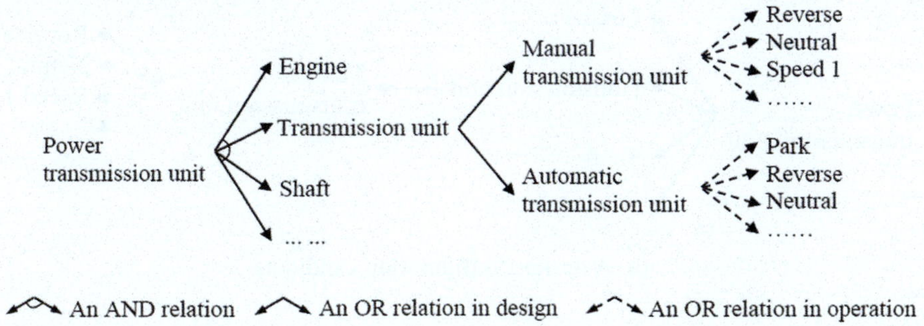

Fig. 4.19 Hybrid AND-OR tree for modeling design configuration candidates and operation configuration states

manual transmission unit or automatic transmission unit, needs to be selected as the transmission unit in the power transmission system, these two transmission units are associated with an OR relation in design. When the super-node is supported by one of its sub-nodes in the operation stage, all these sub-nodes have an OR relation in operations. For example, since only one of the different gear layouts is selected for a specific gear mode in the operation stage, these gear layouts are associated with an OR relation in operations.

From design requirements, three types of AND-OR relations are used to build a hybrid AND-OR tree that can be used to create design configuration candidates and operation configuration states. In this section, the following rules are used to generate different feasible design configuration candidates from the hybrid AND-OR tree:

- The root node has to be selected first.
- When a node is selected and all of its sub-nodes are associated with an AND relation, all these sub-nodes should be selected.
- When a node is selected and all of its sub-nodes are associated with an OR relation in design, only one of these sub-nodes should be selected.
- When a node is selected and all of its sub-nodes are associated with an OR relation in operation, all of these sub-nodes should be selected.

Generally, each feasible design configuration candidate is modeled by an AND-OR tree with AND relations and OR relations in operations. An example of the feasible design configuration candidate created from the hybrid AND-OR tree shown in Fig. 4.19 is given in Fig. 4.20a.

From a feasible design configuration candidate, different operation configuration states in the product operation stage can be generated. Rules for creating product operation configuration states are described as follows:

- The root node in the design configuration candidate AND-OR tree has to be selected first.
- When a node is selected and all of its sub-nodes are associated with an AND relation, all these sub-nodes should be selected.

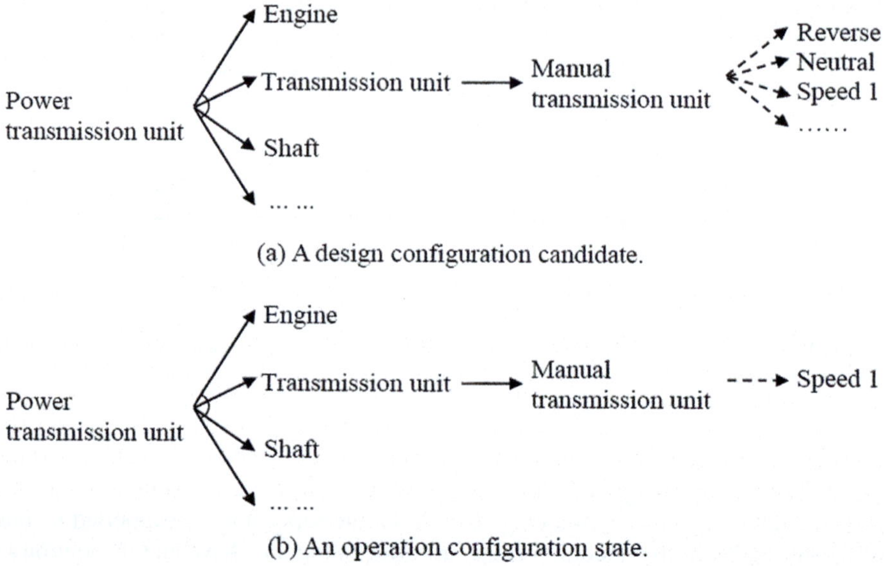

Fig. 4.20 Design configuration candidate and an operation configuration state generated from the hybrid AND-OR tree

- When a node is selected and all of its sub-nodes are associated with an OR relation in operations, only one of these sub-nodes should be selected.

Each AND-OR tree of the design configuration candidate can be used to create several product operation configuration states. Each possible product operation configuration state in the operation stage is modeled by a tree with only AND relationships. An example of the operation configuration state generated from the design configuration candidate shown in Fig. 4.20a is given in Fig. 4.20b.

(2) Modeling of parameters associated with configurations

Design configuration candidates and operation configuration states in adaptable design are further modeled by parameters including product parameters and operating parameters. As discussed in Sect. 4.3.1, the product/operating parameters associated with each configuration node are classified into four categories: un-adaptable design parameters, $\boldsymbol{X}^{\mathrm{D}}$, adaptable design parameters, $\boldsymbol{X}^{\mathrm{A}}$, unchangeable non-design parameters, $\boldsymbol{X}^{\mathrm{U}}$, and changeable non-design parameters, $\boldsymbol{X}^{\mathrm{C}}$. The detailed discussions on the adaptations of parameters in the operation stage considering these four types of parameters and target values of functional performances have been provided in Sect. 4.3.1.

For an adaptable product, in addition to the four categories of product/operating parameters, each configuration node is also associated with a probability, P, representing the percentage of time that this configuration node is used in the product operation stage. Probabilities of configuration nodes are calculated based on the following rules:

- The probability of a configuration node is described by a number between 0 and 1. Probability of the root node is 1.
- When sub-nodes of a super-node are associated with an AND relation or an OR relation in design, probabilities of these sub-nodes should be the same as the probability of the super-node.
- When sub-nodes of a super-node are associated with an OR relation in operations, the sum of the probabilities of these sub-nodes should be equal to the probability of the super-node.

A product operation configuration state can be modeled by a collection of leaf nodes and their associated parameters. If $C^{(i,j)}$ represents the number of leaf nodes of the product operation configuration in the *j*-th adaptation state for the *i*-th design configuration candidate and $P^{(i,j,k)}$ represents the probability of the *k*-th leaf node in this operation configuration state, the probability of the product operation configuration in the *j*-th adaptation state, $P^{(i,j)}$, used in the product operation stage can be calculated by:

$$P^{(i,j)} = \prod_{k=1}^{C^{(i,j)}} P^{(i,j,k)}. \tag{4.29}$$

For an adaptable product, if $C^{(i)}$ represents the number of all possible product operation configurations for the *i*-th design configuration candidate, we have:

$$\sum_{j=1}^{C^{(i)}} P^{(i,j)} = 1. \tag{4.30}$$

The four types of product/operating parameters are associated with configuration nodes. Parameters in the sub-nodes can be used to calculate parameters in the super-node through their relationships. Usually, numerical functions or computer programs can be used to model relationships among the parameters. In adaptable design, two categories of relationships among the parameters are considered as follows:

- A parameter in super-nodes can be defined by a mathematical function of parameters in the sub-nodes when these sub-nodes are associated with an AND relation. For example, in Fig. 4.20, the weight of the power transmission system in an automobile, W, can be calculated from parameters of its sub-nodes including weight of engine, W_E, weight of transmission unit, W_T, weight of shaft, W_S, etc.:

$$W = W_E + W_T + W_S + \cdots$$

- A parameter in the super-node can be defined by a piecewise mathematical function of parameters in sub-nodes when these sub-nodes are associated with an OR relation, either OR relation in design or OR relation in operations. For example, in Fig. 4.20, the weight of the transmission unit of a power transmission system,

W_T, can be calculated by parameters of its sub-nodes including weight of manual transmission unit, W_M, or weight of automatic transmission unit, W_A:

$$W_T = \begin{cases} W_M, & \text{when the manual transmission unit is selected} \\ W_A, & \text{when the automatic transmission unit is selected} \end{cases}. \quad (4.32)$$

By using the two categories of relationships among parameters associated with configuration nodes, relationships among parameters in the root node and parameters in different leaf nodes of a hybrid AND-OR tree can be established.

(3) Evaluation of the robustness

An adaptable design is evaluated by the overall robustness considering different product adaptation states in the operation stage. In order to obtain the overall robustness of an adaptable design, relationships among functional performances and parameters associated with the product operation configuration in a specific adaptation state should be established.

As discussed in Sect. 4.3.1, an adaptable product can be created to satisfy multiple functional requirements. Moreover, two categories of product adaptations can be achieved by an adaptable product: adaptations at the same time period and adaptations at different time periods. To simplify the discussion and avoid the unnecessary complexity, cases with only one functional performance and adaptation tasks at the same time period are considered. Suppose parameters of the *i*-th design configuration candidate can be described by un-adaptable design parameters, $\boldsymbol{X}^{\mathrm{D}}$, adaptable design parameters, $\boldsymbol{X}^{\mathrm{A}}$, unchangeable non-design parameters, $\boldsymbol{X}^{\mathrm{U}}$, and changeable non-design parameters, $\boldsymbol{X}^{\mathrm{C}}$, and relationships between functional performance, F, and parameters can be defined by:

$$F = f\left(\boldsymbol{X}^{\mathrm{D}}, \boldsymbol{X}^{\mathrm{A}}, \boldsymbol{X}^{\mathrm{U}}, \boldsymbol{X}^{\mathrm{C}}\right), \quad (4.33)$$

where $f()$ is the relationship between various parameters and functional performances.

Under influence of uncertainties, parameters can deviate from their target values. Suppose $\Delta\boldsymbol{X}^{\mathrm{D}}$, $\Delta\boldsymbol{X}^{\mathrm{A}}$, $\Delta\boldsymbol{X}^{\mathrm{U}}$ and $\Delta\boldsymbol{X}^{\mathrm{C}}$ are deviations of parameters $\boldsymbol{X}^{\mathrm{D}}$, $\boldsymbol{X}^{\mathrm{A}}$, $\boldsymbol{X}^{\mathrm{U}}$ and $\boldsymbol{X}^{\mathrm{C}}$, respectively, and deviation of the functional performance, ΔF, can be calculated by:

$$\Delta F = f(\boldsymbol{X}^{\mathrm{D}} + \Delta\boldsymbol{X}^{\mathrm{D}}, \mathbf{X}^{\mathrm{A}} + \Delta\boldsymbol{X}^{\mathrm{A}}, \boldsymbol{X}^{\mathrm{U}} + \Delta\boldsymbol{X}^{\mathrm{U}}, \boldsymbol{X}^{\mathrm{C}} + \Delta\boldsymbol{X}^{\mathrm{C}}) - f(\boldsymbol{X}^{\mathrm{D}}, \mathbf{X}^{\mathrm{A}}, \mathbf{X}^{\mathrm{U}}, \boldsymbol{X}^{\mathrm{C}}). \quad (4.34)$$

In the operation stage, values of adaptable design parameters, $\boldsymbol{X}^{\mathrm{A}}$, can be adapted to satisfy the changed functional requirement and changeable non-design parameters. Many measures, such as signal-to-noise ratio (SNR) and variance of functional performance, can be used to evaluate the robustness of a product design. Since the target value of a functional performance for an adaptable product can change within a certain range, the SNR is selected as the robustness measure to consider both the functional performance and its variation. The robustness measure of the *j*-th product operation configuration state for the *i*-th design configuration candidate with the target value of functional performance, F^{R}, and values of changeable non-design parameters, $\boldsymbol{X}^{\mathrm{C}}$, can be calculated by:

$$R^{(i,j)}\left(F^{\mathrm{R}},\ \boldsymbol{X}^{\mathrm{C}}\right) = 10\log_{10}\left(\frac{\mu_{\mathrm{F}}^2}{\sigma_{\mathrm{F}}^2}\right), \tag{4.35}$$

where μ_{F} represents the nominal value of functional performance, and $\sigma_{\mathrm{F}}{}^2$ represents the variance of the functional performance deviation.

Suppose both the target value of functional performance, F^{R}, and values of changeable non-design parameters, $\boldsymbol{X}^{\mathrm{C}}$, are changed with continuous values, and according to Sect. 4.3, the robustness measure of the *j*-th product operation configuration state for the *i*-th design configuration candidate, $R^{(i,j)}$, can be calculated by:

$$R^{(i,j)} = \int_{F^{\mathrm{R}}_{\min}}^{F^{\mathrm{R}}_{\max}} \int_{\boldsymbol{X}^{\mathrm{C}}_{\min}}^{\boldsymbol{X}^{\mathrm{C}}_{\max}} p\left(F^{\mathrm{R}},\ \boldsymbol{X}^{\mathrm{C}}\right) \cdot R^{(i,j)}\left(F^{\mathrm{R}},\ \boldsymbol{X}^{\mathrm{C}}\right) \cdot \mathrm{d}\boldsymbol{X}^{\mathrm{C}}\mathrm{d}F^{\mathrm{R}}, \tag{4.36}$$

where $p(F^{\mathrm{R}}, \boldsymbol{X}^{\mathrm{C}})$ is the probability density function with the required target value of functional performance and changeable non-design parameters. $F^{\mathrm{R}}_{\max}$ and $F^{\mathrm{R}}_{\min}$ are the upper and lower bounds of the required functional performance, respectively. $\boldsymbol{X}^{\mathrm{C}}_{\max}$ and $\boldsymbol{X}^{\mathrm{C}}_{\min}$ are the upper and lower bound vectors of the changeable non-design parameters, respectively.

According to Eq. (4.36), the overall robustness measure of the *i*-th design configuration candidate, $R^{(i)}$, can be calculated by:

$$R^{(i)} = \sum_{j=1}^{C^{(i)}} \left[P^{(i,j)} \cdot R^{(i,j)}\right], \tag{4.37}$$

where $P^{(i,j)}$ represents the probability of the product operation configuration in the *j*-th adaptation state for the *i*-th design configuration candidate, and $C^{(i)}$ represents the number of possible product operation configuration states for the *i*-th design configuration candidate.

According to Eq. (4.37), the best overall robustness measure, *R*, considering all design configuration candidates, can be obtained as:

$$R = R^{(r)} = \mathrm{Max}\left\{R^{(1)}, \ldots, R^{(n)}\right\}, \tag{4.38}$$

where *r* represents the *r*-th design configuration candidate that has been selected as the best design configuration.

(4) A two-level optimization method to identify the optimal configuration and parameters

Both configurations and parameters are considered for robust adaptable design. A two-level optimization method is developed for identifying the optimal configurations and parameters: configuration optimization for identifying the optimal design configuration candidate and parameter optimization for identifying the optimal values of product/operating parameters for each design configuration candidate. The two-level optimization is conducted in a loop. In this loop, first a feasible design configuration candidate is generated, and parameter optimization for this design configuration candidate is then carried out. The optimal design configuration candidate and associated parameter values are obtained based on the following rules:

- A feasible design configuration candidate is created from the hybrid AND-OR tree.
- All possible product operation configuration states are created from a design configuration candidate.
- The optimal parameter values of a design configuration candidate are obtained through the parameter optimization to maximize the robustness considering all changes in both configurations and parameters in the product operation stage.
- Different design configuration candidates are evaluated by their best robustness measures corresponding to the optimal parameter values in order to identify the best design configuration candidate.

The two-level optimization considering product configurations and their associated parameters is formulated as follows:

(1) *Configuration optimization*

The configuration optimization is carried out to identify the optimal design configuration candidate from all the feasible design configuration candidates. According to Eq. (4.38), the optimization model is formulated as follows:

$$\begin{aligned} &\text{Find: the } r\text{-th design configuration candidate} \\ &\text{Maximize} : R = R^{(r)} \\ &\text{Subject to: } 1 \le r \le n \end{aligned} \tag{4.39}$$

where n is the number of all feasible design configuration candidates. For each design configuration candidate, the best robustness corresponding to the optimal parameter values achieved through the parameter optimization is used to evaluate this design configuration candidate.

(2) *Parameter optimization*

The parameter optimization is conducted to identify the optimal parameter values of a design configuration candidate such that the product performances are the least sensitive to the parameter variations caused by uncertainties. The parameter optimization model is formulated as follows:

$$
\begin{aligned}
&\text{Find the values of un-adaptable design parameters } \boldsymbol{X}^{\mathrm{D}}\\
&\text{Maximize: } R^{(i)}\\
&\text{Subject to: } h_v\left[\boldsymbol{X}^{\mathrm{D}}, \boldsymbol{X}^{\mathrm{A}}, \boldsymbol{X}^{\mathrm{U}}, \boldsymbol{X}^{\mathrm{C}}\right] = 0;\\
&\qquad g_w\left[\boldsymbol{X}^{\mathrm{D}}, \boldsymbol{X}^{\mathrm{A}}, \boldsymbol{X}^{\mathrm{U}}, \boldsymbol{X}^{\mathrm{C}}\right] \leq 0;\\
&\qquad v = 1, 2, \ldots; w = 1, 2, \ldots
\end{aligned}
\tag{4.40}
$$

Genetic programming is employed for the configuration optimization. Genetic programming is an evolutionary method to solve an optimization problem when solutions of the optimization problem can be modeled by a tree data structure. Since multiple solutions are considered at the same time, genetic programming can prevent the optimal solution from falling into a local optimal point. In genetic programming, individuals (also called chromosomes) are used to describe multiple solutions in the population of a generation. Reproduction, crossover, and mutation are three operations used to evolve individuals from a generation to the next one with a better average evaluation measure [10]. The optimization based on genetic programming to identify the optimal product design configuration is conducted in the following steps:

Step 1. Creating the initial generation with n individuals. Each individual represents a feasible product design configuration candidate that is created randomly from the hybrid AND-OR tree.

Step 2. Obtaining the robustness measure of each individual in the current generation through the parameter optimization. Each robustness measure is used as the fitness of the corresponding individual.

Step 3. Creating a new generation from the current generation by repeating the three operations, i.e., reproduction, crossover, and mutation, until the number of individuals in the new generation reaches n.

Step 4. Selecting the newly created generation as the current generation to obtain the robustness measure of each individual in the current generation through the parameter optimization.

Step 5. If the average fitness of a generation cannot be significantly improved in the last m generations, or the predefined maximum generation has been reached, the evolution process should be stopped, and the best individual in the current generation is selected as the optimal product configuration. Otherwise, go to Step 3.

4.3.3 *Robust Adaptable Design of Open Architecture Products*

In addition to parameters and configurations, product architecture also provides influence on adaptability of the product. As discussed in Sect. 2.2.3, an open architecture product is an adaptable product with open interfaces that allow the third-party vendors to develop new add-on modules with new functions and connect these add-on modules through the open interfaces [13, 25]. Comparing with a closed architecture product for which all the modules of the product are specified in the product development stage, some modules of an open architecture product can be designed in the product operation stage.

In Sects. 4.3.1 and 4.3.2, the robust adaptable design methods considering changes of configurations and parameter values were developed. However, these robust adaptable design methods cannot be used directly to design an open architecture product. This section aims at developing modeling, evaluation, and optimization methods for the robust adaptable design of open architecture products to identify the optimal design of an open architecture product whose functional performance measures are the least sensitive to variations of product/operating parameters due to uncertainties.

(a) Modeling of open architecture product

An open architecture product (OAP) is composed of a platform, add-on modules, and open interfaces. During the operation stage, different add-on modules are connected to the platform through the open interfaces to form different operation configuration states for different functions. At the design stage, different design configuration candidates are provided such that the best one is selected considering the product robustness. Both operation configuration states and design configuration candidates are further modeled by parameters. The platform and add-on modules are connected through open interfaces.

An OAP usually has several open interfaces. For each interface, different add-on modules can be connected with the platform. For the OAP shown in Fig. 2.13 of Sect. 2.2.3, the platform (M^{P}) has l interfaces. For the i-th (i = 1, ..., l) interface, m_i specific add-on modules $M^{\mathrm{S}}_{i1}, \ldots, M^{\mathrm{S}}_{im_i}$ and the unknown add-on module M^{U}_i are considered.

During the product operation stage, each interface of the platform can be used to connect with different add-on modules. Each combination of add-on modules and the platform for achieving a certain function is called an operation configuration state. When n operation configuration states of an OAP are considered, collection of these operation configuration states, S, is defined by:

$$S = \{S_1, \ldots, S_n\}. \tag{4.41}$$

Each operation configuration state is associated with a probability representing the percentage of time; this operation configuration state is used considering all operation configuration states. Collection of the probabilities for the n operation configuration states, P, is defined by:

$$P = \{P_1, \ldots, P_n\}. \tag{4.42}$$

The i-th operation configuration state, S_i (i = 1, …, n), is modeled by its platform and add-on modules as follows:

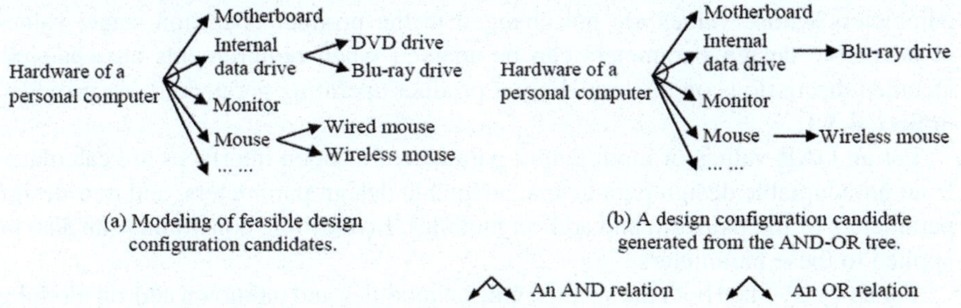

Fig. 4.21 Modeling of design configuration candidates

$$S_i = \{M_1, \ldots, M_l, M^{\mathrm{P}}\}, \tag{4.43}$$

where M_j represents the add-on module connected through the j-th open interface of platform M^{P} ($j = 1, \ldots, l$).

For the design of an OAP, different feasible design configuration candidates can be created from the same design requirements. The different configurations of platform and each of the specific add-on modules can be modeled by an AND-OR tree. In the AND-OR tree, when all the sub-nodes need to be selected to support a super-node, all of these sub-nodes are associated with an AND relation. As shown in Fig. 4.21a, for example, a personal computer is composed of a motherboard, an internal data drive, a monitor, a mouse, etc. When the super-node is supported by one of its sub-nodes in the process of design configuration candidate selection, all these sub-nodes are associated with an OR relation. For example, either the DVD drive or Blu-ray drive needs to be selected for the internal data drive. For each AND-OR tree, a configuration candidate with only AND relations can be created based on the following rules:

- The root node should be selected first.
- When sub-nodes of a selected node are associated with an AND relation, all these sub-nodes should be selected.
- When sub-nodes of a selected node are associated with an OR relation, only one of these sub-nodes should be selected.

Figure 4.21a shows an AND-OR tree and Fig. 4.21b shows a created feasible design configuration candidate with only AND relations. A complete design configuration candidate is modeled by a design configuration candidate for the platform and design configuration candidates for all the add-on modules.

A product operation configuration state or a product design configuration candidate can be further modeled by parameters. As discussed in Sect. 4.3.1, product/operating parameters can be classified into design parameters and non-design parameters. While values of design parameters need to be determined at the design stage, values of non-design parameters are provided as given conditions in design. The design parameters can be further classified into un-adaptable design parameters and adaptable design parameters. Comparing with the un-adaptable design parameters whose values are not changed in the product operation stage, values of adaptable design parameters can be adapted when requirements are changed. Detailed discussions on the modeling of product/operating parameters are provided in Sect. 4.3.1.

For an OAP, values of input/output parameters of open interfaces are calculated from un-adaptable design parameters, adaptable design parameters, and non-design parameters of the platform and add-on modules. In addition, constraints can also be applied to these parameters.

For an OAP, since both the specific add-on modules and unknown add-on modules need to be considered, values of the adaptable design parameters are changed in two different ways in the product operation stage.

- Change of Adaptable Design Parameter Values with Specific Add-on Modules

When all of the add-on modules are specific add-on modules, the relation among product functional performance F, un-adaptable design parameters $\boldsymbol{X}^{\mathrm{D}}$, adaptable design parameters $\boldsymbol{X}^{\mathrm{A}}$, and non-design parameters $\boldsymbol{X}^{\mathrm{N}}$ can be described by $F = f(\boldsymbol{X}^{\mathrm{D}}, \boldsymbol{X}^{\mathrm{A}}, \boldsymbol{X}^{\mathrm{N}})$. In this case, the adaptable design parameters are calculated from the target functional performance, un-adaptable design parameters, and non-design parameters. In other words, values of $\boldsymbol{X}^{\mathrm{A}}$ are calculated from F, $\boldsymbol{X}^{\mathrm{D}}$, and $\boldsymbol{X}^{\mathrm{N}}$. Unique values of the adaptable design parameters should be obtained according to the experience of design engineer.

- Change of Adaptable Design Parameter Values with Unknown Add-on Modules

When some add-on modules are unknown add-on modules, since these unknown add-on modules are only defined by constraints of interface parameters, the relation among product functional performance F, un-adaptable design parameters $\boldsymbol{X}^{\mathrm{D}}$, adaptable design parameters $\boldsymbol{X}^{\mathrm{A}}$, non-design parameters $\boldsymbol{X}^{\mathrm{N}}$, input parameters of unknown add-on modules $\boldsymbol{I}^{\mathrm{U}}$, and output parameters of unknown add-on modules $\boldsymbol{O}^{\mathrm{U}}$ can be described by $F = f(\boldsymbol{X}^{\mathrm{D}}, \boldsymbol{X}^{\mathrm{A}}, \boldsymbol{X}^{\mathrm{N}}, \boldsymbol{I}^{\mathrm{U}}, \boldsymbol{O}^{\mathrm{U}})$. In this case, the adaptable design parameters are calculated from the target functional performances, un-adaptable design parameters, non-design parameters, and input/output interface parameters of the unknown add-on modules. In other words, values of $\boldsymbol{X}^{\mathrm{A}}$ are calculated from F, $\boldsymbol{X}^{\mathrm{D}}$, $\boldsymbol{X}^{\mathrm{N}}$, $\boldsymbol{I}^{\mathrm{U}}$, and $\boldsymbol{O}^{\mathrm{U}}$. Since the input/output interface parameters are usually defined by discrete and continuous parameters, all these possible input/output interface parameter values have to be considered. In addition, experience of design engineers is used to obtain the unique values of adaptable design parameters to avoid the problem of multiple solutions.

(b) Modeling of interfaces for interactions between platform and add-on modules

For an open architecture product, the add-on modules are connected with the platform through open interfaces. Interactions between the platform and add-on modules are defined by input and output parameters of the open interfaces. Figure 4.22 shows interactions between a platform and add-on modules through interfaces. Values of input parameters of an interface for an add-on module are determined by values of

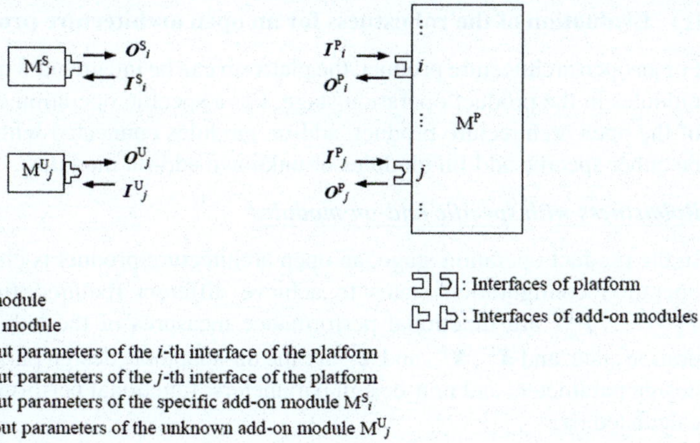

Fig. 4.22 Interactions between platform and add-on modules

the corresponding output parameters of the interface for the platform, while values of input parameters of an interface for the platform are determined by values of the corresponding output parameters of the interface for the add-on module.

When a specific add-on module is connected with the platform, such as M_i^S for the *i*-th interface shown in Fig. 4.22, values of input and output parameters of the interfaces for the platform and specific add-on module are calculated from parameters of the platform and add-on module.

When an unknown module is considered, such as M_j^U for the *j*-th interface shown in Fig. 4.22, constraints are used to define input and output parameters of the interface for the unknown add-on module. The input and output parameters of the interface for the unknown add-on module are defined by continuous and discrete parameters.

When the interface input/output parameters are continuous parameters, they are defined by:

$$I_j^U \in \left[L_j^I,\ U_j^I\right], \text{ and } O_j^U \in \left[L_j^O,\ U_j^O\right], \tag{4.44}$$

where I_j^U and O_j^U are input and output parameters of the unknown add-on module, and $[L_j^I, U_j^I]$ and $[L_j^O, U_j^O]$ are the lower/upper boundaries of I_j^U and O_j^U, respectively.

When the interface input/output parameters are discrete parameters, they are defined by:

$$I_j^U \in \{A_j^I, B_j^I, \ldots\}, \text{ and } O_j^U \in \{A_j^O, B_j^O, \ldots\}, \tag{4.45}$$

where $\{A_j^I, B_j^I, \ldots\}$ and $\{A_j^O, B_j^O, \ldots\}$ are the feasible values of I_j^U and O_j^U, respectively.

(c) Evaluation of the robustness for an open architecture product

For an open architecture product, the platform can be integrated with different add-on modules in the product operation stage. For a specific operation configuration state of the open architecture product, add-on modules connected with the platform can be either specific add-on modules or unknown add-on modules.

Robustness with specific add-on modules

In the product operation stage, an open architecture product is changed to different operation configuration states to achieve different required functions. Suppose $F_1^{(s)}, \ldots, F_{n_s}^{(s)}$ are functional performance measures of the *s*-th operation configuration state, and $\boldsymbol{X}^{\mathrm{D}}$, $\boldsymbol{X}^{\mathrm{A}}$, and $\boldsymbol{X}^{\mathrm{N}}$ are the un-adaptable design parameters, adaptable design parameters and non-design parameters; functional performance, $F_k^{(s)}$, can be calculated by:

$$F_k^{(s)} = f_k^{(s)}\left(\boldsymbol{X}^{\mathrm{D}}, \boldsymbol{X}^{\mathrm{A}}, \boldsymbol{X}^{\mathrm{N}}\right), k = 1, \ldots, n_s,\ s = 1, \ldots, n, \tag{4.46}$$

where $f_k^{(s)}()$ represents the relation among parameters $\boldsymbol{X}^{\mathrm{D}}$, $\boldsymbol{X}^{\mathrm{A}}$, $\boldsymbol{X}^{\mathrm{N}}$ and $F_k^{(s)}$, n_s is the number of functional performance measures for the *s*-th operation configuration state, and *n* is the total number of operation configuration states.

Under influence of uncertainties, parameters can deviate from their target values. Suppose $\Delta\boldsymbol{X}^{\mathrm{D}}$, $\Delta\boldsymbol{X}^{\mathrm{A}}$, and $\Delta\boldsymbol{X}^{\mathrm{N}}$ are deviations of parameters $\boldsymbol{X}^{\mathrm{D}}$, $\boldsymbol{X}^{\mathrm{A}}$, and $\boldsymbol{X}^{\mathrm{N}}$, respectively, and deviation of the functional performance, $\Delta F_k^{(s)}$, can be calculated by:

$$\Delta F_k^{(s)} = f_k^{(s)}(\boldsymbol{X}^{\mathrm{D}} + \Delta\boldsymbol{X}^{\mathrm{D}}, \boldsymbol{X}^{\mathrm{A}} + \Delta\boldsymbol{X}^{\mathrm{A}}, \boldsymbol{X}^{\mathrm{N}} + \Delta\boldsymbol{X}^{\mathrm{N}}) - F_k^{(s)}. \tag{4.47}$$

Since both functional performance and variations of the performance are considered in the signal-to-noise ratio (SNR), the SNR has been selected as the robustness evaluation measure. Therefore, the robustness measure of the *s*-th operation configuration state for the *i*-th design configuration candidate considering parameters $\boldsymbol{X}^{\mathrm{D}}$ can be calculated by:

$$R^{(i,s)}(\boldsymbol{X}^{\mathrm{D}}) = \sum_{k=1}^{n_s}\left[\omega_k^{(s)} 10 \log_{10}\left(\frac{\mu_k^2}{\sigma_k^2}\right)\right], \tag{4.48}$$

where μ_k represents the nominal value of functional performance which can be calculated using Eq. (4.46), σ_k^2 represents the variance of the functional performance deviation which can be calculated using Eq. (4.47), $\omega_k^{(s)}$ represents the weighting factor for functional performance $F_k^{(s)}$, and n_s is the number of functional performance for the *s*-th operation configuration state.

The overall robustness of the *i*-th design configuration candidate considering all the *n* operation configuration states with the un-adaptable design parameters $\boldsymbol{X}^{\mathrm{D}}$ can be calculated by:

$$R^{(i)}(\boldsymbol{X}^{\mathrm{D}}) = \sum_{s=1}^{n} [\mathrm{P}_s \cdot R^{(i,s)}(\boldsymbol{X}^{\mathrm{D}})], \tag{4.49}$$

where P_s represents the probability of use for the s-th operation configuration state.

From Eq. (4.49), the optimal parameters of $\boldsymbol{X}^{\mathrm{D}}$ to achieve the best overall robustness measure for the i-th design configuration candidate can be obtained through optimization:

$$R^{(i)} = \max_{\text{w.r.t. } \boldsymbol{X}^{\mathrm{D}}} : R^{(i)}(\boldsymbol{X}^{\mathrm{D}}). \tag{4.50}$$

Robustness with unknown add-on modules

For an unknown add-on module, since values of the interface input/output parameters can be varied as continuous values and discrete values, the product performance measures are also influenced by the changes of these interface input/output parameters. Suppose $\boldsymbol{I}^{\mathrm{U}}$ and $\boldsymbol{O}^{\mathrm{U}}$ represent input and output parameters of unknown add-on modules, and functional performance $F_k^{(s)}$ can be calculated by:

$$F_k^{(s)} = f_k^{(s)}\left(\boldsymbol{X}^{\mathrm{D}}, \boldsymbol{X}^{\mathrm{A}}, \boldsymbol{X}^{\mathrm{N}}, \boldsymbol{I}^{\mathrm{U}}, \boldsymbol{O}^{\mathrm{U}}\right), k = 1, \ldots, n_s,\ s = 1, \ldots, n, \tag{4.51}$$

where $f_k^{(s)}$ () represents the relation among parameters $\boldsymbol{X}^{\mathrm{D}}$, $\boldsymbol{X}^{\mathrm{A}}$, $\boldsymbol{X}^{\mathrm{N}}$, $\boldsymbol{I}^{\mathrm{U}}$, $\boldsymbol{O}^{\mathrm{U}}$ and $F_k^{(s)}$, n_s is the number of functional performance measures for the s-th operation configuration state, and n is the total number of operation configuration states.

Suppose $\Delta\boldsymbol{I}^{\mathrm{U}}$ and $\Delta\boldsymbol{O}^{\mathrm{U}}$ are deviations of parameters $\boldsymbol{I}^{\mathrm{U}}$ and $\boldsymbol{O}^{\mathrm{U}}$, respectively, and deviation of the functional performance, $\Delta F_k^{(s)}$, can be calculated by:

$$\Delta F_k^{(s)} = f_k^{(s)}(\boldsymbol{X}^{\mathrm{D}} + \Delta\boldsymbol{X}^{\mathrm{D}}, \boldsymbol{X}^{\mathrm{A}} + \Delta\boldsymbol{X}^{\mathrm{A}}, \boldsymbol{X}^{\mathrm{N}} + \Delta\boldsymbol{X}^{\mathrm{N}}, \boldsymbol{I}^{\mathrm{U}} + \Delta\boldsymbol{I}^{\mathrm{U}}, \boldsymbol{O}^{\mathrm{U}} + \Delta\boldsymbol{O}^{\mathrm{U}}) - F_k^{(s)}. \tag{4.52}$$

Robustness measure of the s-th operation configuration state for the i-th design configuration candidate considering parameters $\boldsymbol{X}^{\mathrm{D}}$, $\boldsymbol{I}^{\mathrm{U}}$, and $\boldsymbol{O}^{\mathrm{U}}$ can be calculated by:

$$R^{(i,s)}(\boldsymbol{X}^{\mathrm{D}}, \boldsymbol{I}^{\mathrm{U}}, \boldsymbol{O}^{\mathrm{U}}) = \sum_{k=1}^{n_s} \left[\omega_k^{(s)} 10 \log_{10}\left(\frac{\mu_k^2}{\sigma_k^2}\right)\right], \tag{4.53}$$

where μ_k represents the nominal value of functional performance which can be calculated using Eq. (4.51), σ_k^2 represents the variance of the functional performance deviation which can be calculated using Eq. (4.52), $\omega_k^{(s)}$ represents the weighting factor for functional performance $F_k^{(s)}$, and n_s is the number of functional performance for the s-th operation configuration state.

Since values of interface input/output parameters for unknown add-on modules are varied as continuous and discrete parameters, the influence of the different values of interface input/output parameters for the unknown add-on modules on product performance measures has to be considered. Both statistical and worst-case methods are developed considering possible changes of interface parameter values for unknown add-on modules as follows.

- The statistical method for the calculation of robustness

Changes of input/output interface parameters can be continuous and discrete. If $\boldsymbol{I}^{\mathrm{U}}$ and $\boldsymbol{O}^{\mathrm{U}}$ are continuous parameters, $\mathrm{P}(\boldsymbol{I}^{\mathrm{U}})$ and $\mathrm{P}(\boldsymbol{O}^{\mathrm{U}})$ are the probability density functions of $\boldsymbol{I}^{\mathrm{U}}$ and $\boldsymbol{O}^{\mathrm{U}}$ for the s-th operation configuration state, and $\boldsymbol{I}_{\mathrm{L}}^{\mathrm{U}}$, $\boldsymbol{I}_{\mathrm{U}}^{\mathrm{U}}$, $\boldsymbol{O}_{\mathrm{L}}^{\mathrm{U}}$, and $\boldsymbol{O}_{\mathrm{U}}^{\mathrm{U}}$ are the lower/upper boundaries of the $\boldsymbol{I}^{\mathrm{U}}$ and $\boldsymbol{O}^{\mathrm{U}}$, the robustness measure of the s-th operation configuration state for the i-th design configuration candidate considering the un-adaptable design parameters $\boldsymbol{X}^{\mathrm{D}}$ is calculated by:

$$R^{(i,s)}(\boldsymbol{X}^{\mathrm{D}}) = \int_{\boldsymbol{I}_{\mathrm{L}}^{\mathrm{U}}}^{\boldsymbol{I}_{\mathrm{U}}^{\mathrm{U}}} \int_{\mathbf{O}_{\mathrm{L}}^{\mathrm{U}}}^{\mathbf{O}_{\mathrm{U}}^{\mathrm{U}}} \mathrm{P}(\boldsymbol{I}^{\mathrm{U}})\mathrm{P}(\boldsymbol{O}^{\mathrm{U}})R^{(i,s)}(\boldsymbol{X}^{\mathrm{D}}, \boldsymbol{O}^{\mathrm{U}}, \boldsymbol{I}^{\mathrm{U}})\mathrm{d}\boldsymbol{O}^{\mathrm{U}}\mathrm{d}\boldsymbol{I}^{\mathrm{U}}. \tag{4.54}$$

If $\boldsymbol{I}^{\mathrm{U}}$ and $\boldsymbol{O}^{\mathrm{U}}$ are discrete parameters for the s-th operation configuration state, the robustness measure of the s-th operation configuration state for the i-th design configuration candidate considering the un-adaptable design parameters $\boldsymbol{X}^{\mathrm{D}}$ is calculated by:

$$R^{(i,s)}(\boldsymbol{X}^{\mathrm{D}}) = \sum\sum \mathrm{P}(\boldsymbol{I}^{\mathrm{U}})\mathrm{P}(\boldsymbol{O}^{\mathrm{U}})R^{(i,s)}(\boldsymbol{X}^{\mathrm{D}}, \boldsymbol{I}^{\mathrm{U}}, \boldsymbol{O}^{\mathrm{U}}). \tag{4.55}$$

- The worst-case method for calculation of robustness

The worst-case robustness considering changes of input/output interface parameters is calculated. The robustness measure of the s-th operation configuration state for the i-th design configuration candidate considering the un-adaptable design parameters $\boldsymbol{X}^{\mathrm{D}}$ is calculated by:

$$R^{(i,s)}(\boldsymbol{X}^{\mathrm{D}}) = \min_{w.r.t.\ \boldsymbol{I}^{\mathrm{U}},\, \boldsymbol{O}^{\mathrm{U}}} : \; R^{(i,s)}(\boldsymbol{X}^{\mathrm{D}}, \boldsymbol{I}^{\mathrm{U}}, \boldsymbol{O}^{\mathrm{U}}). \tag{4.56}$$

The best robustness measure for the i-th design configuration candidate, $R^{(i)}$, considering un-adaptable design parameters $\boldsymbol{X}^{\mathrm{D}}$ can be calculated using Eq. (4.50) in which the $R^{(i,s)}$ is calculated from Eqs. (4.54–4.55), or Eq. (4.56).

(d) A multi-level optimization model to identify the optimal design

A multi-level optimization model is developed to identify the best design configuration candidate and its un-adaptable design parameter values considering robustness. The multi-level optimization is formulated as follows:

The first-level optimization:

The first-level optimization is used to identify the optimal design configuration candidate from all the feasible design configuration candidates. The first-level optimization model is formulated as follows:

$$\begin{aligned} &\text{Find: the } i\text{-th design configuration candidate} \\ &\text{Maximize: } R = R^{(i)} \\ &\text{Subject to: } 1 \le i \le p \end{aligned} \tag{4.57}$$

where i represents the i-th design configuration candidate, and p is the number of all feasible design configuration candidates.

The second-level optimization:

The second-level optimization is used to identify the optimal un-adaptable design parameter values of a design configuration candidate. The second-level optimization model is formulated as follows:

$$\begin{aligned} &\text{Find: the un-adaptable design parameters } \boldsymbol{X}^{\mathrm{D}} \\ &\text{Maximize: } R^{(i)} = \sum_{s=1}^{n} \left[P_s \cdot R^{(i,s)}(\boldsymbol{X}^{\mathrm{D}}) \right] \\ &\text{Subject to: } \boldsymbol{X}_{\mathrm{L}}^{\mathrm{D}} \le \boldsymbol{X}^{\mathrm{D}} \le \boldsymbol{X}_{\mathrm{U}}^{\mathrm{D}} \end{aligned} \tag{4.58}$$

where $\boldsymbol{X}_{\mathrm{L}}^{\mathrm{D}}$ and $\boldsymbol{X}_{\mathrm{U}}^{\mathrm{D}}$ represent the lower boundaries and upper boundaries of $\boldsymbol{X}^{\mathrm{D}}$, respectively. The robustness measure $R^{(i,s)}(\boldsymbol{X}^{\mathrm{D}})$ in Eq. (4.58) is obtained from Eqs. (4.49, 4.54–4.55) or Eq. (4.56).

The third-level optimization:

When the worst-case method is used to evaluate the product robustness, the third-level optimization is needed. The third-level optimization is used to calculate the worst-case robustness considering changes of input/output interface parameters. According to Eq. (4.56), the third-level optimization model is formulated as follows:

$$\begin{aligned} &\text{Find: the input and output parameters } \boldsymbol{I}^{\mathrm{U}}, \boldsymbol{O}^{\mathrm{U}} \\ &\text{Minimize: } R^{(i,s)}(\boldsymbol{X}^{\mathrm{D}}) = R^{(i,s)}(\boldsymbol{X}^{\mathrm{D}}, \boldsymbol{I}^{\mathrm{U}}, \boldsymbol{O}^{\mathrm{U}}) \\ &\text{Subject to: } \boldsymbol{I}_{\mathrm{L}}^{\mathrm{U}} \le \boldsymbol{I}^{\mathrm{U}} \le \boldsymbol{I}_{\mathrm{U}}^{\mathrm{U}};\ \boldsymbol{O}_{\mathrm{L}}^{\mathrm{U}} \le \boldsymbol{O}^{\mathrm{U}} \le \boldsymbol{O}_{\mathrm{U}}^{\mathrm{U}} \end{aligned} \tag{4.59}$$

where $\boldsymbol{I}_{\mathrm{L}}^{\mathrm{U}}$, $\boldsymbol{I}_{\mathrm{U}}^{\mathrm{U}}$, $\boldsymbol{O}_{\mathrm{L}}^{\mathrm{U}}$, and $\boldsymbol{O}_{\mathrm{U}}^{\mathrm{U}}$ are the lower/upper boundaries of $\boldsymbol{I}^{\mathrm{U}}$ and $\boldsymbol{O}^{\mathrm{U}}$, respectively.

Parameter optimization is conducted by using the constrained nonlinear optimization function provided in the MATLAB optimization toolbox, while the configuration optimization is conducted by using genetic programming [16]. The basic steps for the optimization based on genetic programming to identify the optimal product design configuration have been explained in Sect. 4.3.2.

4.4 Adaptable Design Considering Adaptation Operations

4.4.1 Assembly Operation Sequences for Adaptable Design

An adaptable product must meet diversified requirements of product users. Functions of an adaptable product should be easily upgraded in its lifespan, which can be implemented by replacing or updating function modules of the adaptable product. An efficient method is required to meet this need [20].

4.4.1.1 Serial Assembly and Parallel Assembly

Serial assembly and parallel assembly are introduced for product assembly sequence planning (ASP). Based on the assembly operation modeling scheme introduced in Sect. 2.4, a serial assembly is defined based on following Condition 1, and a parallel assembly is defined based on Condition 2.

Condition 1: $|AC_{i(i+1)}|=|AC_{(i+1)i}|=1 \vee 2$, $|AC_{ij}|=|AC_{ji}|=0 \vee 9$, $|j-i|>1$, $i = 1, 2, \ldots, n_{con-1}$, $j = 1, 2, \ldots, n_{con}$,

where $|AC_{i(i+1)}|=|AC_{(i+1)i}|=1 \vee 2$ shows that there are connections between P_i and P_{i+1}. $|AC_{ij}|=|AC_{ji}|=0 \vee 9$ means that there are no connections between P_i and P_{i+1}. n_{con} is the total number of parts in a serial assembly. The structural diagram of a serial assembly is shown in Fig. 4.23a. Equation (4.60) is an example of the assembly constraint matrix of a serial assembly.

Condition 2: $|AC_{ij}|=|AC_{ji}|=1 \vee 2$, $|AC_{jl}|=0 \vee 9$, $j \neq l \neq i$, $i = 1$, $j = 2, \ldots, n_{para}$, $l = 3, \ldots, n_{para}$,

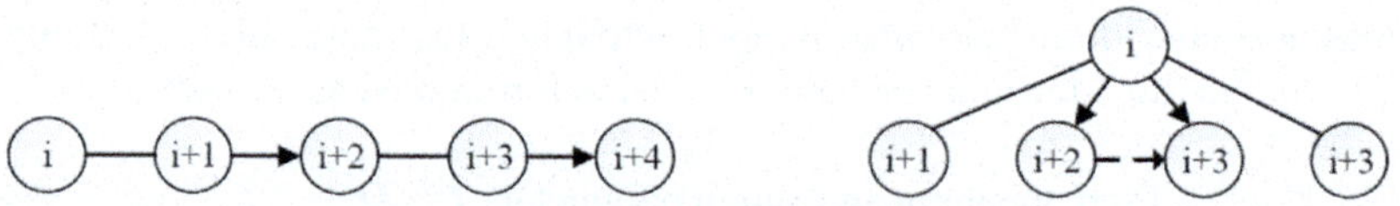

(a) Structure diagram of serial assembly. (b) Structure diagram of parallel assembly.

Fig. 4.23 Structure diagram of serial and parallel assemblies

where $|AC_{ij}|$=$|AC_{ji}|$=$1 \vee 2$ shows that there are connections between P_i and P_j. $|AC_{ji}|$=$0 \vee 9$ shows that there are not connections between P_i and P_j. n_{para} is the total number of parts in a parallel assembly. The structure graph of a parallel assembly is shown in Fig. 4.23b. Equation (4.61) is an example of the assembly constraint matrix of a parallel assembly.

$$AC^{\text{con}} = \begin{matrix} & i\ i_1\ i_2\ i_3\ i_4 \\ \begin{matrix} i \\ i_1 \\ i_2 \\ i_3 \\ i_4 \end{matrix} & \begin{bmatrix} 0 & 2 & 0 & 0 & 0 \\ 2 & 0 & 1 & 0 & 0 \\ 0 & -1 & 0 & 2 & 0 \\ 0 & 0 & 2 & 0 & 1 \\ 0 & 0 & 0 & -1 & 0 \end{bmatrix} \end{matrix}, \tag{4.60}$$

$$AC^{para} = \begin{matrix} & i\ i_1\ i_2\ i_3\ i_4 \\ \begin{matrix} i \\ i_1 \\ i_2 \\ i_3 \\ i_4 \end{matrix} & \begin{bmatrix} 0 & 2 & 1 & 1 & 2 \\ 2 & 0 & 0 & 0 & 0 \\ -1 & 0 & 0 & 9 & 0 \\ -1 & 0 & -9 & 0 & 0 \\ 2 & 0 & 0 & 0 & 0 \end{bmatrix} \end{matrix}. \tag{4.61}$$

Search steps of a serial or parallel assembly are as follows:

(1) If an assembly constraint matrix meets both Conditions 1 and 2 at the same time, it is a mixed serial and parallel assembly. If not, moving to (2).
(2) If an assembly constraint matrix meets Condition 1, it is a serial assembly. If not, go to 3).
(3) If an assembly constraint matrix meets Condition 2, it is a parallel assembly.

4.4.1.2 ASP for Components Inside a Module

Assembly sequences for components in a module are generated through a process of resetting the assembly constraint matrix following changes of constraint relations of components in the assembly, such as assembly constraint matrix AC_k of module M_k. The process resets the matrix to make all nonzero elements on row i into positive to assemble component P_i. The processes are shown in Fig. 4.24 as follows:

(1) The selection of base part: A component with a large and stable surface is selected as a base part for other components to install. Priority is given to the part meeting serial assembly, then parallel assembly to reduce the operation time. The row and column of the base part are moved into the first row and column in matrix AC_k to get AC_k^1, and then, the first row and column of AC_k^1 are set to 0 to get AC_k^{1-1}.
(2) Search for the parallel assembly: If there are elements in their rows with values of $0, 9$, or -9 in AC_k^{1-1}, they meet the parallel assembly (h_{para1}) condition. Actions

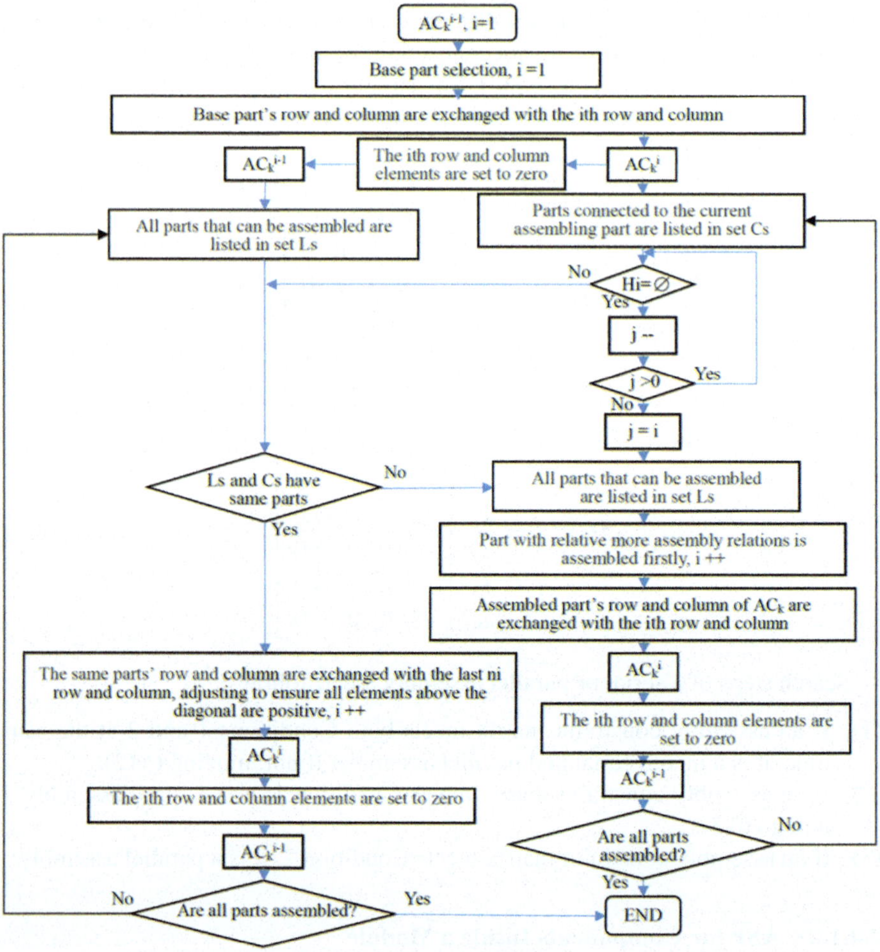

Fig. 4.24 Flowchart of the assembly for parts in a module

will move these elements into the last rows and columns in the matrix to get AC_k^2, and adjust their values in the diagonal into positive, and assign 0 to their rows and columns to get AC_k^{2-1}. There may be different sequences of components in the parallel assembly. Otherwise, $AC_k^2 = AC_k^1$, $AC_k^{2-1} = AC_k^{1-1}$.

(3) Search for the serial assembly: If there are elements in their rows with values of 1, without -1, -9 in AC_k^{1-1}, they meet the serial assembly. A component with more relations with other components is assembled first. The row and column of this element are exchanged with the second row and column in AC_k^2 to get AC_k^3, and assign 0 to its row and column to get AC_k^{3-1}.

(4) Analysis of AC_k^{3-1}: If elements of the part in row of AC_k^{3-1} are zero, the process is same as (2) for the parallel assembly. If there is no negative value in the row

of element searched in (3), the process is same as (3) for the serial assembly. If there is a negative value, elements with no-negative values will be selected to assemble in AC_k^{3-1}. A component with more relations with other components is assembled first.AC_k^{3-1} is then adjusted to get AC_k^{4-1}. The process continues until all components are assembled to form the final assembly matrix.

4.4.1.3 ASP for Modules of a Product

The module sequence can be similarly searched in the module level of the structure graph. As shown in Fig. 4.25, WT is used to represent a transform matrix of the sequence generation. The support platform module is selected as the base in the module assembly. Rows and columns of WT are exchanged based on the selected support module, followed the assembled modules gradually until all the modules and components in the module level are assembled. Modules are assembled following a sequence of common modules, customized modules, and personalized modules in a product.

4.4.2 Disassembly Operation Sequences for Adaptable Design

The total disassembly time is used as a criterion to evaluate the disassembly sequence for the disassembly cost. Disassembly time varies in different disassembly operations. When a target component is selected, sub-assemblies containing the target component can be identified. These sub-assemblies will be considered to be separated from the product. A disassembly sequence searching process starts from the target component to the end product. Two types of matrices are used to generate the disassembly sequence, namely the multi-level constraint matrices and fastener-component matrix. Multi-level constraint matrices are used to examine constraints between components and to skip infeasible sequences. Fastener-component matrix is used to find fasteners that should be removed in advance before the component can be disassembled. For those fasteners or constraints that can only be removed by destructive disassembly operations, they will be identified and removed by specific tools [31].

Based on the multi-level constraint matrices, the searching algorithm starts from the end product to the target component. For example, if a target component T is selected for recycling, the "leaf nodes" or sub-assemblies with target component T can be identified. Only for those sub-assemblies that contain target component T will be searched. The searching algorithm will stop when the target component is reached.

The search algorithm is performed to get the target component from the lowest level of sub-assembly based on the product bill of material. For example, if a sub-assembly including the target component has n components in total, a permutation

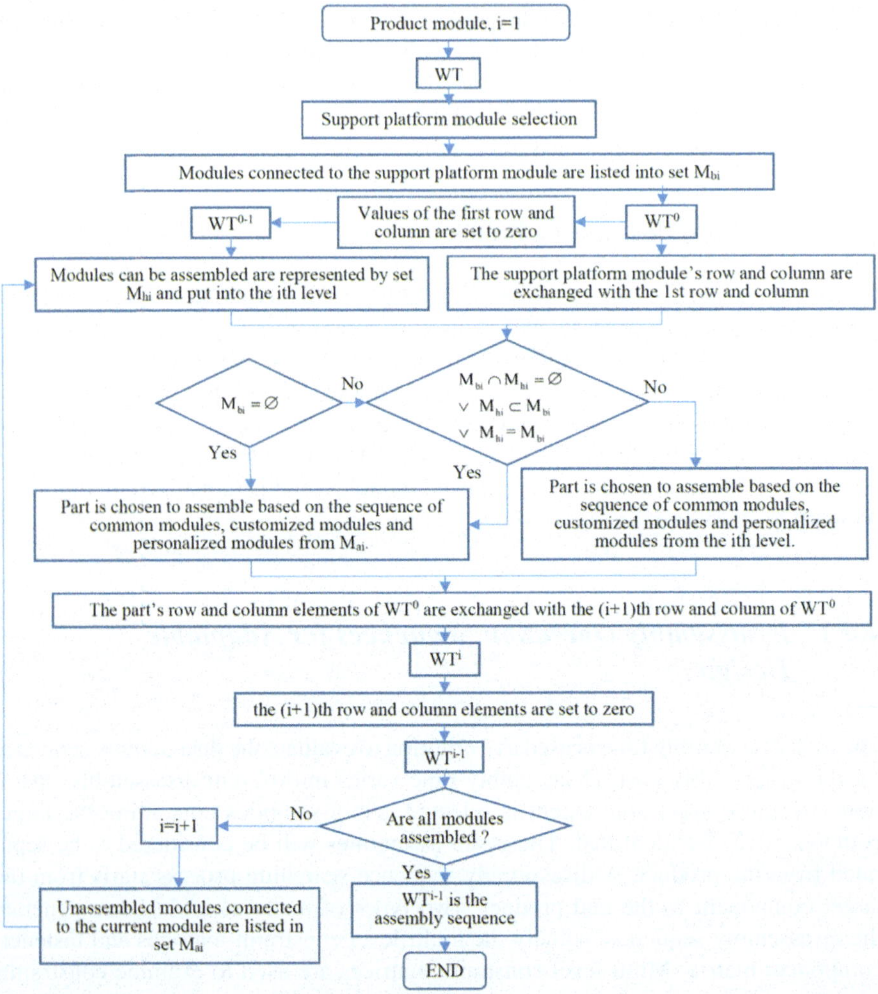

Fig. 4.25 Flowchart of ASP for modules of a part

will be made to form sequences. All sequences are examined to determine the disassembly feasibility. For example, the algorithm checks disassembly sequence S(i). The first component S(i,1) in sequence S(i) will be examined using the corresponding constraint matrix. If component S(i,1) can be removed, a following check with the fastener matrix is applied to find fasteners that should be removed in advance. Then, the second component S(i,2) will be checked. If component S(i,1) fails the feasibility check, it means that the component cannot be removed, and hence, sequence S(i) is infeasible and will be skipped. All sequences are ended up by the target component T and all feasible disassembly sequences are recorded in the "All_Arrays". The disassembly cost comparison is made among all feasible sequences and the optimal one will be selected.

After searching operations of each level's matrices, sequences of disassembling each level's sub-assemblies will be combined. By evaluating all feasible disassembly sequences, an optimal sequence can be decided. The disassembly sequence planning (DSP) process can be summarized in following steps:

(1) Determining a target component.
(2) Search for the lowest level of sub-assemblies which contains the target component.
(3) Generating all feasible disassembly sequences using the multi-level constraint matrices and fastener-component matrix.
(4) Comparing feasible sequences using predefined criteria, such as the disassembly time.
(5) Selecting the optimal disassembly sequence.

4.5 Adaptable Design Considering Mass Personalization

A product that meets the maximum user requirement will have the highly competitive ability in the market. A product to meet the individual need is called the personalized product. The personalized product that can be made using mass production is called the mass personalization, which is highly pursued by industry to gain the maximum market share using the existing production capacity.

One of the key measures of the personalized product is the interaction between the user requirement for the product and the way of the product to respond to the user need. As there is a variety of users with different needs for a product, the mass personalization is a challenge to industrial implementation using the traditional design and manufacturing method. An effective and cognitive method is needed for product personalization [24].

Open architecture products (OAPs) can effectively support the implementation of product personalization. However, the method to form the open architecture product is still under research. Following parts of this section introduce the adaptable design of OAPs for mass personalization.

4.5.1 Module Planning for OAPs

Module types of OAPs can be planned by the integration of extended quality function deployment (QFD) and axiomatic design methods [37]. The degree of variety (DV) is expanded as a quantitative measure for various components to decide the difference of OAP function modules. The weighted coefficient of variance index is used as a quantitative description of the variance measure to decide the module type. The index of weighted coefficients of variances is used as a quantitative description of the variance measure as follows:

$$\mathrm{VC}_i = \sum_{j=1}^{n} \mathrm{TCR}_j \times \mathrm{EC}_j, \tag{4.62}$$

$$\mathrm{rTIW}_i\% = \overline{\mathrm{TIW}_i} / \sum_{i=1}^{m} \mathrm{TIW}_i \times 100\%, \tag{4.63}$$

$$\mathrm{wCOV}_i = \mathrm{rTIW}_i \times \mathrm{VC}_i, \tag{4.64}$$

where i is a component number in the design matrix; VC_i is the variance coefficient of technical requirements; TCR_j is the correlation coefficient between the customer requirement and technical requirement; EC_j is the expected requirement change; TIW_i is the technical importance weight; rTIW is the relative importance of the average of technical requirements; and wCOV is the weighted coefficient of variance.

The function variance is then decided by values of wCOVs for OAPs' modules either meeting constant or variant function needs (FNs). The constant FNs form common platform modules. Variant FNs determine customized and personalized modules. Designers can specify FNs based on the function variant in product applications. DV_i is used to assess components' contributions in different OAPs' modules as follows:

$$x_{ci} = \sum_{q=1}^{lc} (\mathrm{RW}_q \times C_{qi}) \text{ for constant TVs}, \tag{4.65}$$

$$x_{vi} = \sum_{r=1}^{lv} (\mathrm{RW}_r \times C_{ri}) \text{ for variant TVs}, \tag{4.66}$$

$$\mathrm{DV}_i(\%) = X_{Vi} / X_{Ci} \times 100(\%), \tag{4.67}$$

where x_{ci} is constant target values TVs*;* x_{vi} is variant TVs; RW *is* the weight of technical requirements; C_{qi} is relationship values between components and FNs, decided by TVs(C); l_c is numbers of the total constant TVs(C)*;* C_{ri} is relationship values between components and FNs, decided by TVs(V); and l_v is numbers of the total variant TVs(V).

The design structure matrix (DSM) is finally formed to cluster product components to indicate degrees of the component variable. According to DVs, components are divided into three types of modules based on two thresholds of the variant. Thresholds are set based on variant FNs over constant FNs. From the assumption of need changes, different values of thresholds can be applied using manufacturers' data and user requirement changes. Following relationships of components and their *DV* values, components can be clustered into modules for an OAP.

4.5.2 Detail Design of OAPs

The detail design transforms module concepts into physical structures of an OAP, which is conducted by converting engineering recourses into physical utilizable formats. Mechanical components are designed to support energy/force operations. For example, an electric vehicle can be driven by the electrical power through energy/force operations as shown in Fig. 4.26.

Therefore, finding mechanical components for the energy/force action is critical in the OAP detail design. Therefore, a process is suggested in this research to transform the design concept into detailed mechanical elements through the energy/force transformation. Figure 4.27 shows a process of the energy/force transformation.

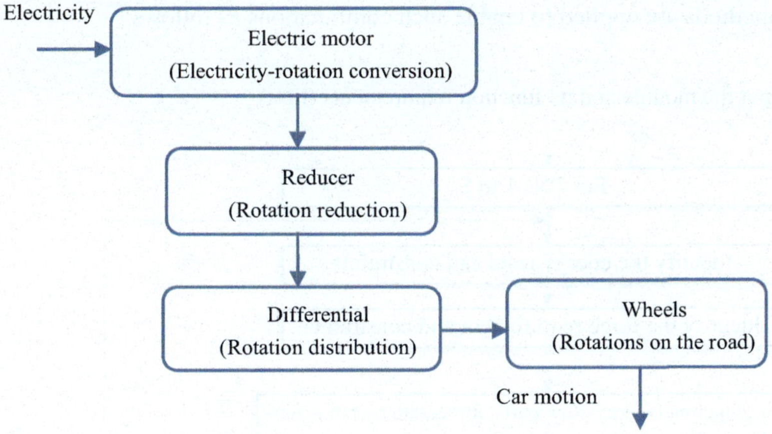

Fig. 4.26 Driving of an electric vehicle

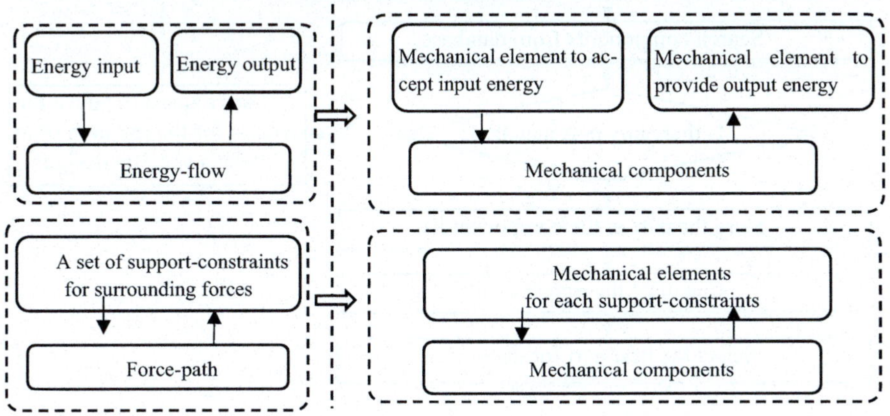

Fig. 4.27 Transformation in the energy/force flow system

Design for the transformation first searches available components to meet function needs from the part database formed based on data in handbooks of engineering design. The component search uses criteria including functions to transfer energy and force; geometry to meet needs of the component shape, size, dimension and tolerance, raw material; position to support and link other parts; features in manufacturing, assembly, and disassembly. The special design is considered only if there is no required component available to reduce manufacturing cost. A flowchart of the search process is shown in Fig. 4.28 [24].

Therefore, in the detail design, mechanical components are first searched based on the need of transferring energy/force to generate the power/motion required in the conceptual design. These components are then combined to form the function module. When integrating components together, limitations of materials and manufacturing methods may cause conflicts, such as the interference and un-matching. Two methods are applied to enable such combinations as follows:

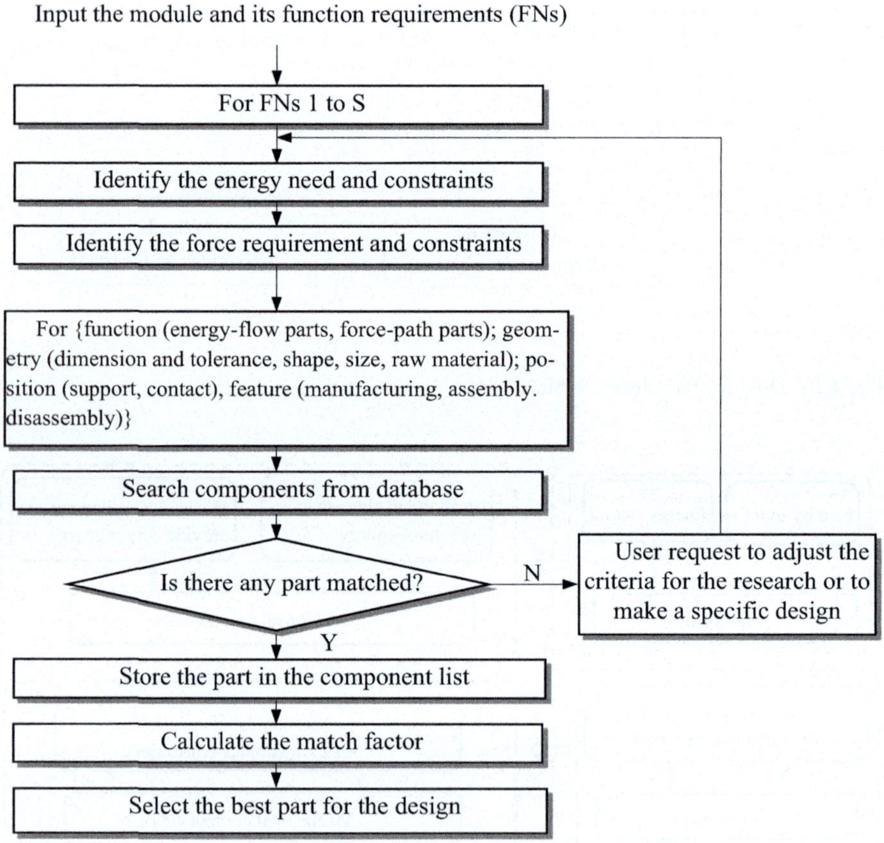

Fig. 4.28 Component search for design alternatives

(1) Each energy-force operation may have different mechanical components. It is possible to select a proper one so that all mechanical components can fit together. For example, a bearing seat can have different profiles for different fixtures.
(2) Energy flows and force paths can be adjusted to enlarge the selection of mechanical components. For example, a straight tube can be bent to avoid a spatial interference.

In both situations, an initial structure is adjusted to best meet the overall solution of a product.

4.5.3 Product Personalization

Product functions are normally processed as constant factors in the product design. However, they can vary significantly for an OAP that will be adjusted by product users in the product lifespan. This requires multiple configurations of product functions when users' requirement changes. As an OAP, some of its functions are commonly required, some are alternatively selected, and some are personal options. These functions are met using common, customized, and personal function modules connected by interfaces [12]. Therefore, following criteria are considered for OAPs to meet product personalization [24]:

(1) Adaptability: Customized and personalized function modules are designed to satisfy different requirements of users. The impact of customized and personalized function modules on the common function modules should be reduced to the minimum.
(2) Openness: Personal function modules are designed to be opened to the third-party vendors through open interfaces.

The personalization is the product change ability opened to the public to meet users' requirement changes. The product openness relies on the product interface efficiency [11]. Interface efficacy (IE) can be used as a measure of the interface efficiency based on key factors in interface design and operations. It evaluates the effectiveness of interfaces for connecting modules considering the product structure, interface connection, and module handling in the operation based on the design for assembly and disassembly. IE suggests that the interface should be designed simply in structure and easy in operation for the product function and reliability.

IE is proposed based on the geometrical and operational complex of the connection operation for modules and interfaces. Criteria are measuring factors including interface connector attributes such as the number, size, and weight; positioning attributes such as the ease of position, and easy of handling; and operation attributes such as accessibility, ease of assembly, tool applications. Geometry complexity includes parts' size, shape, and weight. Operation complexity considers fasteners' operation and position. Values of these elements are assigned by weighting factors for their importance in interface operations, including parts connected, connectors used, geometry complexity, operation complexity, tools used in the operation, and operation accessibility. For commonly used various fasteners, including bolt-nut-washer, screw, screw-washer, pin fit, taper fit, key-key way, and spline fit, their weighting factors are assigned considering the number of parts and tools used in the operation.

4.6 Summary

In this chapter, basic processes are introduced for adaptable design considering product and design adaptabilities. Main elements in the adaptable design processes, including definition of design requirements, generation of design candidates, evaluation of design candidates, and identification of the optimal adaptable design, are summarized. To ensure the quality and performance of adaptable product, robust adaptable design methods are introduced by considering: (1) changes of parameter values, (2) changes of product configurations, and (3) product with an open architecture. In addition, methods for adaptable design considering adaptation operations and mass personalization are also provided, respectively. Tools and techniques that can facilitate adaptable design will be discussed in the next chapter.

References

1. Y. Akao, *Quality Function Deployment: Integrating Customer Requirements into Product Design* (Productivity Press, Cambridge, MA, USA, 1990)
2. J.S. Arora, *Introduction to Optimum Design* (McGraw-Hill, New York, NY, USA, 1989)
3. A. Brem, P.A. Nylund, G. Schuster, Innovation and de facto standardization: the influence of dominant design on innovative performance, radical innovation, and process innovation. Technovation **50–52**, 79–88 (2016)
4. K.M. Chen, R.J. Liu, Interface strategies in modular product innovation. Technovation **25**(7), 771–782 (2005)
5. Q. Cheng, W. Li, D. Xue, Z. Liu, P. Gu, K. Li, Design of adaptable product platform for heavy-duty gantry milling machines based on sensitivity design structure matrix. J. Mech. Eng. Sci. **231**(24), 4495–4511 (2017)
6. K. Dubey, V. Yadava, Robust parameter design and multi-objective optimization of laser beam cutting for aluminium alloy sheet. Int. J. Adv. Manuf. Technol. **38**(3–4), 268–277 (2008)
7. M. Gadalla, D. Xue, An approach to identify the optimal configurations and reconfiguration processes for design of reconfigurable machine tools. Int. J. Prod. Res. **56**(11), 3880–3900 (2018)
8. P. Gu, Adaptable design with flexible interface systems. Trans. Soc. Des. Process Sci. (SDPS) **8**(3), 61–74 (2004)
9. M. Hashemian, Design for adaptability. Ph.D. dissertation (Department of Mechanical Engineering, University of Saskatchewan, Saskatoon, Canada, 2005)
10. G. Hong, L. Hu, D. Xue, Y.L. Tu, Y.L. Xiong, Identification of the optimal product configuration and parameters based on individual customer requirements on performance and costs in one-of-a-kind production. Int. J. Prod. Res. **46**(12), 3297–3326 (2008)
11. C. Hu, Q. Peng, P. Gu, Interface adaptability for an industrial painting machine. Comput.-Aided Des. Appl. **11**(2), 182–192 (2014)

12. C. Hu, Q. Peng, P. Gu, Adaptable interface design for open-architecture product. Comput.-Aided Des. Appl. **12**(2), 1–10 (2015)
13. E.K.R.E. Huizingh, Open innovation: state of the art and future perspectives. Technovation **31**(1), 2–9 (2011)
14. Y.T. Ko, Modeling a hybrid-compact design matrix for new product innovation. Comput. Ind. Eng. **107**, 345–359 (2017)
15. Y. Koren, S.J. Hu, P. Gu, M. Shpitalni, Open-architecture products. Ann. CIRP **62**(2), 719–729
16. J.R. Koza, *Genetic Programming: On the Programming of Computers by Means of Natural Selection* (The MIT Press, Cambridge, MA, USA, 1992)
17. Y. Li, Design for product adaptability. M.Sc. thesis (Department of Mechanical and Manufacturing Engineering. University of Calgary, Calgary, Canada, 2007)
18. Y. Li, P. Gu, D. Xue, Design process for product adaptability: method and application, in *Proceedings of the 16th CIRP International Design Seminar, Kananaskis*. Alberta, Canada (2006), pp. 448–455
19. L. Ma, B. Forouraghi, A hyperspherical particle swarm optimizer for robust engineering design. Int. J. Adv. Manuf. Technol. **67**(5–8), 1091–1102 (2013)
20. H. Ma, Q. Peng, J. Zhang, P. Gu, Assembly sequence planning for open-architecture products. Int. J. Adv. Manuf. Technol. **94**, 1551–1564 (2018)
21. J. MacQueen, Some methods for classification and analysis of multivariate observations, in *Proceedings of the Fifth Berkeley Symposium on Mathematical Statistics and Probability*, Berkeley, CA, USA (1967), pp. 281–297
22. M. Martinez, D. Xue, A modular design approach for modeling and optimization of adaptable products considering the whole product life-cycle spans. J. Mech. Eng. Sci. **232**(7), 1146–1164 (2018)
23. M. Orihata, C. Watanabe, Evolutional dynamics of product innovation: the case of consumer electronics. Technovation **20**(8), 437–449 (2000)
24. Q. Peng, Y. Liu, J. Zhang, Personalization for massive product innovation using open architecture. Chin. J. Mech. Eng. **31**(1), 34 (2018)
25. Q. Peng, Y. Liu, P. Gu, Development of an open-architecture electric vehicle using adaptable design, in ed. by A. Azevedo, *Advances in Sustainable and Competitive Manufacturing Systems* (2013), pp. 79–90.
26. Q. Peng, Y. Liu, P. Gu, Z. Fan, Development of an open-architecture electric vehicle using adaptable design, in *Lecture Notes in Mechanical Engineering* (Springer, Switzerland, 2013)
27. Q. Peng, Y. Liu, P. Gu, Improvement of product adaptability by efficient module interactions, in *Proceedings of the ASME Design Engineering Technical Conference*. DETC2014-35183 (2014)
28. S. Russell, P. Norvig, *Artificial Intelligence: A Modern Approach* (Prentice Hall, Upper Saddle River, NJ, USA, 2002)
29. K.C. Sih, *The Theory of Grey Information Relation* (Chun-Hwa Publishing Co., Taipei, 1997)
30. N. Suh, *Axiomatic Design: Advances and Applications* (Oxford University Press, New York, 2001)
31. H. Wang, Q. Peng, J. Zhang, P. Gu, Selective disassembly planning for the end-of-life product. Procedia CIRP 512–517 (2017)
32. D. Xue, A multi-level optimization approach considering product realization process alternatives and parameters for improving manufacturability. J. Manuf. Syst. **16**(5), 337–351 (1997)
33. D. Xue, G. Hua, V. Mehrad, P. Gu, Optimal adaptable design for creating the changeable product based on changeable requirements considering the whole product life-cycle. J. Manuf. Syst. **31**(1), 59–68 (2012)
34. H. Yang, D. Xue, Y.L. Tu, Modeling of the non-linear relations among different design and manufacturing evaluation measures for multi-objective optimal concurrent design. Concur. Eng.: Res. Appl. **14**(1), 43–53 (2006)
35. J. Zhang, Y. Chen, D. Xue, P. Gu, Robust design of configurations and parameters of adaptable products. Front. Mech. Eng. **9**(1), 1–14 (2014)

36. J. Zhang, D. Xue, P. Gu, Adaptable design of open architecture products with robust performance. J. Eng. Des. **26**(1–3), 1–23 (2015)
37. C. Zhao, Q. Peng, Development of a paper-bag-folding machine using open architecture for adaptability. Proc. Inst. Mech. Eng. Part B J. Eng. Manuf. **229**(S1), 155–169 (2015)
38. F. Zhou, Y. Ji, R.J. Jiao, Affective and cognitive design for mass personalization. J. Intell. Manuf. **24**(5), 1047–1069 (2013)

Chapter 5
Tool and Techniques for Adaptable Design

5.1 Modular Design Tools

5.1.1 Modularization for Adaptable Design Using Matrix Approach

5.1.1.1 Introduction

Modular structure allows a product to have independent function units that can be added or replaced on the product to meet requirement changes. Modularity is one of the most important features of adaptable products (APs). APs are built based on the modular structure to meet user requirements using different function modules. This section introduces methods for modules planning of APs. Modules form AP's independent functions to enable the product extendable and upgradable abilities. AP components are clustered to form function modules for appropriate configurations of the structure to adapt users' requirement changes in the product lifespan.

Modular design integrates extended Quality Function Deployment (QFD) and axiomatic design principles for planning AP structures to meet the product requirement change. The degree of variety is used as a quantitative measure for variability of components to decide different function modules. Industrial products are developed using the introduced method.

QFD is a common method to guide the product concept design based on user requirements. It builds relations of design contents and customer needs in product concept development by mapping customer needs into function demands to decide design targets and technical parameters. QFD uses a tool of House of Quality (HOQ) to connect customers' needs and product functions in the process of design concept development.

Much research has been conducted using QFD in the early stage of product development. A dynamic QFD structure can be defined for updating customer data and

P. Gu et al., *Adaptable Design*, Research on Intelligent Manufacturing,
https://doi.org/10.1007/978-981-99-5869-6_5

evolving requirements of product. The QFD-based method has been used in product platform development for improving the product family design. QFD combined with the Kano model can increase understanding of customer needs and user satisfactions. QFD has also been used to accomplish the design of modular products in the exploration of design requirements and modular product analysis. The final configuration of products can be established using QFD through the linear integer programming. The general variety index and coupling index can measure redesign effort required for future changes of products based on QFD.

Axiomatic design was proposed for systematic understanding of the product design [26]. It provides a common guideline for QFD applications with two axioms. One is the independent axiom for independence of function requirements (FRs) to be maintained. FRs are defined as the minimum number of independent requirements to meet the design goal. The other is information axiom to satisfy the independence axiom. Different methods have been proposed for product design based on axiomatic design. As the axiomatic design recommends product structure using the one-to-one match of independent functional requirements, modular design is commonly applied in planning product structure and configuration.

Therefore, most of research on the modularity originates from Suh's independence axiom for the complexity reduction of customization. Modularization has been used as a tool to break the product structure into smaller and manageable units. Modular products use the similarity of physical and functional architecture for the minimum incidental interaction between physical components. A measure of modularity can be defined based on the independence of functional requirements and their sensitivity in design parameters. A design incidence matrix can be formed to build the product platform based on axiomatic design. Methods for modules planning need to capture all aspects of modularity for adaptable product design, especially for planning personalized function modules.

In summary, there are many methods for modules planning to improve product structures and applications. These methods aim at either design of a product family, or design of the product platform. They are mainly proposed for the product design under the traditional production mode, which benefits designers or manufacturers, not users for requirement changes during the product application. A traditional product has to be redesigned and remanufactured to meet the requirement change. Adaptable product (AP) is proposed to meet the individual need by upgrading or replacing of personal modules by users in the working stage of the product. Similarly, as modular products, APs require relatively independent product modules and common platforms. But the AP objective is achieved by personal modules for the product to be extended and upgraded in the product lifespan. The existing design methods, however, cannot meet these requirements.

This section introduces methods to plan functional modules of APs by integrating extended QFD and expanded axiomatic design.

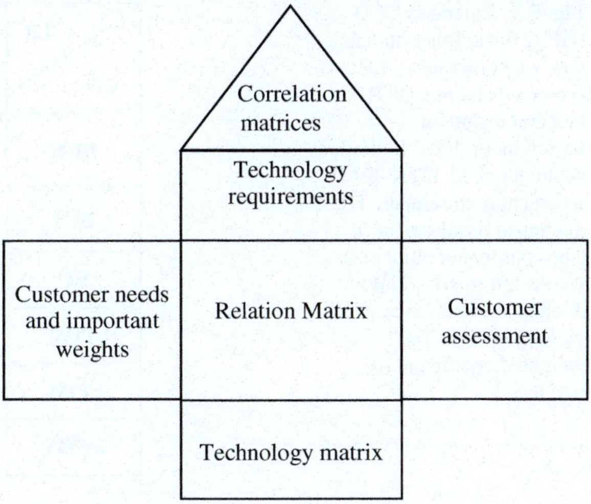

Fig. 5.1 House of quality

5.1.1.2 Modular Planning

Extended QFD

Quality Function Deployment (QFD) is a tool of product design to analyze user needs and decide design targets and technical requirements [24]. The core of QFD is "House of Quality" (HOQ) including customer requirements, importance rates, and technical methods required to meet the requirements in a matrix form as shown in Fig. 5.1 [10]. Each element in the matrix is related to others through the link of needs and ways to meet the needs. The bottom of the "house" is for designers to set the target values and rank for importance. The right part of the "house" lists the assessment of the requirements.

QFD is one of the most used tools in product design to transfer customer needs into design components. As increasing customer demands for individual featured products, it is not enough for considering only general functional needs in product design. Individual user's need should also be considered. An adaptable product (AP) uses personal function modules to meet requirements in the product lifespan. By adding or upgrading function modules to an existing AP, different function products can be formed in cost-effective solutions. In order to plan different modules of APs to decide common modules as platforms, most required functions as customized modules, and potential functions for personal modules, QFD can be extended to include not only general or basic product function requirements known in the design stage, but potential requirements in the product lifespan as well. Product requirements are divided into two parts: one is basic function needs (BFN), the other is individual customer needs (ICN). Corresponding technical requirements are divided into general technical requirements (GTR) and AP's technical requirements (ATR). The product modules for BFN are made by mass production. Function modules for

Fig. 5.2 Extended QFD (BFN: Basic function need; CN: Customer need; EC: Expected change; GTR: General technical requirement; ICN: Individual customer need; OTR: OAP technical requirement; TR: Technical requirement; *i*: The component number in the design matrix; TIW: Technical importance weight; wCOV: The weighted coefficient of variance.)

TR / **CN**	**GTR**	**OTR**	**Weight**	**Expected change**
BFN	*Relations*		W_i	*ECi*
ICN				
VC	*VCi*			
TIW	*TIWi*			
rTIW	*iTIWi*			
wCOV	*wCOV*			

ICN have small or potential needs, these are personal modules. Figure 5.2 shows contents of an extended QFD with the possible AP's technical importance weight (TIW) and target product value (TPV).

To decide module types of APs, a weighted coefficient of variance (wCOV) is introduced to assess technical requirements (TRs) for closeness to meet customer needs. The wCOV is then scaled for ranking of TRs' contributions to form product function modules into AP's common modules, customized function modules, and personal function modules [35].

Extended Axiomatic Design

Axiomatic design is used to maintain the module independence for function requirements (FRs). Design for mapping of FRs and design parameters (DPs) of technical requirements (TRs) carries processes to search solutions of FRs until one-to-one DPs matches are completed as shown in Fig. 2.12. The degree of variety (DV) can be decided using the method of module planning introduced in Sect. 4.5.1 and Eq. (4.67). DV_i is then used to assess components' contributions in different APs' modules.

Design Structure Matrix (DSM) for Module Clustering

A DSM is formed based on relations of components to combine components into modules. DV is used as a measure of the component variety to cluster components into blocks for forming three types of modules of APs. Any interaction outside the clustered blocks is considered as interfaces. The diagonal of the DSM is the DV used to indicate the component variable.

Fig. 5.3 Clustered DSM without using DV

	C_1	C_2	C_5	C_3	C_4	C_6
C_1		X				
C_2	X		X			
C_5		X		X	X	
C_3			X		X	
C_4			X	X		X
C_6					X	

Fig. 5.4 Clustered DSM using DV

	C_1	C_2	C_5	C_3	C_4	C_6
C_1	DV_1	X				
C_2	X	DV_2	X			
C_5		X	DV_5	X	X	
C_3			X	DV_2	X	
C_4			X	X	DV_4	X
C_6					X	DV_6

According to DVs, components are divided into three types of modules based on two thresholds of $f1$ and $f2$. Thresholds are set based on predicted requirement changes in the product life cycle. The threshold values are used to divide modules into common platforms, customized modules, and personal modules. Component C_i with $DV_i < f1$ are defined as common platform modules, component C_i with $DV_i >$ $f2$ are to form personal modules. Customized modules are structured by components with DV values between $f1$ and $f2$. It is obviously that a greater $f1$ will generate more platform components, and a smaller $f2$ will form more personal modules. A manufacturer can also set the thresholds based on expected product changes.

Figures 5.3 and 5.4 show the difference of clustered modules without DV and with DV in the module development process. Figure 5.3 shows product modules after clustering based on the relationship of components only. Using DVs and thresholds, components with similar DV values are clustered into one module as shown in Fig. 5.4 where DSM can be clustered to form three types of modules for APs.

After types of modules are decided based on the DV of components and thresholds, the modules can be planned for the product configuration. Different types of modules have different requirements in the product application.

5.1.1.3 Modular Design for a Paper-Bag Folding Machine

Paper bags are commonly used as containers in groceries. There are different types of paper bags in the market. Figure 5.5 shows a machine used to fold and bond the bottom of the paper bag. The original paper-bag folding machine was designed as an integral structure. One machine can only be used for one type of the paper bag. Different machines are required to fold different sizes or shapes of bags, which increases cost of the machine application. It is expected from the machine users that the machine can meet production requirements of different bags. An adaptable paper-bag folding machine is developed to meet the requirement.

Extended QFD of the Paper-Bag Folding Machine

Change needs of the machine function are based on the user survey. The change requirements in the machine application are called the expected change of user requirements. A column is added in the matrix of Table 5.1 to estimate qualitative values of the user requirement change, such as high = 3, medium = 2, and low = 1. Therefore, an extended QFD of the paper-bag folding machine can be built based on basic function needs and future requirements in the machine lifespan as shown in Table 5.1.

Using data of the extended QFD in Table 5.1 and Eqs. (5.1)–(5.3), the index of weighted coefficient of variance is calculated. Results are listed in Table 5.1 that are

Fig. 5.5 A paper-bag folding machine

Table 5.1 Function requirements of the paper-bag folding machine

	storage	Taking bag	Sending bag	bonding	Open bag	Folding bottom	Stereotypes	Pull rod	Folding side	arrange	Lateral pressure side	flattening	Rotating	adjustment	replacement	Variable station	weight	Expect change
Automatic bag	●	●	●										◎				3	L
Bearing ability				●						○							5	M
Folding bottom					◎	●											5	M
Folded neatly							◎		○	◎	◎	○					5	H
Easy to store								◎	◎		◎	◎					3	L
efficiency				○		◎	○						◎			◎	3	H
Machine dimensions		○	○										●			◎	1	L
Variable paper size														●	●		3	H
Variable function																●	3	H
VC	9	10	10	21	6	27	12	1	6	11	12	6	21	27	27	39		
TIW	87	90	90	155	48	174	58	29	45	65	77	45	87	87	87	126		
rTIW	6.4%	6.7%	6.7%	11.5%	3.6%	12.9%	4.3%	2.1%	3.3%	4.8%	5.7%	3.3%	6.4%	6.4%	6.4%	9.3%		
wCOV	0.58	0.67	0.67	2.41	0.21	3.48	0.52	0.02	0.2	0.53	0.68	0.2	1.35	1.74	1.74	3.64		
Percent	16%	18%	18%	66%	6%	96%	14%	1%	5%	15%	19%	5%	37%	48%	48%	100%		
result	c	c	c	v	c	v	c	c	c	c	c	c	v	v	v	v		

Relationships [Strong (●), Medium (◎), Weak (○)]
Expected changes of customer needs [high ($H = 3$), medium ($M = 2$), low ($L = 1$)]

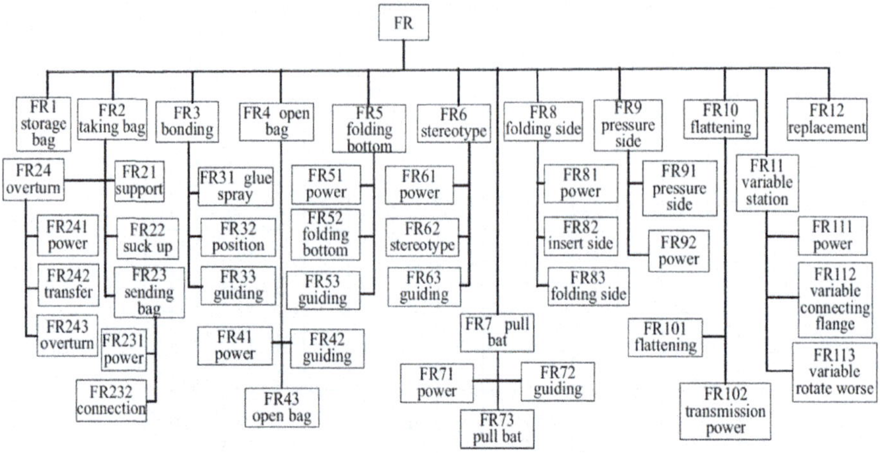

Fig. 5.6 Function requirements of the paper-bag folding machine

used to plan components for modules based on the change probability of function requirements.

Extended Axiomatic Design for Function Decomposition

According to axiomatic design, the function decomposition is achieved through the mapping of FRs and TPs using methods in section "Extended Axiomatic Design". Figure 5.6 shows FRs of the bag folding machine based on user requirements. Figure 5.7 lists corresponding DPs mapped from FRs. One-to-one relations are listed in Table 5.2 for function requirements and design parameters of the machine. The design matrix of the machine between FRs and DPs can then be established. Numerical values, such as 1, 3, and 9, are assigned to FRs of components for their abilities to meet the FRs, which are used for the degree of variety calculation in the design matrix shown in Table 5.3.

DSM for Planning of Module Types

DSM is formed to cluster parts. The diagonal in the DSM is DV that indicates degrees of the component variation. According to DV, components are divided into three types based on two thresholds of the variant. Thresholds are set based on variant TRs over constant TRs. Based on the assumption of need changes, different values of thresholds can be applied based on manufacturer data of the requirement change. Table 5.4 shows an example of the paper-bag folding machine using 1% and 10% for f_1 and f_2 thresholds, respectively. According to the relationship of components and

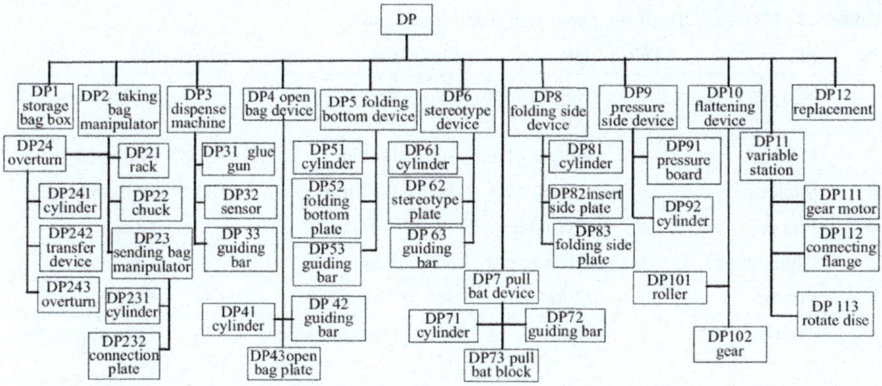

Fig. 5.7 Design parameters of the paper-bag folding machine based on FRs in Fig. 5.6

their DV values, components are clustered into modules for the adaptable machine as shown in Table 5.4.

Following the process, modules of the paper-bag folding machine are divided into common platform modules, personal modules and customized modules as shown in Figs. 5.8, 5.9 and 5.10, respectively. Figure 5.8 shows ten common modules used to form the platform of the paper-bag folding machine. Paper-bag folding machine's personal function modules are shown in Fig. 5.9, those modules are predicted in the design stage but will be used or updated during the machine application. Figure 5.10 lists different customized modules made by the machine producer and chosen by users during the purchasing process. Users can also make or purchase their personal modules from other suppliers to meet their needs. Based on these three types of modules, an adaptable paper-bag folding machine is formed to meet user's individual need when personal modules are installed with common and customized modules in the machine.

5.1.1.4 Modular Formation of the Paper-Bag Folding Machine

To test feasibility of the proposed machine architecture, potential change requirements of the paper-bag folding machine are proposed in Table 5.5 based on the manufacturer's suggestion, user survey and market forecast. It is required to use eight to ten working processes for the machine to complete operations listed in Table 5.5. The proposed AP machine can be changed from eight to ten workstations to meet these needs through changing personal modules. When the size and/or shape of the paper bag are changed, personal function modules can be replaced in the machine to meet different sized and/or shaped paper-bag requirements.

Figure 5.11 shows an example of the personal paper-bag folding machine with eight workstations based on the process requirement 2 in Table 5.5. The machine includes one common platform; seven customized modules: bag storage module 9,

Table 5.2 FRs and DPs of the paper-bag folding machine

No.	FR	No.	DP	No.	FR	No.	DP
1	Storage	1	Bag storage box	29	Power	29	Cylinder
2	Support	2	Chuck plane	30	Guide	30	Guide bar
3	Suck up	3	Chuck	31	Pull rod	31	Pull on block
4	Power	4	Cylinder	32	Adjust	32	Gear screw
5	Connection	5	Even board	33	Power	33	Cylinder
6	Power	6	Cylinder	34	Plug side	34	Side board
7	Transmit	7	Shaft	35	Pressure side	35	Lateral pressure crank
8	Flip	8	Flip sat	36	Adjust	36	Gear scram
9	Power	9	Cylinder	37	Power	37	Cylinder
10	Guide	10	Guide bar	38	Pressure bag	38	Pressure bags pole
11	Connection	11	Sliding plate	39	Finishing side	39	Lateral pressure
12	Hang bag	12	Flapper	40	Finishing bottom	40	Bag equivalent plate
13	Adjust	13	Adjusting rod	41	Adjust	41	Gear screw
14	Glue spray	14	Glue gun	42	Power	42	Cylinder
15	Guide	15	Guide bar	43	Pressure side	43	Pressure side plate
16	Position	16	Sensor	44	Adjust	44	Gear screw
17	Adjust	17	Nut screw	45	Flatten	45	Roller
18	Power	18	Cylinder	46	Power	46	DC geared motor
19	Guide	19	Guide bar	47	Transmit	47	Gear
20	Hang bag	20	Hang bag board	48	Transmit	48	Shaft
21	Power	21	Cylinder	49	Power	49	Gear motor
22	Folding bottom	22	Folding board	50	Position	50	CAM divider
23	Guide	23	Guide bar	51	Clip bag	51	Clip bag board
24	Adjust	24	Gear screw	52	Adjust	52	Adjusting device
25	Power	25	Cylinder	53	Replacement	53	Rotating disk
26	Stereotypes	26	Stereotypes board	54	Variable station	54	Fixed chassis
27	Guide	27	Guide bar				
28	Adjust	28	Gear screw				

FR function requirements; *DP* design parameters

Table 5.3 Design matrix of the paper-bag folding machine

	1	2	3	4	5	6	7	8	9	10	11	12	13	14	15	16	17	18	19	20	21	22	23	24	25	26	27	28	29	30	31	32	33	34	35	36	37	38	39	40	41	42	43	44	45	46	47	48	49	50	51	52	53	54	R	RW
1	9																																																						c	16
2		9																																																					c	
3			9																																																				c	
4				9																																																			c	
5		3		3	9																																																		c	18
6						9																																																	c	
7						3	9	3																																															c	
8							3	9																																															c	
9									9																																														c	
10										9																																													c	
11											9																																												c	18
12												9																																											v	
13												3	9																																										c	
14														9																																									c	
15															9																																								c	66
16																9																																							v	
17																	9																																						c	
18																		9																																					c	
19																			9	3																																			c	6
20																				9																																			v	
21																					9																																		c	
22																						9																																	v	96
23																							9																																c	
24																						1		9																															c	
25																									9																														c	
26																										9	3																												c	14
27																											9																												c	
28																												9																											c	
29																													9																										c	
30																														9																									c	1
31																														3	9																								c	
32																																9																							c	
33																																	9																						c	
34																																		9																					c	5
35																																			9																				c	
36																																		3		9																			c	
37																																					9																		c	
38																																						9																	c	
39																																							9																c	15
40																																								9															c	
41																																									9														c	
42																																										9													c	
43																																											9												c	19
44																																												9											c	
45																																													9										c	
46																																														9									c	5
47																																														3	9								c	
48																																																9							c	
49																																																3	9	3			3	3	c	37
50																				1						1								1									1							9					v	
51																																																			9				c	
52													1				1							1				1				1				1					1			1								3			c	48
53	3	3	3									3								3		3				1																	1								3	3			v	
54																																																		1			3	3	39	100
Xv	144	144	144	0	0	0	0	0	0	0	0	306	0	0	0	0	0	0	0	235	0	1008	0	0	0	85	0	0	0	0	0	0	0	37	0	0	0	0	0	0	0	0	85	0	0	0	0	0	0	433	144	144	300	300		
Xc	144	180	162	216	162	216	216	216	162	162	162	54	210	594	594	594	642	54	72	18	864	96	864	174	126	126	168	174	9	12	9	57	45	60	45	93	135	135	135	135	183	171	171	219	45	60	45	444	333	111	333	144	111	111		
DV	1	0.8	0.89	0	0	0	0	0	0	0	0	5.67	0	0	0	0	0	0	0	13.1	0	10.5	0	0	0	0.67	0	0	0	0	0	0	0	0.62	0	0	0	0	0	0	0	0	0.5	0	0	0	0	0	0	3.9	0.4	1	2.7	2.7	44	
	0.02	0.02	0.02	0	0	0	0	0	0	0	0	0.13	0	0	0	0	0	0	0	0.30	0	0.24	0	0	0	0.02	0	0	0	0	0	0	0	0.01	0	0	0	0	0	0	0	0	0.01	0	0	0	0	0	0	0.09	0.01	0.02	0.06	0.06		

Table 5.4 Clustered modules of the paper-bag folding machine

	4	5	8	6	7	9	10	13	11	15	16	14	17	18	19	21	23	24	25	27	28	29	30	31	32	33	35	36	38	39	37	40	41	42	44	45	46	47	49	48	1	2	3	26	43	34	51	52	20	12	22	50	53	54
4	0	X	X																																																			
5	X	0																																							X												
8	X		0	X																																																		
6			X	0	X																																																	
7				X	0																																																	
9						0																																																
10							0		X																																								X					
13								0																																														
11							X		0																																								X				X	
15										0		X																																										
16											0																																											
14										X		0	X																																									
17												X	0																																								X	
18														0	X																																							
19														X	0	X																																		X				
21															X	0																																						
23																	0	X																																				
24																	X	0																																				
25																			0	X																																		
27																			X	0																								X										
28																					0																																X	
29																						0		X																														
30																							0	X																														
31																						X	X	0	X																													
32																								X	0																												X	
33																										0	X																											
35																										X	0																											
36																												0																									X	
38																													0																									
39																														0		X																						
37																															0	X	X																					
40																														X	X	0																						
41																															X		0																				X	
42																																		0																				
44																																			0																		X	
45																																				0		X																
46																																					0	X																
47																																				X	X	0																
49																																							0	X														
48																																							X	0												X		
1																																									0.02												X	
2		X																																								0.02	X										X	
3																																										X	0.02										X	
26																				X																								0								X		
43																																													0.01							X		
34																																														0.01						X		
51																																															0.01							
52																																																0.02						
20							X		X																																								0.0.3				X	
12															X																																			0.13			X	
22																																																			0.24		X	
50																																								X				X	X	X						0.09		X
53									X				X						X		X				X			X					X		X						X	X	X						X	X	X		0.06	X
54																																																				X	X	0.06

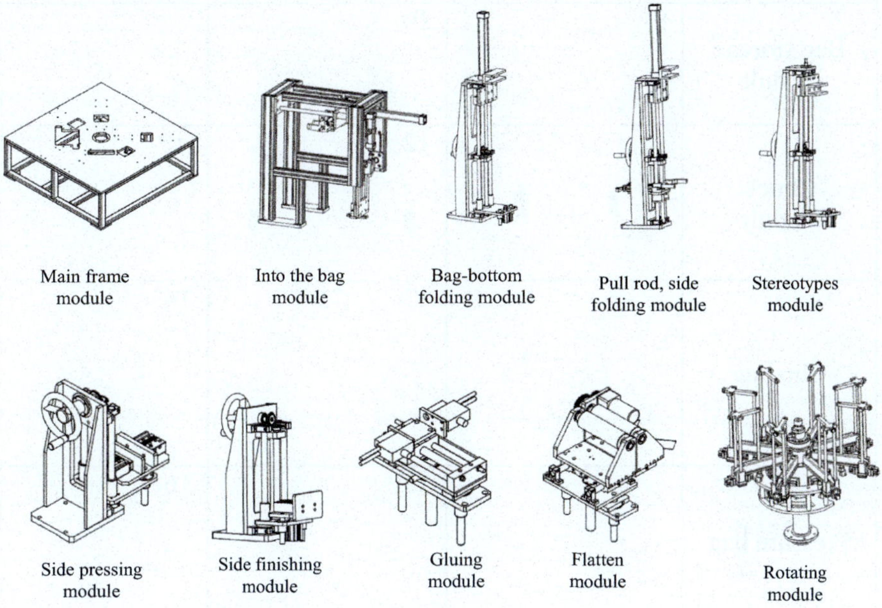

Fig. 5.8 Paper-bag folding machine's platform modules

Different folding bottom board	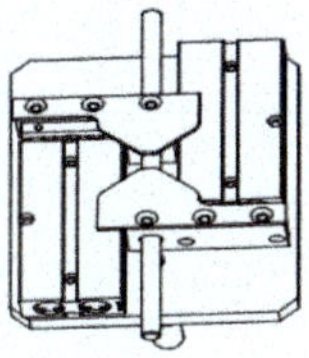	2 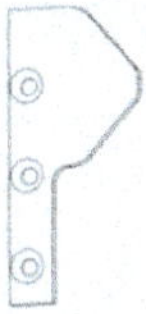	3 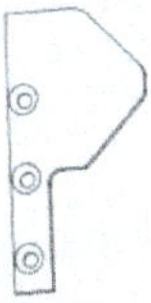
Different size hang bags board	4 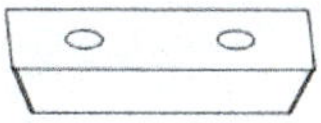	5 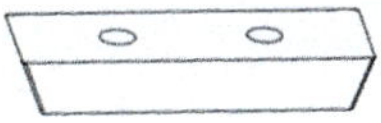	6 
Different support bag board	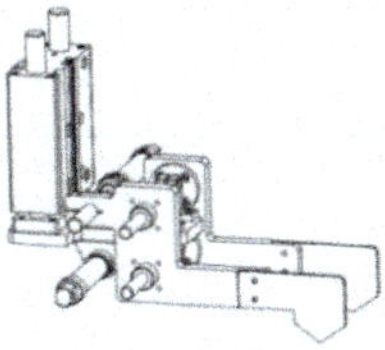	7 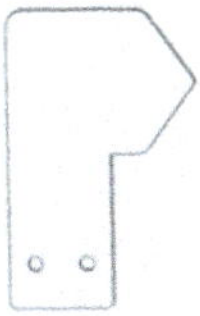	8 

Fig. 5.9 Paper-bag folding machine's personal function modules

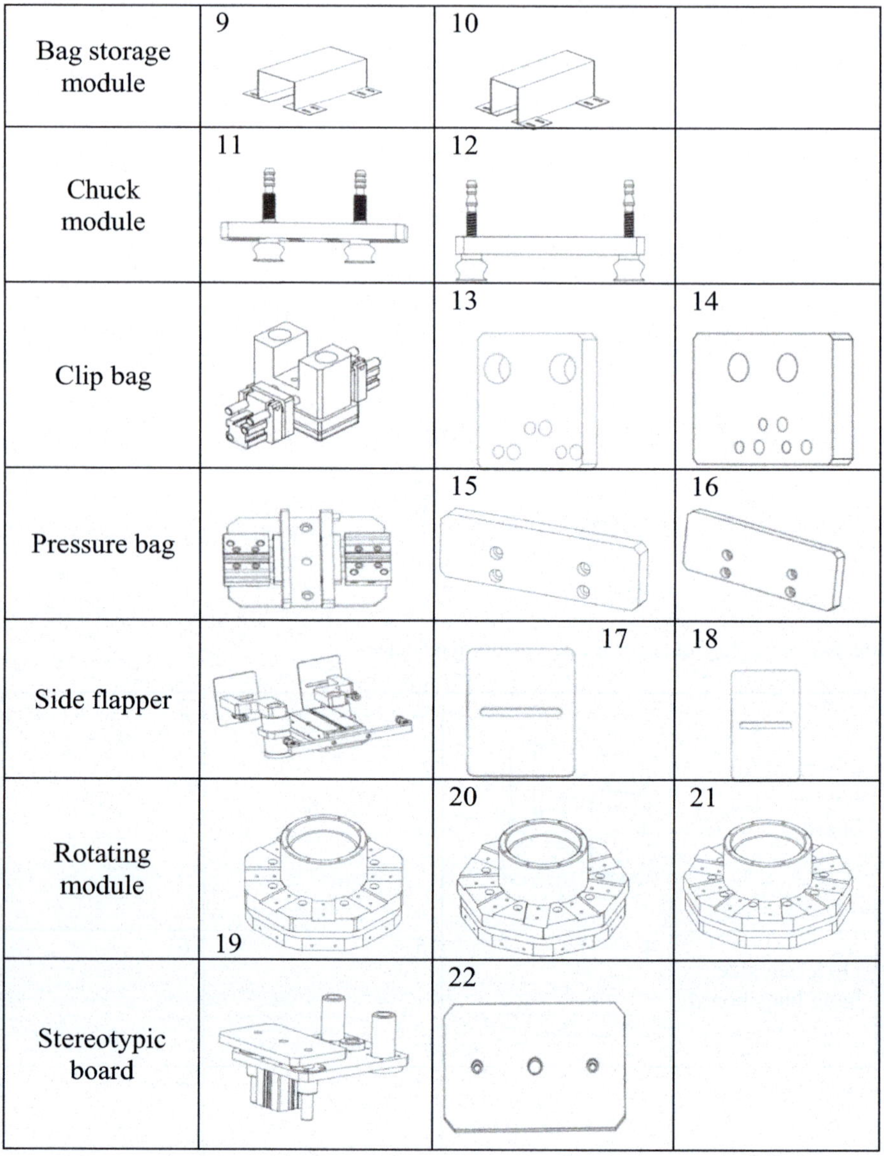

Fig. 5.10 Paper-bag folding machine's customized modules

chuck module 11, clip bag plate 13, pressure bagplate 15, side flapper 17, rotating module 20, and stereotypic board 22; and three personal modules: offolding bottom board 2, hang bags board 5, and support bag board 7. The machine is used to meet the bag specifications $D = \{(\text{Length, Wide, Side})|145 \le L \le 300, W = 130, S = 40\}$.

Table 5.5 Predicted paper-bag folding operations

No.	Different requirements for the paper-bag folding machine
1	Bag setting manipulator → Gluing → Bag-bottom folding → Shape forming → Side folding → Side pressing → Code printing → Bag removal
2	Bag setting manipulator → Gluing → Bag-bottom folding → Shape forming → Side folding → Side finishing → Side pressing → Bag removal
3	Bag setting manipulator → Gluing → Bag-bottom folding → Shape forming → Side folding → Side finishing → Side pressing → Code printing → Bag removal
4	Punch → Bag setting manipulator → Gluing → Bag-bottom folding → Shape forming → Side folding → Side finishing → Side pressing → Bag removal
5	Punch → Bag setting manipulator → Gluing → Bag-bottom folding → Shape forming → Side folding → Side finishing → Side pressing → Code printing → Bag removal

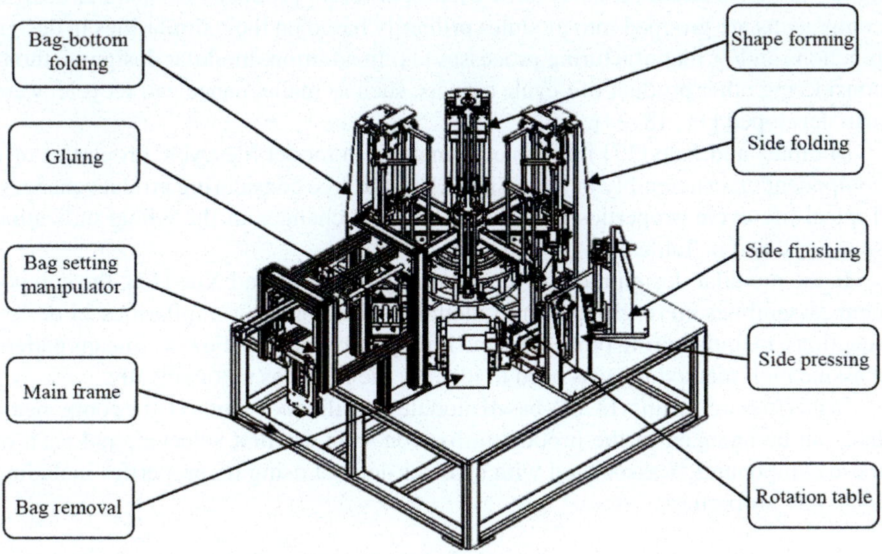

Fig. 5.11 Paper-bag folding machine with eight workstations

If the requirement is changed to the need of process 3 in Table 5.5 for adding a function of code printing, it is required to change only the rotating module and to adjust the position of common modules. When the size of the paper-bag changes to specifications $D = \{(\text{Length, Wide, Side})|145 \leq L \leq 300, W = 80, S = 50\}$, it will need to replace the existing modules using three personal modules: the folding bottom board 3, hang bag board 6, and support bag board 8. The machine can then meet the requirement.

In summary, the paper-bag folding machine with the adaptable design provides a cost-effective solution for users working for different jobs. The solution allows users using one machine to meet different job requirements by changing personal modules.

The machine can be upgraded by users in their working place, which reduces cost of using multi-machines.

5.1.2 Pattern Clustering Approach

Modular design approach has been widely used for providing product variety and reducing production costs. In module design, similar components of the product are grouped into relatively independent modules such that these modules can be easily assembled in manufacturing and disassembled in maintenance [1]. Since the modules are relatively independent, they can be designed and manufactured separately. In addition, manufacturing cost can be reduced and product variety can be increased by designing standard modules to be used in different products. In modular design, components are grouped into modules primarily based on their similarities in design functions and/or manufacturing processes [1]. In addition, modular design methods considering other product life-cycle aspects, such as maintenance and recycle, were also developed [11, 18, 34].

Martinez and Xue [17] introduced a method to model life-cycle properties of a component or an assembly in the design AND-OR tree considering product changes. Typical life-cycle properties considering product changes in the whole utilization span are listed in Table 5.6.

In the modular design method developed by Martinez and Xue [17], the components/assemblies were grouped into modules based on their similarities in design functions, manufacturing processes, and life-cycle properties. Fuzzy c-means pattern classification method [4] was used to identify the modules automatically.

In this fuzzy c-means (FCM)-based module classification method, the components that can be changed in the product utilization span are first selected, and each of these components is associated with a vector defined using its properties including life-cycle properties:

$$D_j = \left(D_{j1}, D_{j2}, \ldots, D_{jp}\right), \quad j = 1, 2, \ldots, n \tag{5.1}$$

Table 5.6 Examples of typical life-cycle properties considering product changes

Category	Life-cycle property	Value
Quantitative	Maintenance frequency for a module	18 months
	Lifespan of automobile battery	4 years
	Start time for a module	3rd year
	Time span with reliability over 95%	1.5 years
Qualitative	Degradation of performance	Low
	Chance in technology advancement	High

where n is the total number of selected components, and p is the total number of selected property measures. Suppose the number of modules is selected as c, the module classification is conducted through the following steps:

Step 1. Initialize the membership μ_{ij} of D_j for cluster i ($i = 1, 2, \ldots, c$) using a random number such that

$$\sum_{i=1}^{c} \mu_{ij} = 1, \quad j = 1, 2, \ldots, n \tag{5.2}$$

Step 2. Compute the fuzzy centroid vector V_i for $i = 1, 2, \ldots, c$ using

$$V_i = \frac{\sum_{j=1}^{n} (\mu_{ij})^m D_j}{\sum_{j=1}^{n} (\mu_{ij})^m}, \quad i = 1, 2, \ldots, c \tag{5.3}$$

where the fuzziness index m is a real number greater than 1.

Step 3. Update the fuzzy membership μ_{ij} by

$$\mu_{ij} = \frac{\left(\frac{1}{d^2(D_j, V_i)}\right)^{\frac{1}{m-1}}}{\sum_{i=1}^{c} \left(\frac{1}{d^2(D_j, V_i)}\right)^{\frac{1}{m-1}}}, \quad i = 1, 2, \ldots, c; \; j = 1, 2, \ldots, n \tag{5.4}$$

where

$$d^2(D_j, V_i) = \sum_{k=1}^{p} (D_{jk} - V_{ik})^2, \quad i = 1, 2, \ldots, c; \; j = 1, 2, \ldots, n \tag{5.5}$$

Step 4. Calculate

$$J_m = \sum_{i=1}^{c} \sum_{j=1}^{n} (\mu_{ij})^m d^2(D_j, V_i) \tag{5.6}$$

Steps 2, 3, and 4 are repeated continuously until the value of J_m reaches a minimum. The fuzzy c-means clustering method always converges to a strict minimum of J_m although different initial values may lead to a different local minimum. In this method, the fuzziness index m is selected as 2. To improve the quality of classification, the elements in the vector D_j are first normalized to the values between 0 and 1.

In the fuzzy c-means clustering method, the number of clusters has to be selected first. For identification of modules in design of an adaptable product, the proper number of modules is not known. To avoid the excessive computation effort by trying clustering with different cluster numbers, a method to identify the optimal number of clusters [31] is employed for identification of the optimal module number in adaptable product design.

First the concentration level of the data clustering is analyzed. The concentration level represents the average distance of the data points to their cluster centers. Suppose the distance between a p-dimensional data point, D_j, and its cluster center, V_i, is calculated by:

$$\overline{d}_{ij} = \overline{d}_{ij}(D_j, V_i) = \sqrt{\frac{\sum_{k=1}^{p}(D_{jk} - V_{ik})^2}{p}} \tag{5.7}$$

The cluster concentration level, $\overline{D}_c$, is defined by:

$$\overline{D}_c = \frac{\sum_{i=1}^{c}\sum_{j=1}^{N_i}\overline{d}_{ij}}{N} \tag{5.8}$$

where N_i is the number of data points in the i-th cluster, N the number of all data points, and c the cluster number. The cluster concentration level, $\overline{D}_c$, decreases when the cluster number, c, increases. The relative decrease of cluster concentration level, $\lambda(\overline{D}_c)$, is used as the measure to identify the optimal cluster number c:

$$\lambda(\overline{D}_c) = \frac{\overline{D}_{c-1} - \overline{D}_c}{\overline{D}_c} \tag{5.9}$$

For each clustering with a specific number of clusters, the concentration level, $\overline{D}_c$, is calculated and compared against other concentration levels obtained with different numbers of clusters. The optimal number of clusters is identified as the one for which the relative decrease of cluster concentration level, $\lambda(\overline{D}_c)$, has reached a peak.

In the design of the adaptable equipment to test two newly designed aircraft pumps developed by Martinez and Xue [17], 13 components were selected and each of the 13 components was associated with 19 properties as shown in Table 5.7.

The optimal number of modules was then identified. The cluster concentration levels and relative changes of cluster concentration levels considering different numbers of clusters are shown in Table 5.8. From this table, the best number of clusters was identified as 7 (shown in bold font).

All the components were then clustered into the 7 modules using the FCM. The result of FCM was a matrix as shown in Table 5.9 to describe the relations between the components and the clusters in the form of membership functions. The membership function for a component represented the degree that component belonged to that cluster. In this example, each component belonged to the cluster with the highest membership value (shown in bold font). The seven modules and components in these seven modules are shown in Table 5.9 and Fig. 5.12.

Table 5.7 Components and their properties [17]

Component	Component Properties																		
	Operation phase			Design function			Assembly in operation									Assembly in manufacturing			
	Phase I	Phase II	Phase III	Large chamber	Small chamber	Refrigeration system	Top panel	Side panel (1 m^3)	Side panel (0.5 m^3)	Door with window	Door without window	Small door	Circulation fan	Cryogenic system	Mechanical refrigeration system	Circulation fan	Door with window	Door without window	Small door
Top panel	1	1	1	1	1	0	1	1	1	1	1	1	1	0	0	1	0	0	0
Side panel 1 for 1 m^3 chamber	1	1	0	1	1	0	1	1	0	1	1	0	0	0	0	0	1	1	0
Side panel 2 for 1 m^3 chamber	1	1	0	1	1	0	1	1	0	1	1	0	0	0	0	0	1	1	0
Side panel 3 for 1 m^3 chamber	1	1	0	1	1	0	1	1	0	1	1	0	0	0	0	0	1	1	0
Side panel 1 for 0.5 m^3 chamber	0	0	1	0	1	0	1	0	1	0	0	1	0	0	0	0	0	0	1
Side panel 2 for 0.5 m^3 chamber	0	0	1	0	1	0	1	0	1	0	0	1	0	0	0	0	0	0	1
Side panel 3 for 0.5 m^3 chamber	0	0	1	0	1	0	1	0	1	0	0	1	0	0	0	0	0	0	1
Door with a window for 1 m^3 chamber	1	0	0	1	0	0	1	1	0	1	0	0	0	0	0	0	1	0	0
Door without a window for 1 m^3 chamber	0	1	0	1	0	0	1	1	0	0	1	0	0	0	0	0	0	1	0
Door for 0.5 m^3 chamber	0	0	1	0	1	0	1	0	1	0	0	1	0	0	0	0	0	0	1
Circulation fan	1	1	1	1	1	0	1	0	0	0	0	0	1	0	0	1	0	0	0
Cryogenic system	1	0	0	0	0	1	0	0	0	0	0	0	0	1	0	0	0	0	0
Mechanical refrigeration system	0	1	1	0	0	1	0	0	0	0	0	0	0	0	1	0	0	0	0

Table 5.8 Evaluation to clustering results with different numbers of clusters [17]

Number of clusters	Cluster concentration level $\overline{D}_c$	Relative change measure $\lambda(\overline{D}_c)$
3	0.0987	0.0719
4	0.0973	0.0143
5	0.0950	0.0242
6	0.0896	0.0603
7	**0.0598**	**0.4983**
8	0.0549	0.0893
9	0.0531	0.0339
10	0.0503	0.0557
11	0.0467	0.0771
12	0.0421	0.1093
13	0.0397	0.0605

Table 5.9 Membership functions achieved using FCM [17]

Cluster	Components												
	Top panel	Side panel 1 (1 m^3)	Side panel 2 (1 m^3)	Side panel 3 (1 m^3)	Side panel 1 (0.5 m^3)	Side panel 2 (0.5 m^3)	Side panel 3 (0.5 m^3)	Door with window	Door without window	Small door	Circulation fan	Cryogenic system	Mechanical refrigeration
1	**2.65 E−01**	7.93 E−05	7.93 E−05	7.93 E−05	2.00 E−05	2.00 E−05	2.00 E−05	3.49 E−04	3.49 E−04	2.00 E−05	**9.83 E−01**	7.43 E−05	1.05 E−04
2	1.68 E−01	**9.99 E−01**	**9.99 E−01**	**9.99 E−01**	1.30 E−05	1.30 E−05	1.30 E−05	6.79 E−04	6.79 E−04	1.30 E−05	2.84 E−03	6.11 E−05	7.06 E−05
3	1.33 E−01	1.55 E−04	1.55 E−04	1.55 E−04	1.57 E−05	1.57 E−05	1.57 E−05	**9.98 E−01**	4.58 E−04	1.57 E−05	2.89 E−03	9.62 E−05	8.52 E−05
4	1.33 E−01	1.55 E−04	1.55 E−04	1.55 E−04	1.57 E−05	1.57 E−05	1.57 E−05	4.58 E−04	**9.98 E−01**	1.57 E−05	2.89 E−03	7.50 E−05	1.06 E−04
5	9.00 E−02	5.09 E−05	5.09 E−05	5.09 E−05	1.96 E−05	1.96 E−05	1.96 E−05	2.73 E−04	3.42 E−04	1.96 E−05	2.86 E−03	1.35 E−04	**9.99 E−01**
6	1.29 E−01	5.06 E−05	5.06 E−05	5.06 E−05	**1.00 E + 00**	**1.00 E + 00**	**1.00 E + 00**	2.71 E−04	2.71 E−04	**1.00 E + 00**	2.83 E−03	7.46 E−05	1.06 E−04
7	8.34 E−02	5.55 E−05	5.55 E−05	5.55 E−05	1.74 E−05	1.74 E−05	1.74 E−05	3.91 E−04	3.03 E−04	1.74 E−05	2.53 E−03	**9.99 E−01**	1.70 E−04
Module	1	2	2	2	6	6	6	3	4	6	1	7	5

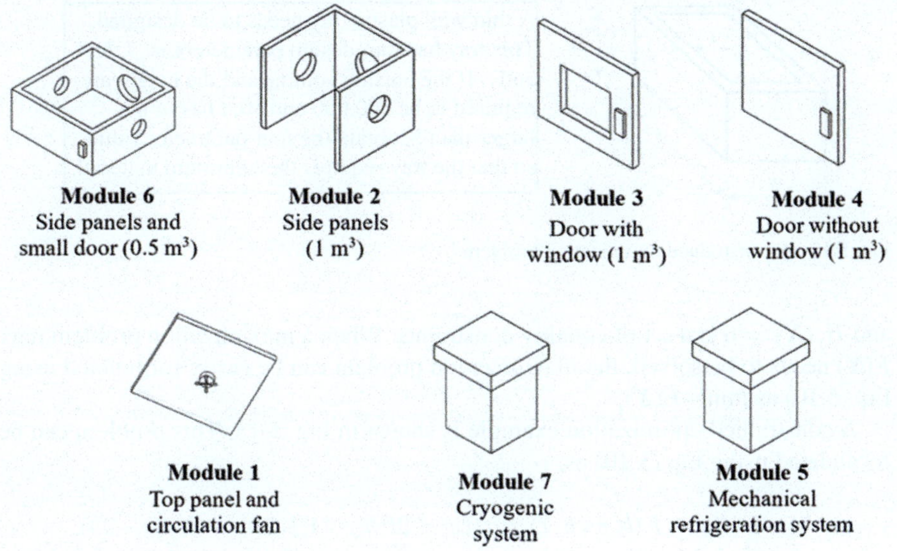

Fig. 5.12 Modules for the adaptable product [17]

5.2 Optimization Tools

As discussed in Sect. 4.2.4, optimization plays an important role to achieve the optimal adaptable design from all feasible design solutions. In this section, various optimization approaches, methods, and tools for adaptable design are provided.

5.2.1 *Constrained Optimization*

Optimization was first developed as a numerical method to identify the optimal values of a numerical function [2]. A constrained optimization problem is defined by:

$$
\begin{aligned}
&\min_{\text{w.r.t. } \boldsymbol{X}} F(\boldsymbol{X}) \\
&\text{Subject to:} \\
&\qquad \boldsymbol{X}_L \le \boldsymbol{X} \le \boldsymbol{X}_U \\
&\qquad G_i(\boldsymbol{X}) \ge 0, i = 1, 2, \ldots, m \\
&\qquad H_j(\boldsymbol{X}) = 0, j = 1, 2, \ldots, q
\end{aligned}
\tag{5.10}
$$

In Eq. (5.10), $\boldsymbol{X}$ is a vector of variables $\boldsymbol{X} = (X_1, X_2, \ldots, X_n)$ to be optimized, $F(\boldsymbol{X})$ is the optimization objective function to be minimized, $\boldsymbol{X}_L$ and $\boldsymbol{X}_U$ are vectors of lower and upper boundaries for $\boldsymbol{X}$, $G_i(\boldsymbol{X}) \ge 0$ is a set of inequality constraints,

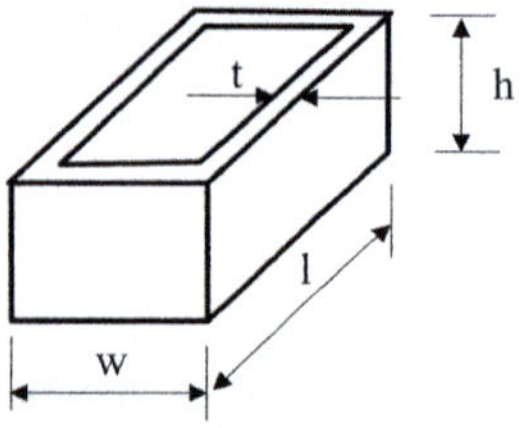

A thin wall plastic tray needs to be designed. This tray has four design parameters, w, l, h, and t. If the outside volume and the height are required to be 500,000 and 50, t has to be larger than 5, obtain the four parameter values, so that the tray requires the minimum material.

Fig. 5.13 A constrained optimization problem

and $H_j(\boldsymbol{X}) = 0$ is a set of equality constraints. When a maximization problem max $F(\boldsymbol{X})$ needs to be solved, the maximization problem can be easily formulated using Eq. (5.10) as min – $F(\boldsymbol{X})$.

A constrained optimization example is shown in Fig. 5.13. This problem can be formulated using Eq. (5.10) as:

$$\begin{aligned}
&\text{Min } F(\boldsymbol{X}) = F(l, w, h, t) = t(lw + 2l(h - t) + 2(w - 2t)(h - t)) \\
&\text{Subject to:} \\
&\qquad h - 50 = 0 \\
&\qquad wlh - 500{,}000 = 0 \\
&\qquad t - 5 \geq 0
\end{aligned} \tag{5.11}$$

Optimization problems through numerical search are primarily classified into two categories:

- Unconstrained optimization problems.
- Constrained optimization problems.

Numerous optimization search methods (e.g., steepest descent method and Gauss–Newton method) have been developed for solving various types of unconstrained optimization problems considering optimization quality and efficiency. The numerical search process for optimization is illustrated in Fig. 5.14. The search is started from an initial solution selected through best guess or randomly. In each iteration of optimization search, search direction and search step (i.e., distance) are decided to find the next location to be evaluated. When the next point is better than the currently achieved optimal solution, the next point is then selected as the optimal solution. When the next point is worse than the currently achieved optimal solution, the next point is then discarded. At early stage in the optimization process, large search steps are selected to improve optimization efficiency. In late stage in the area closer to the optimal solution, small search steps are selected. The final optimal solution is identified when optimization objective function cannot be further reduced in a number of iterations.

When constraints are considered as shown in Fig. 5.15, the solutions only in the feasible region satisfying all the constraints are searched. The optimal solution in Fig. 5.15 considering constraints is different from the optimal solution given in Fig. 5.14 without considering constraints.

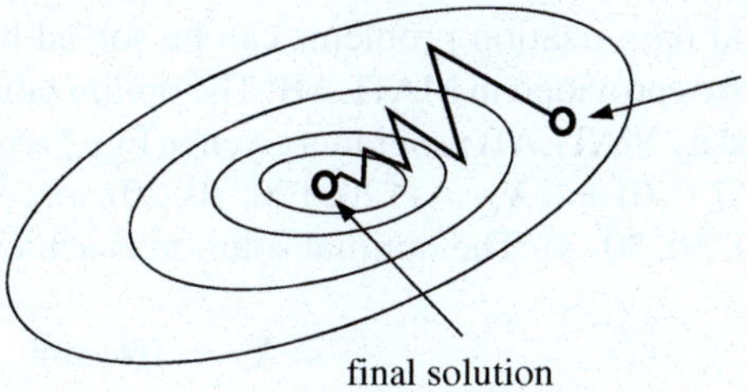

Fig. 5.14 Numerical search process in optimization

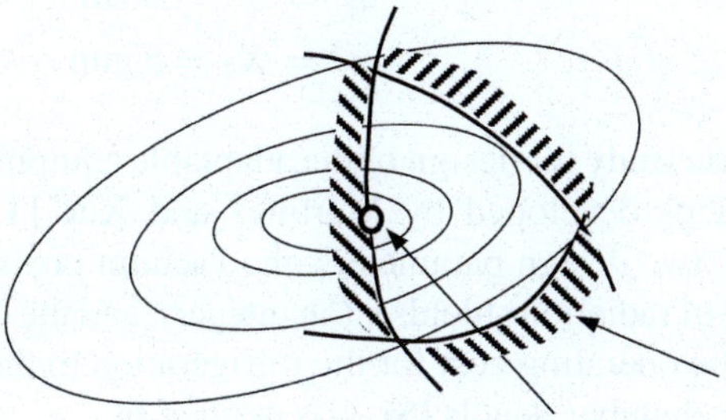

Fig. 5.15 Optimization considering constraints

A constrained optimization problem is usually converted into an unconstrained optimization problem by introducing a penalty function $S(\boldsymbol{X})$ modeled by constraints and adding this penalty function to the original optimization objective function.

$$\min_{\text{w.r.t. } \boldsymbol{X}} D(\boldsymbol{X}, \rho) = F(\boldsymbol{X}) + \frac{1}{\rho} S(\boldsymbol{X}) \tag{5.12}$$

where $F(\boldsymbol{X})$ is the original optimization objective function and ρ is a positive number.

The penalty function $S(\boldsymbol{X})$ is defined by:

$$S(\boldsymbol{X}) = \sum_i \delta_i |G_i(\boldsymbol{X})|^\alpha + \sum_j |H_j(\boldsymbol{X})|^\beta \tag{5.13}$$

where $G_i(\boldsymbol{X})$ and $H_j(\boldsymbol{X})$ are inequality constraints and equality constraints given in Eq. (5.10), and α and β are constants usually having values of 1 and 2. δ_i in Eq. (5.13) is defined by:

$$\delta_i = \begin{cases} 0, \text{ if } G_i(\boldsymbol{X}) \geq 0 \\ 1, \text{ if } G_i(\boldsymbol{X}) < 0 \end{cases} \tag{5.14}$$

When a search solution falls in a location that violates constraints, the value of the penalty function $S(\boldsymbol{X})$ is then calculated as a very large number, leading to a large value in $D(\boldsymbol{X}, \rho)$ in Eq. (5.12). The search location is subsequently moved to the feasible region that satisfies the constraints with a very small value of $S(\boldsymbol{X})$.

Numerical optimization problems can be solved by computer programs such as those toolboxes provided in MATLAB. The optimization problem given in Eq. (5.11) can be solved by MATLAB programs given in Fig. 5.16. Boundaries of $\boldsymbol{X}$ are assigned as $\boldsymbol{X}_L = (0, 0, 0, 0)$ and $\boldsymbol{X}_U = (120, 120, 70, 20)$, and initial values of $\boldsymbol{X}$ are assigned as $\boldsymbol{X}_0 = (30, 30, 30, 3)$. The optimal solution is achieved as:

$$\begin{aligned} l^* &= X_1 = 100\,\text{mm} \\ w^* &= X_2 = 100\,\text{mm} \\ h^* &= X_3 = 50\,\text{mm} \\ t^* &= X_4 = 5\,\text{mm} \end{aligned}$$

In the case study for design of the adaptable equipment to test two newly designed aircraft pumps developed by Martinez and Xue [17], the optimization problem considering two design parameters, the vacuum pressure P (Pa, a real number) and the number of radiation shields n (an integer), and the objective function of total cost, including the operating cost for the refrigeration to the chamber ($) and the product cost for the chamber panels ($), was defined by:

$$\min C(P, n)$$

The optimal values were achieved as $P = 0.1$ (Pa) and $n = 5$ as shown in Fig. 5.17.

RunOpt.m

```
%% container optimization
%% x(1): l, x(2): w, x(3): h, x(4): t

x0=[30;30;30;3]; % Set the initial values
lb=[0;0;0;0]; % Set the lower boundaries
ub=[120;120;70;20]; % Set the upper boundaries
[x,fval,exitflag]=fmincon(@myfun,x0,[],[],[],[],lb,ub,@mycon)
```

myfun.m

```
function f=myfun(x)
%%f(l,w,h,t) = t(lw + 2l(h-t) + 2(w-2t)(h-t))
f=x(4)*(x(1)*x(2)+2*x(1)*(x(3)-x(4))+2*(x(2)-2*x(4))*(x(3)-x(4)));
```

mycon.m

```
function [c,ceq]=mycon(x)
c(1)=5.0-x(4); %% 5.0 - t <= 0
ceq(1)=x(3)-50.0; %% h - 50.0 = 0
ceq(2)=x(1)*x(2)*x(3)-500000.0; %% lwh - 500000.0 = 0
```

Fig. 5.16 MATLAB programs for constrained optimization

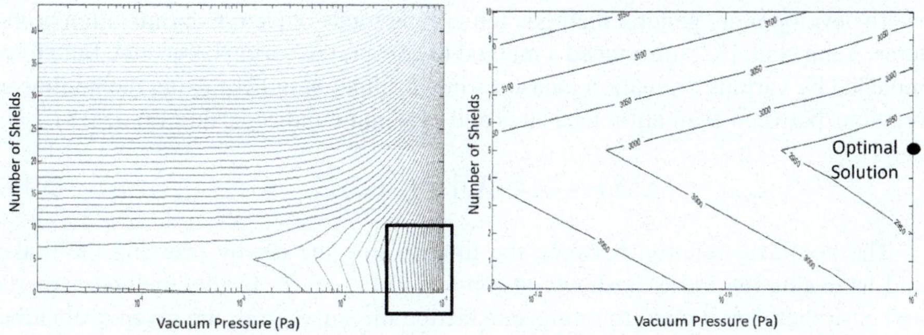

(a). Whole optimization space **(b).** Partial optimization space

Fig. 5.17 Optimization of two design parameters [17]

5.2.2 Multi-objective Optimization

Since the adaptable product is evaluated by adaptability and other evaluation measures in different life-cycle phases, identification of the optimal design is a typical multi-objective optimization problem.

A multi-objective optimization problem is defined by:

$$\min_{\text{w.r.t. } \boldsymbol{X}} (F_1(\boldsymbol{X}), F_2(\boldsymbol{X}), \ldots, F_m(\boldsymbol{X})) \tag{5.15}$$

where $F_1(\boldsymbol{X}), F_2(\boldsymbol{X}), \ldots, F_m(\boldsymbol{X})$ are m objective functions to be minimized.

A simple way to solve this multi-objective optimization problem is to converting these different objective functions, usually in different units and scales, into comparable values by multiplying scaling factors and adding these scaled objective functions to build the new objective function as shown in Eq. (5.16).

$$\min_{\text{w.r.t. } \boldsymbol{X}} W_1 F_1(\boldsymbol{X}) + W_2 F_2(\boldsymbol{X}) + \cdots + W_m F_m(\boldsymbol{X}) \tag{5.16}$$

where $W_1, W_2, \ldots, W_m$ are m scaling factors. Selection of proper values for these scaling factors, however, is a challenging task.

To develop more meaningful scaling factors, Xue et al. [30] introduced a method to combine different objective functions together by comparing their improvement from a selected basic design using Eq. (5.17).

$$\min_{\text{w.r.t. } \boldsymbol{X}} \sum_{i=1}^{m} \frac{F_i(\boldsymbol{X}) - F_i(\boldsymbol{X}_0)}{F_i(\boldsymbol{X}_0)} \tag{5.17}$$

where $\boldsymbol{X}_0$ is the selected basic design. The optimization results are different when different basic designs are selected.

To develop more general methods for solving multi-objective optimization problems, Yang et al. [32] introduced a method to convert the various objective functions modeled by various evaluation measures into evaluation indices using the nonlinear relations between evaluation measures and evaluation indices using Eq. (5.18).

$$I_i(\boldsymbol{X}) = I_i[F_i(\boldsymbol{X})], i = 1, 2, \ldots, m \tag{5.18}$$

The nonlinear relation between the time to stop the car by pressing the brake and its evaluation index is shown at top left in Fig. 5.18. In this method, experts and customers are invited to rating satisfaction measures from the given evaluation measures. Cubic polynomial curve-fitting is then conducted to model the nonlinear relation.

The final objective function is defined by:

$$\max_{\text{w.r.t. } \boldsymbol{X}} I(\boldsymbol{X}) = W_1 I_1(\boldsymbol{X}) + W_2 I_2(\boldsymbol{X}) + \cdots + W_m I_m(\boldsymbol{X}) \tag{5.19}$$

where $W_1, W_2, \ldots, W_m$ are m weighting factors between 0 and 1 representing importance of these evaluation indices.

In the design of the automobile caliper disc brake developed by Yang et al. [32], four evaluation measures were selected:

- T: stopping time (s).
- t_f: final temperature of disc (°F).

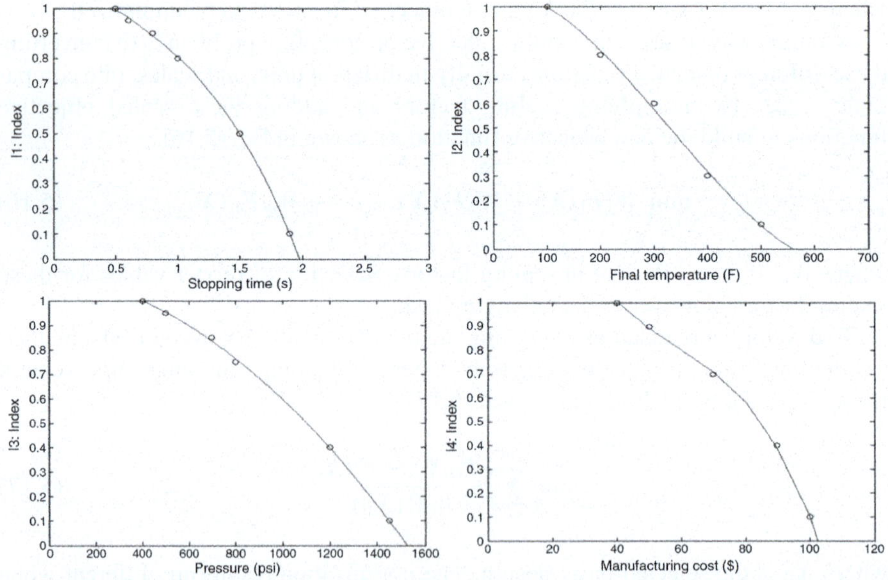

Fig. 5.18 Nonlinear relations between evaluation measures and evaluation indices [32]

- p: pressure on disc (psi).
- C: manufacturing cost ($).

The nonlinear relations between these evaluation measures and their evaluation indices were achieved through curve-fitting:

- $I_1 = 1.1734 - 0.4163T + 0.1885T^2 - 0.1404T^3$
- $I_2 = 1.0600 + 0.9524 \times 10^{-4} t_f - 0.8214 \times 10^{-5} t_f^2 + 0.8333 \times 10^{-8} t_f^3$
- $I_3 = 1.1045 - 0.9788 \times 10^{-4} p - 0.4095 \times 10^{-6} p^2 - 0.2100 \times 10^{-12} p^3$
- $I_4 = 2.310 - 0.05970C + 0.8765C^2 - 0.5 \times 10^{-5} C^3$

The evaluation measures and their evaluation indices are given in Fig. 5.18. The objective function with 6 design variables $\boldsymbol{X}$ was defined by:

$$\max_{\text{w.r.t. } \boldsymbol{X}} I(\boldsymbol{X}) = 0.551 I_1(\boldsymbol{X}) + 0.062 I_2(\boldsymbol{X}) + 0.088 I_3(\boldsymbol{X}) + 0.299 I_4(\boldsymbol{X}) \quad (5.20)$$

5.2.3 *Multi-level Optimization*

The constrained optimization methods and multi-objective optimization methods discussed in Sects. 5.2.1 and 5.2.2 are numerical methods to search for the best values of parameters from the given objective functions. Since selection of design configurations also plays an important role in adaptable design, optimization to select the best design configuration has to be considered when many feasible design configurations can be selected.

Figure 5.19 shows an AND-OR tree for design of an adaptable vibratory feeder [33]. In this design, the adaptable vibratory feeder served as both the linear vibratory feeder and the bowl vibratory feeder. The linear vibratory feeder had a rectangular container. In the operation stage, the rectangular container vibrated to move small parts in a specific direction. The bowl vibratory feeder had a bowl container. In the operation stage, the bowl container vibrated to move the small parts along the spiral track of the bowl container.

From this hybrid AND-OR tree shown in Fig. 5.19, 1296 feasible design configuration candidates can be created based on the rules for the generation of the design configuration candidates. Since each of these design configuration candidates should be further developed, the traditional exhaustive method to generate all design configuration candidates, and evaluate these candidates for selection of the best one is not practical. To improve design efficiency, optimization at configuration level should also be conducted.

Identification of the optimal design configuration and parameters can be accomplished through multi-level optimization model [29]. The optimization is conducted at two different levels: configuration level and parameter optimization level. First

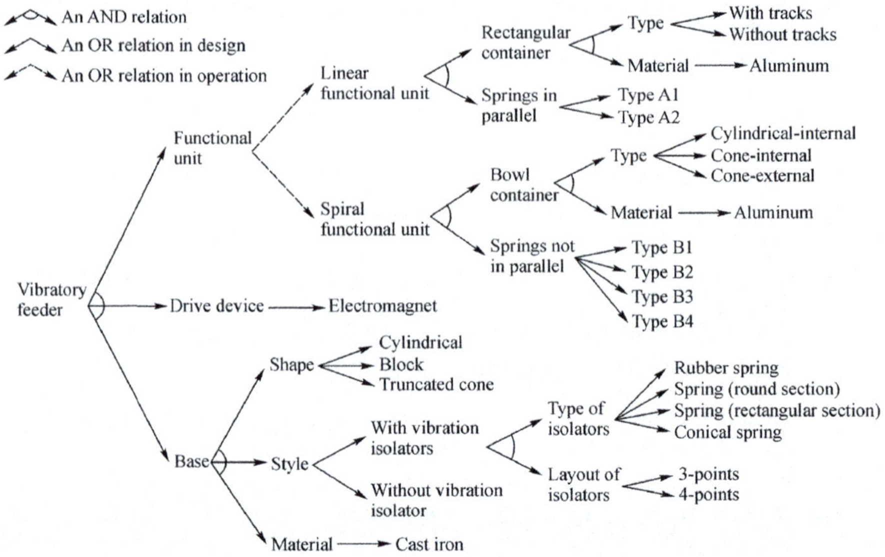

Fig. 5.19 Modeling of different configurations of the adaptable vibratory feeder [33]

parameter optimization for a selected configuration is conducted to obtain the optimal parameter values for this configuration. The best evaluation measure or evaluation index corresponding to the optimal parameter values is then used as the evaluation measure for this design configuration. Among all feasible design configurations, the optimal one is subsequently obtained through configuration optimization.

In this optimization model, the optimal design parameter values for the i-th design configuration solution candidate are achieved through parameter optimization.

$$\begin{aligned}&\text{Find: parameters } \boldsymbol{X}_i\\&\text{Optimize: } I^{(i)}\\&\text{Subject to: } \boldsymbol{X}_i^{(\mathrm{L})} \le \boldsymbol{X}_i \le \boldsymbol{X}_i^{(\mathrm{U})}\\&\qquad h_j(\boldsymbol{X}_i) = 0,\ j = 1, 2, \ldots\\&\qquad g_j(\boldsymbol{X}_i) \le 0,\ j = 1, 2, \ldots\end{aligned} \tag{5.21}$$

where $I^{(i)}$ is the selected evaluation index, $\boldsymbol{X}_i^{(\mathrm{L})}$ and $\boldsymbol{X}_i^{(\mathrm{U})}$ represent the lower boundaries and upper boundaries of $\boldsymbol{X}_i$, respectively, and $h_j(\boldsymbol{X}_i)$ and $g_j(\boldsymbol{X}_i)$ are equality constraints and inequality constraints, respectively.

Among all the p feasible product design configuration solution candidates, the optimal design solution is obtained through configuration optimization.

$$\begin{aligned}&\text{Find: the } i\text{-th design configuration candidate}\\&\text{Maximize: } I = I^{(i)}\end{aligned}$$

$$\text{Subject to: } 1 \leq i \leq p \tag{5.22}$$

where i represents the i-th design configuration candidate, and p is the number of all feasible design configuration candidates.

Parameter optimization can be conducted through numerical search [2]. The methods discussed in Sects. 5.2.1 and 5.2.2 are used for parameter optimization. Configuration optimization is conducted through genetic programming [16]. Since genetic programming is a global optimization method, this method will be explained in Sect. 5.2.4 in details.

In the design of an adaptable vibratory feeder developed by Zhang et al. [33], the two-level optimization problem was defined by:

Configuration Optimization

$$\begin{aligned} &\text{Find: the } r\text{-th design configuration candidate} \\ &\text{Maximize: } S = S_{(r)} = \text{Max}\{S_{(1)}, \ldots, S_{(1296)}\} \\ &\text{Subject to: } 1 \leq r \leq 1296 \end{aligned}$$

Parameter Optimization

$$\begin{aligned} &\text{Find: the values of un-adaptable design parameters } K_p, M_m, \text{ and } a \\ &\text{Maximize: } S_{(i)} \\ &\text{Subject to: } 0.15 \times 10^5 \leq K_p \leq 3.5 \times 10^5; 0.3 \leq M_m \leq 1.5; 0.7 \leq \alpha \leq 1.4 \end{aligned}$$

where S is the objective function considering robustness measures of both the linear vibratory feeder and the bowl vibratory feeder functions of the adaptable product, and K_p, M_m, and α are the design parameters of spring constant of each plate spring, mass of materials stored in the container, and angle between the plate spring and the base. Among the 1296 design configuration candidates, the optimal one as shown in Fig. 5.20 was achieved using genetic programming.

The optimal parameter values were identified as:

$$\begin{aligned} K_p &= 9.9 \times 104 \text{ (N/m)} \\ M_m &= 0.3 \text{ (kg)} \\ \alpha &= 1.32 \text{ (rad)} \end{aligned}$$

5.2.4 Global Optimization

In the traditional optimization methods, the numerical search is conducted to find better solution in each optimization iteration. When the objective function has multiple local optimal points such as the one shown in Fig. 5.21, the search may

(a) Configuration of liner vibratory feeder

- Rectangular container type: With tracks
- Rectangular container material: Aluminum
- Springs in parallel: Type A2
- Bowl container type: Cylindrical-internal
- Bowl container material: Aluminum
- Springs not in parallel: Type B4

(b) Configuration of bowl vibratory feeder

- Drive device: Electromagnet
- Base shape: Block
- Base style: Without vibration isolator
- Base material: Cast iron

Fig. 5.20 Optimal configurations of the adaptable vibratory feeder [33]

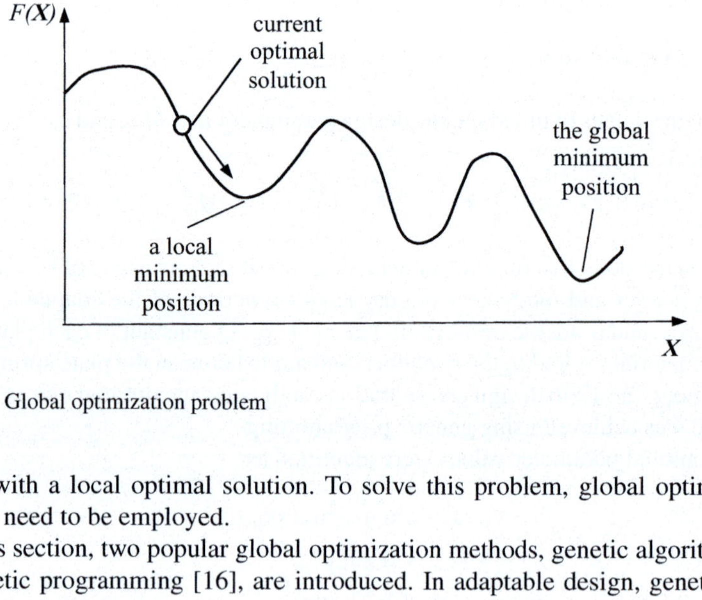

Fig. 5.21 Global optimization problem

end up with a local optimal solution. To solve this problem, global optimization methods need to be employed.

In this section, two popular global optimization methods, genetic algorithm [12] and genetic programming [16], are introduced. In adaptable design, genetic algorithm is used for parameter optimization, while genetic programming is used for configuration optimization.

Both genetic algorithm and genetic programming belong to evolutional computation approach that is based on mechanics of genetics and nature selection as shown in Fig. 5.22. In evolutionary computation, an initial set of n candidate solutions called a population is first generated. The candidate solutions, also called individuals, are evaluated using a pre-defined fitness function. Optimization objective function is used as the fitness function. The average fitness of the individuals in a generation is

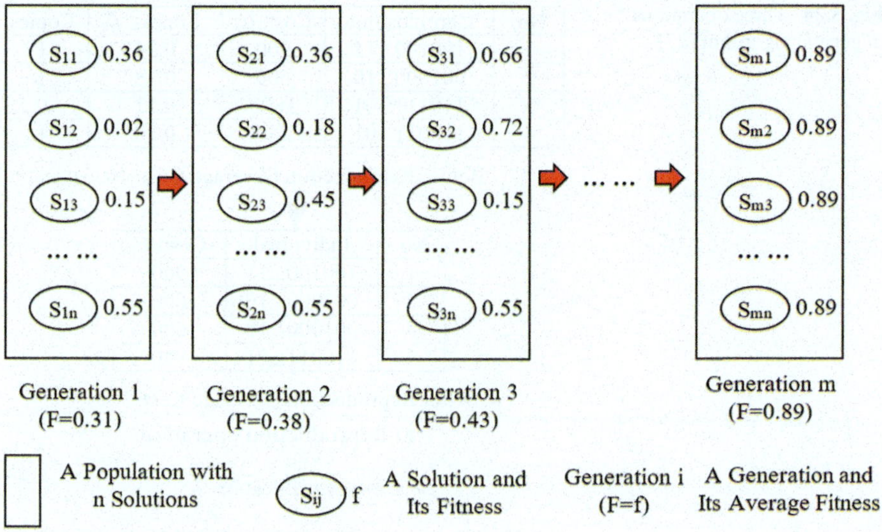

Fig. 5.22 Evolutionary computation approach

used to evaluate the quality of a population. A population is evolved to the next one by duplicating the individuals with the numbers of offspring based on fitness measures of these individuals. Stronger individuals produce more offspring individuals, and weaker individuals produce less offspring individuals or die in the evolution process. Small random changes are subsequently applied to the newly created individuals in the new population. The quality of the population is improved from a generation to the next one until the generation cannot be further improved. The best individual in the final generation is selected as the final optimization result.

Genetic Algorithm

In most genetic algorithm schemes, a binary chromosome of ones and zeros, as shown in Fig. 5.23, is used to describe a candidate solution such as a set of optimization variables. Generally, a number of chromosomes, called a generation of populations, are used for representing multiple optimization variable sets at a certain optimization search stage.

Genetic algorithm-based optimization search starts with initializing the first generation of chromosomes. Three operators, *reproduction, crossover*, and *mutation*, are used to evolve the populations of chromosomes toward the optimal solution as shown in Fig. 5.24.

Fig. 5.23 A *n*-bit chromosome to defined a solution

Fig. 5.24 Three operations in genetic algorithm [27]

No.	Chromosome	Cost: C_i	Fitness: f_i	Copies
1	100100...11	900	0.24	1
2	001000...10	600	0.38	2
3	110111...01	1800	0.12	0
4	110011...01	840	0.26	1

Before Reproduction (Average Cost: 1035)

No.	Individual	Cost: C_i
1	100100...11	900
2	001000...10	600
3	001000...10	600
4	110011...01	840

After Reproduction (Average Cost: 735)

(a) Reproduction operation.

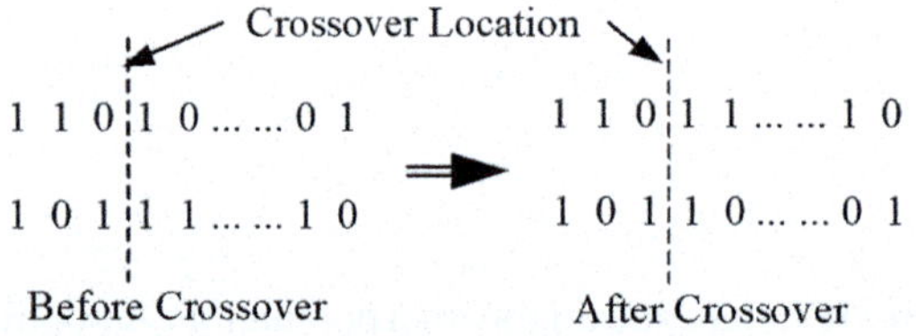

(b) Crossover operation.

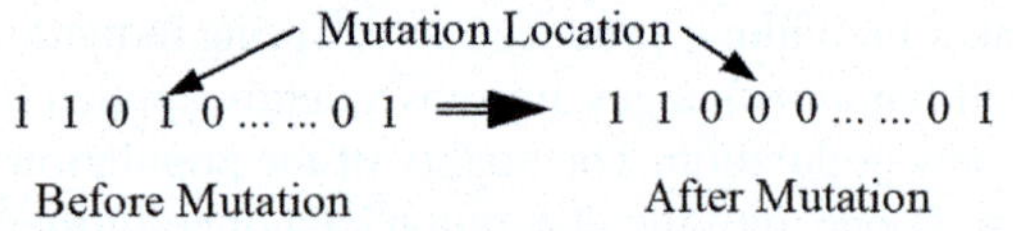

(c) Mutation operation.

Reproduction operator allows the chromosomes with better objective function evaluation measures producing more descendant chromosomes in the next generation. The chromosomes with poor evaluation measures may extinct in the evolution process. The total number of chromosomes in one generation remains the same. In the example shown in Fig. 5.24, each chromosome was used to model a solution to deliver the ordered products to different customers [27]. A delivery solution was evaluated by delivery cost, and costs of the individuals were further converted into fitness measures, representing the percentages of these individuals to be duplicated in the next generation as shown in Fig. 5.24a.

Crossover operator is used to pick up two chromosomes with a specific crossover probability, split these two chromosomes at a random location, and join the first part of one chromosome with the second part of another chromosome as shown in Fig. 5.24b.

Mutation operator is used to pick up a chromosome with a specific mutation probability, select a location randomly, and change its code as shown in Fig. 5.24c.

The search stops when the pre-defined maximum number of generations is reached or the stopping criteria are satisfied.

Since genetic algorithm selects a number of points (chromosomes) to conduct parallel search, it can reduce the chance of being trapped into a local optimum. In addition, genetic algorithm uses probabilistic transition rules instead of the deterministic transition rules to improve the quality of solution.

Genetic Programming

Genetic programming was initiated from genetic algorithm [16]. Compared with genetic algorithm in which a solution is modeled by a fixed-length binary string, a solution in genetic programming can be described by a data structure such as a tree [13]. During the evolution process, elements and lengths of the solutions are changed dynamically. Although genetic programming and genetic algorithm are different in modeling problem solutions, they share the same principles of evolution through natural selections.

Figure 5.25a shows an AND-OR tree developed by Hong et al. [13] in the design of customized windows. From the AND-OR tree, many design solutions can be created, and Fig. 5.25b shows one of these design solutions.

Evolution in genetic programming is also conducted through three types of operations: reproduction, crossover, and mutation.

For reproduction, all individuals in a generation are first evaluated to obtain their fitness measures. An individual is selected for duplication in the next generation based on its fitness using the roulette wheel selection method. In this method, the size of the section in the roulette wheel is proportional to the value of the fitness function of every individual—the bigger the value is, the larger the section is. A marble is thrown in the roulette wheel and the individual where the marble stops is selected. Clearly, the individual with bigger fitness value will be selected more times. Table 5.10 shows the 16 individuals in a generation, and the numbers of individuals to be duplicated in the next generation in the reproduction process.

Crossover is the operation to be applied to the two selected parent individuals. Since each individual is described by a tree, the crossover is conducted by selecting a crossover position randomly for each of the two parent individuals and swapping the two sub-trees with the selected positions as the root nodes of the sub-trees. The crossover operation applied to two individuals is shown in Fig. 5.26.

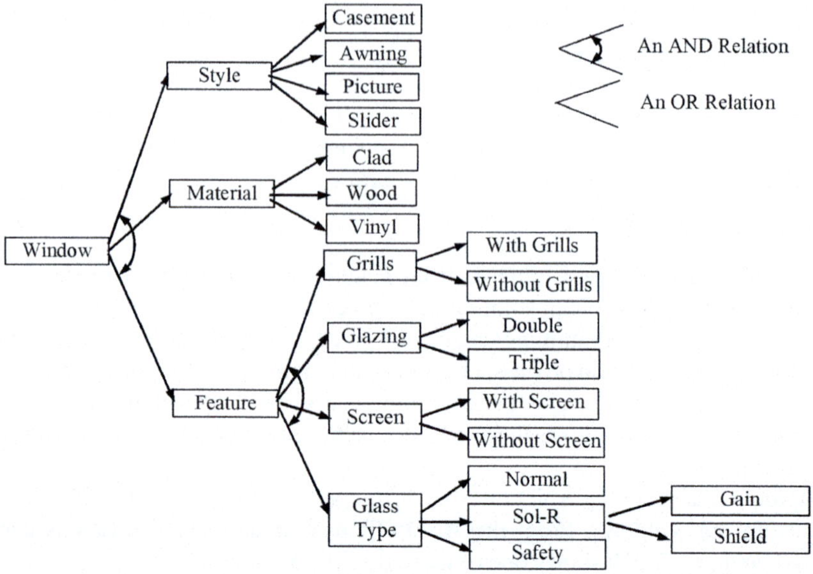

(a). An AND-OR tree for modeling a family of customized windows

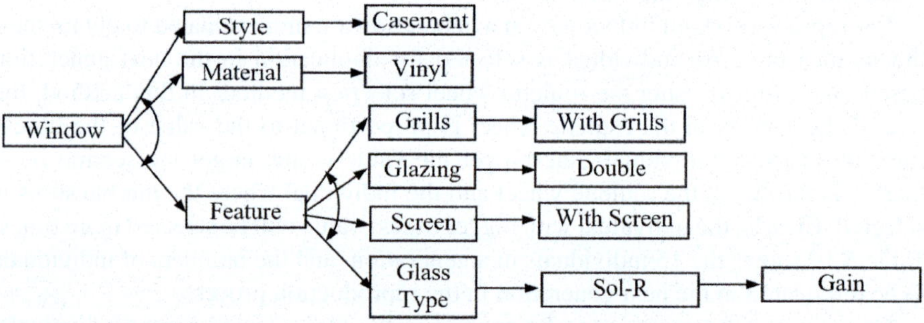

(b). An AND tree for modeling of a specific design solution

Fig. 5.25 Design of customized windows in genetic programming [13]

The crossover position should satisfy the following conditions:

- The node at the selected location should not be a root node in the AND/OR tree.
- The two nodes at the selected two locations of the parent individuals for crossover should have an OR relation.

The crossover operation should be conducted with small probability to avoid significant random changes of individuals in the next generation.

Mutation operation is applied to a selected parent individual. First a mutation position in the individual that matches with an OR node in the AND-OR tree is selected randomly. Then the sub-tree with the selected mutation node as the root

Table 5.10 A generation in genetic programming [13]

Individual Number	Configuration of Window	Parameters (m)		Fitness	Number of Copies
		W	H		
1	Vinyl, Awning, Grills, Double Glazing, No Screen, Normal Glass	2.94	0.9	0.856	1
2	Wood, Picture, Grills, Triple Glazing, Screen, Normal Glass	1.75	2.5	0.817	0
3	Vinyl, Picture, No Grills, Triple Glazing No Screen, Normal Glass	1.75	2.99	0.85	1
4	Wood, Casement, Grills, Triple Glazing, No Screen, Sol-R Shield Glass	0.81	1.45	0.71	1
5	Wood, Awning, No Grills, Double Glazing, No Screen, Sol-R Shield Glass	1.5	1.76	0.861	1
6	Wood, Slider, Grills, Double Glazing, Screen, Normal Glass	1.11	1.56	0.64	0
7	Wood, Awning, Grills, Triple Glazing, No Screen, Normal Glass	1.53	1.78	0.848	2
8	Vinyl, Slider, No Grills, Triple Glazing, No Screen, Normal Glass	0.89	2.03	0.781	2
9	Vinyl, Picture, Grills, Double Glazing, No Screen, Sol-R Shield Glass	2.3	2.17	0.835	1
10	Wood, Picture, Grills, Double Glazing, No Screen, Normal Glass	1.53	2.77	0.813	2
11	Wood, Picture, No Grills, Double Glazing, No Screen, Normal Glass	2.49	1.73	0.82	1
12	Clad, Picture, No Grills, Triple Glazing, No Screen, Sol-R Gain Glass	1.8	1.76	0.797	2
13	Wood, Slider, No Grills, Triple Glazing, No Screen, Sol-R Gain Glass	0.75	2.41	0.659	0
14	Vinyl, Picture, Grills, Triple Glazing, No Screen, Sol-R Shield Glass	2.99	1.73	0.839	1
15	Wood, Awning, Grills, Triple Glazing, Screen, Normal Glass	2.25	1.59	0.833	1
16	Vinyl, Picture, No Grills, Double Glazing, Screen, Sol-R Shield Glass	1.91	2.63	0.841	0
Average Fitness of the Generation				0.798	

node is removed from the individual. Another position in the AND-OR tree that has an OR relation with the previously selected mutation node is then identified as the root node to create a new sub-tree to replace the removed sub-tree. The mutation operation applied to an individual is given in Fig. 5.27.

The mutation operation should be conducted with small probability to avoid significant random changes of individuals in the next generation.

Fig. 5.26 Crossover operation [13]

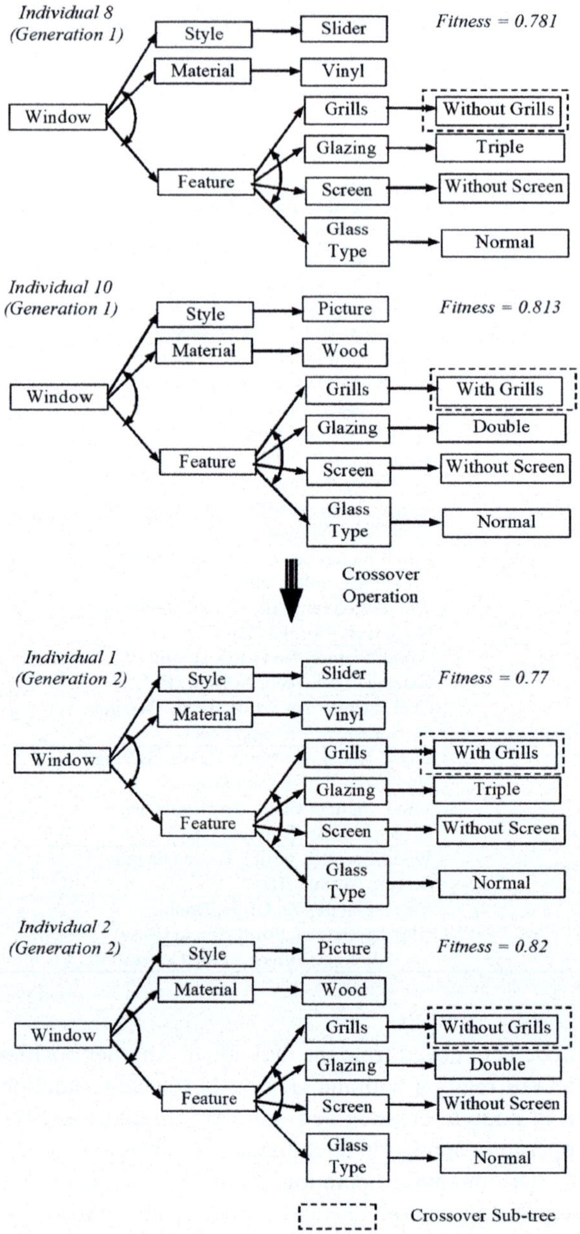

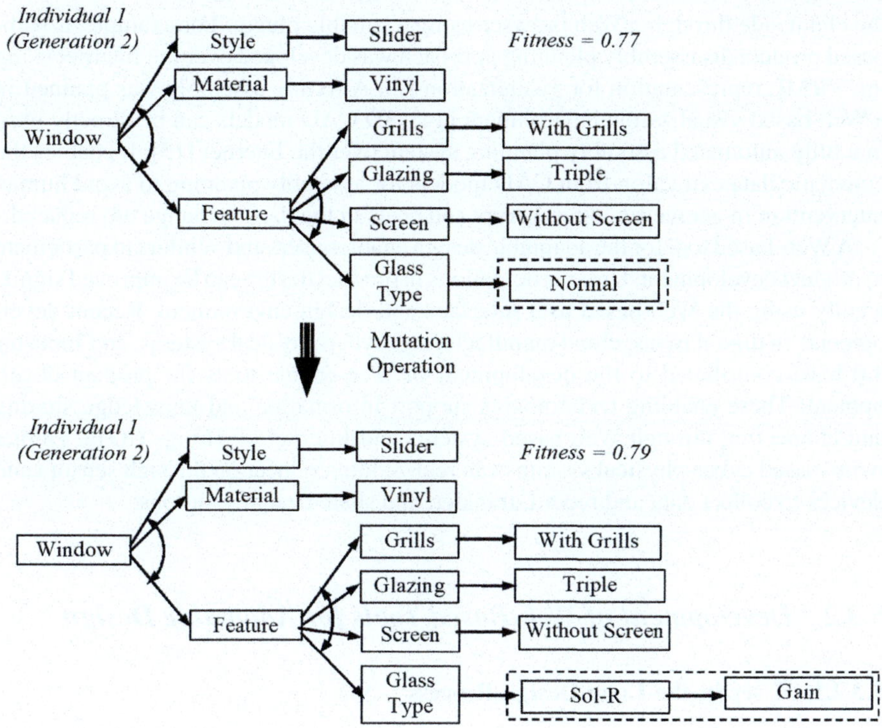

Fig. 5.27 Mutation operation [13]

5.3 Web-Based Tools for Adaptable Design

5.3.1 Introduction

This section introduces web-based tools for adaptable design [5]. Web-based tools connect a physical product and its digital model to form digital twins. The digital model is a dynamic virtual product to simulate and depict attributes and behaviors of the physical product in the real world. Web-based tools support users to collaboratively develop product in a distributed environment of product development.

Web-based tools provide 3D visualized environments for collaborative product design interactions between physical and information domains. Different methods have been proposed for the product representation in Web-based environments. A digital model can describe different layers of product details for scenarios, components, and interfaces. A production line model can be formed for interactions between physical and digital domains to represent operations of the model. A product model can be built in the CAD software system using a standard format such as STEP or WRL to store 3D models in a database server. 3D product models can be viewed on the client side through a Web browser using a suitable plugin. For example, a Web-based product disassembly planning approach was developed over the Internet using the VRML representation for visualization [7]. A fixture assembly was planned in a Web-based visualization environment [14]. 3D CAD models can be directly used in a fully automated assembly planning system over the Internet [15]. It enables the automatic data extraction from CAD models for assembly planning to avoid human intervention in geometric input. Errors and costs in the data exchange are reduced.

A Web-based tool for the adaptable design enables users and vendors to participate in product development. User participations in product design can be improved significantly using the Web-based user interface and design environment. Recent developments in digital twins, cloud manufacturing, cyber-physical systems, and Industry 4.0 have contributed to the development of Web-enable tools for product development. These enabling technologies support information and knowledge sharing and connecting through Web-based systems and Internet of Things (IOT). Digital twins-based cyber-physical systems can realize interconnection through sensors and devices to collect data and record, transfer, and share design solutions.

5.3.2 Development of Web-Based Tools for Adaptable Design

5.3.2.1 User-in-the-Loop Design Process

Adaptable design can be improved using the user-in-the-loop design through Internet. The process and component of a Web-based tool for the user-in-the-loop design are shown in Fig. 5.28. For adaptable products, users select a product platform to add modules to meet their needs from the module base or library. Modules in the library from different sources are shared for different functions of user needs. Details are as follows:

Platform selection and function matching: When users submit product requirements through an online interface, the system searches appropriate platforms from the database through decision-making mechanisms to match the user needs. If there are more alternatives than one solution, users can provide more details to refine the design requirement for a unique solution.

Module component search and product configuration: Product manufacturers and module suppliers provide common functional and customized modules for users to select. All related information is shown on the interface for users to choose the corresponding modules. As there are a large variety of modules and component alternatives, a heuristic search engine is provided to automatically configure modules in products based on constraints defined by the user and manufacturer. When there are not enough modules to meet some special personalized need, the user can contact a third-party module supplier to develop a new personalized module.

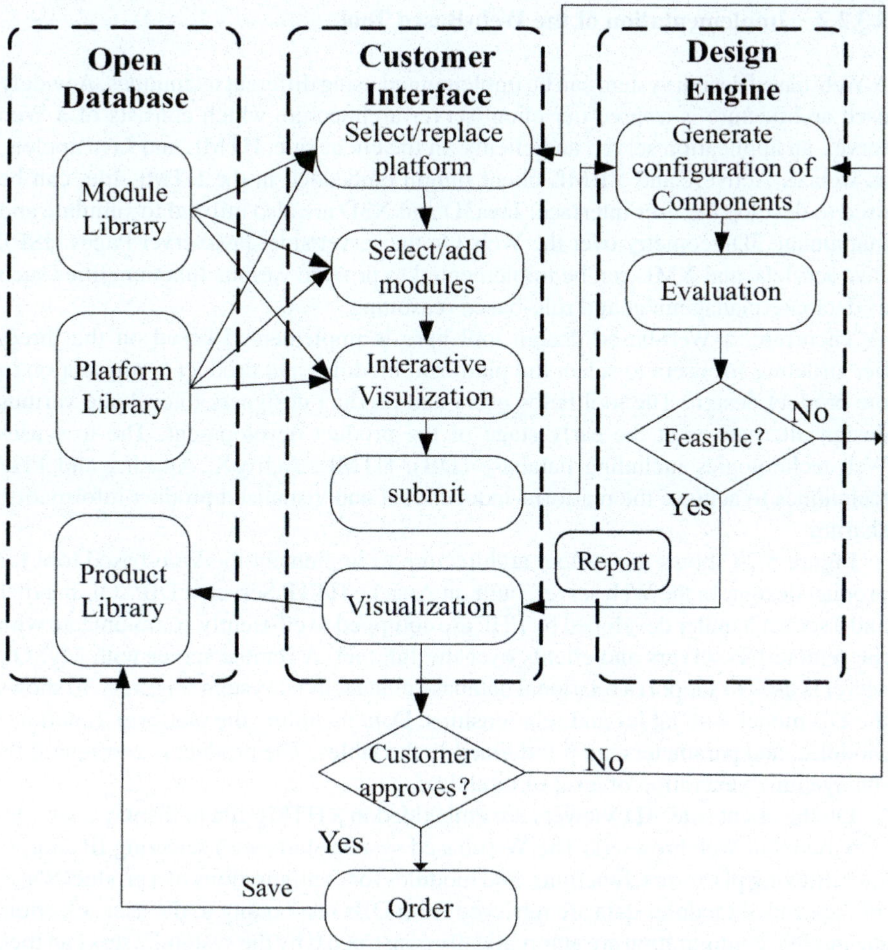

Fig. 5.28 Web-based tool for the user-in-the-loop design

When an initial design is ready, the user submits the selection to the configuration design system that builds the product structure for the user selection. The configuration is evaluated based on various perspectives. The optimal configuration is searched based on design rules formed by the user and manufacturer.

Design results are visually depicted to the user. If an initial design proposed by the user fails to pass the analysis stage, a problem report is generated for the user. The user can change the requirement or modify the selection by using the interactive visualization tool in the interface. After the change is made, the user resubmits the design proposal. Once the proposed structure and solution successfully pass the analysis stage, the system waits for the user confirmation. At this stage the user can either accept and order the product or change it even further and resubmit an alternative proposal. The process is iterative and ends only when the user is satisfied.

5.3.2.2 Implementation of the Web-Based Tool

A Web-based design system can be implemented using different techniques. A widely used architecture is a three-tier client/server architecture which consists of a Web server, an application server, and clients. In the client side, HTML and Java applets, as well as ActiveX and VRML client plugin embedded in the HTML file, can be used to develop the user interface. Java3D and X3D are also utilized to simulate and manipulate 3D geometry over the Web. On the server side, JavaServer Pages (JSP), Java servlets, and XML can be implemented to provide various functionalities such as database management and rule-based reasoning.

Therefore, a Web-based design tool here is implemented based on the three-tier structure for users to select the platform, modules and product configuration in the product design. The tool helps users and product designers to evaluate various design alternatives at the early stage of the product development. The tool uses Web technologies including database (DB), HTML, ActiveX, Apache, and PHP techniques to achieve the platform-independent and visualized product information sharing.

Figure 5.29 shows the system architecture to implement the Web-based tool for product design. In the Web server, built on Apache HTTP Server, a DB/SQL handler and a socket handler developed by PHP are equipped to efficiently communicate with the remote DB servers and clients over the Internet. A remote server with MySQL server is used to support a relational database management system. Figure 5.30 shows the DB model with tables and relationships. Data including the platform, functions, modules, and parameters are pre-defined in the tables. The product is configured by the dynamic generation of a set of modules.

On the client side, 3D Viewers are embedded in a HTML file to display a simulation model in Web browsers. The Web-based system starts with selecting IP address for retrieving platforms, functions, and modules for configurations of a product. Once all the needed modules data are retrieved from DBs responding to the user selection, the module configuration are automatically constructed by the system. Users can then perform the module configuration using the interactive selection or heuristic algorithms on a Web browser. Using the fully interactive functions, users can examine details of the product configuration.

5.3.3 Applications of Web-Based Tools in Adaptable Design

The main user interface of the Web-based tool for the adaptable design is shown in Fig. 5.31a. The product architecture, modules and interfaces as well as associated parameters and documents templates are defined by a Web-based graphic user interface (GUI) as shown in Fig. 5.31b–d, respectively.

All the data, parameters, models, and other useful information of modules, including platform modules, add-on modules, new developed modules, customized and individualized modules, as well as existing history products and related design

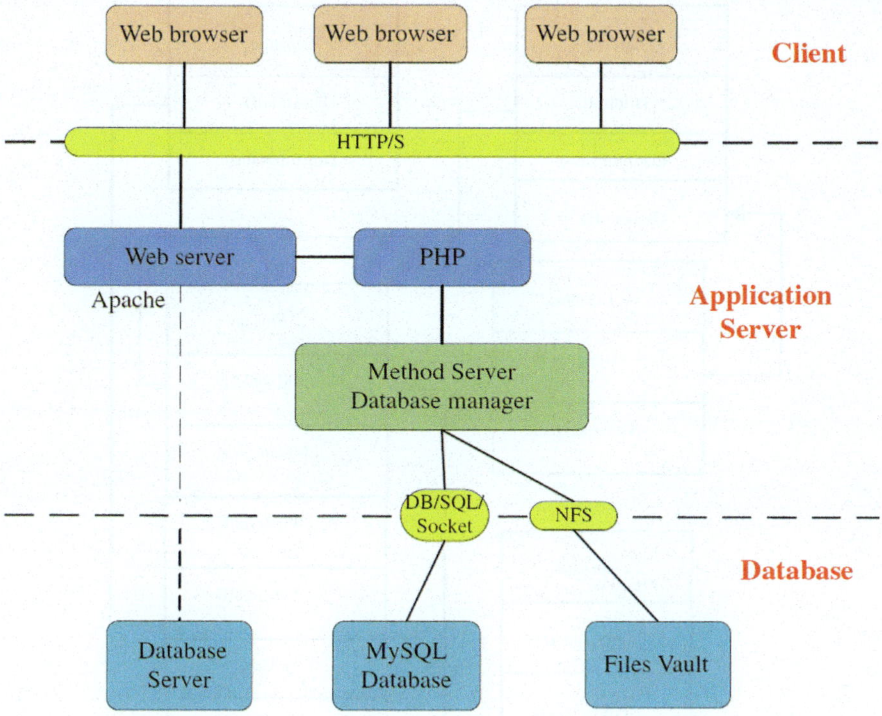

Fig. 5.29 Architecture of the web-based tool

requirements, are stored in shared databases. The functional and performance attributes of each module such as the volume, weight, center of the gravity, cost, and material texture are also stored in the database. Interfaces of the module database and product database are shown in Fig. 5.31e,f, respectively. New modules can be defined and stored into the database via interface as shown in Fig. 5.31g. The cost and adaptability for the existing product configurations can also be measured and displayed on the interface as shown in Fig. 5.31h.

For design of a new product to meet different requirements of individual users, users can use the Web-based interface to select the functional modules. For example, a new configuration is developed for the new need as shown in Fig. 5.32.

As modules in the database are designed independently, users can select them based on the product need. After a platform module is selected, other modules can be chosen. The Web-based user interface helps users in the module selection as shown in Fig. 5.32b. Each selection button in the webpage is related to an available module in the database. The functional attributes of modules are used as a search index or measure index of modules. For a developed adaptable product, modules can be selected as add-on modules. When different functions are required, the user only needs to replace add-on modules by disconnecting and connecting them to adaptable interfaces. Figure 5.33 shows the final assembly of a product to the user with all the modules.

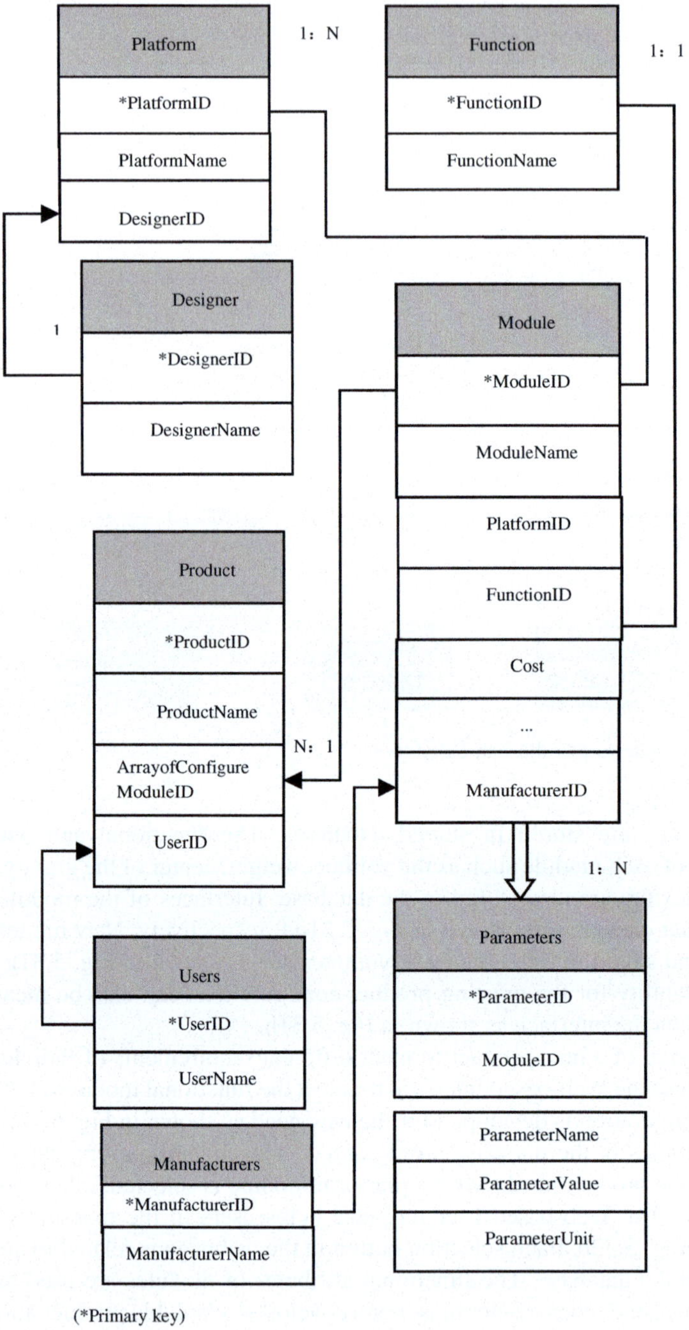

Fig. 5.30 Relational database

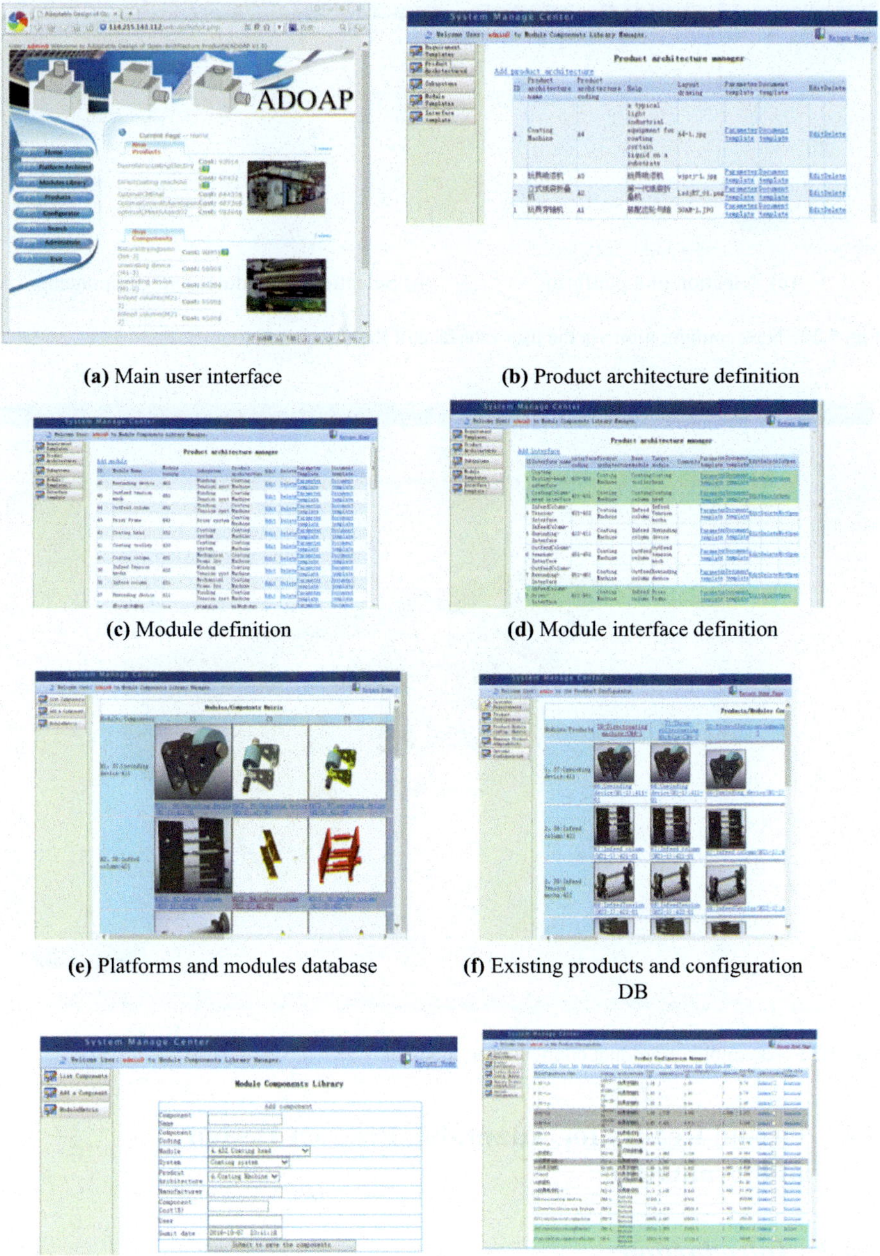

(a) Main user interface

(b) Product architecture definition

(c) Module definition

(d) Module interface definition

(e) Platforms and modules database

(f) Existing products and configuration DB

(g) Adding new modules

(h) Cost, adaptability and openness for configurations

Fig. 5.31 User interactions using the web-based tool in adaptable design

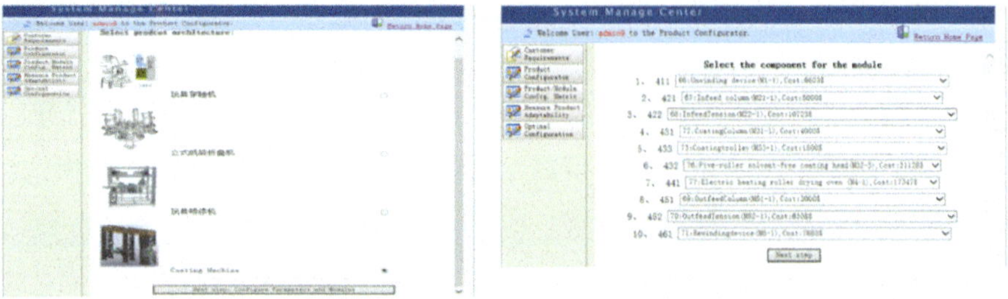

(a) Selection of a platform (b) Selection of modules and components

Fig. 5.32 New configuration via the user interface of the web-based tool

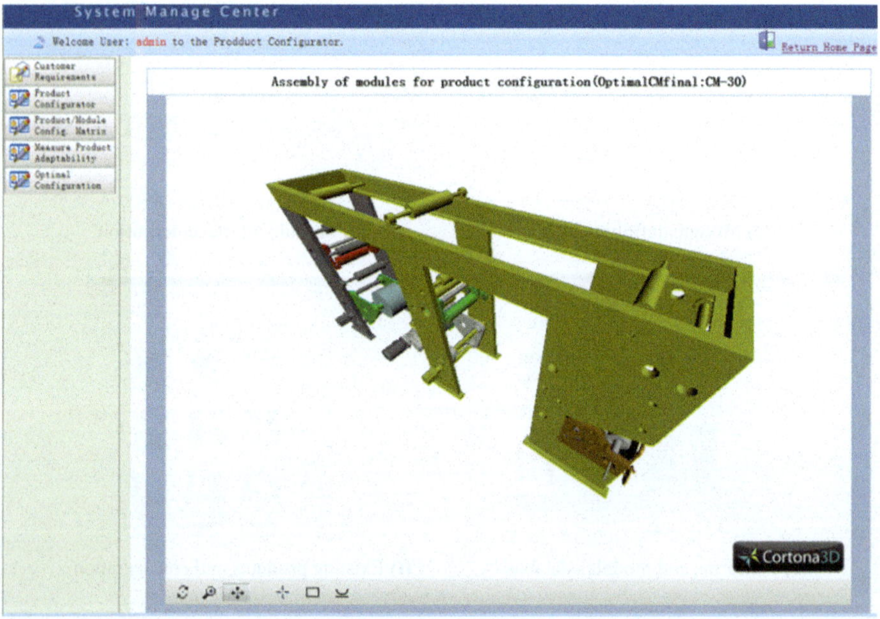

Fig. 5.33 Assembly of product configuration modules

5.4 Virtual Reality for Adaptable Product Design and Evaluation

5.4.1 Introduction

Virtual reality (VR) provides a special virtual environment (VE) with features of immersion, interaction, and information integration to enrich the capacity of product development tools [19]. VR has been known for years, but VR technologies take recent leaps in the power and affordability, a whole new wave of professionals have begun to employ VR as a tool for solving the intractable problems.

VR integrates technologies of the computer simulation, 3D graphics, human–machine interface, multimedia, sensor, and networks. VR uses enhanced computer input/output (I/O) devices to form a 3D immersive environment. Human–computer interactions are conducted in VEs to mimic product operations in the real world. Users can observe and feel products through operations of the products using different VR devices such as the haptic device, data gloves, and motion tracking sensors to capture and control the operation process. User feedbacks can be collected in real-time for product improvement. Different scaled VEs can be formed based on simulation needs of product using different configurations of VR systems from the large sized CAVE systems to personal head-mounted devices (HMD).

VR application-oriented research has developed different methods in engineering fields. A lot of efforts have been made by industries and researchers to apply VR in product development [20]. VR has been applied in product design [28], process planning [21], and manufacturing operations [23]. VR technologies have also been used in the interior design, real estate development, and training [36]. VR applications in manufacturing are called virtual manufacturing to investigate the dynamic process of manufacturing systems [6]. VR can form an immersive environment to simulate details of the product production without the manufacturing cost occurred in traditional methods.

For product design, VR provides tools for the product analysis and validation to test and verify performance of products using digital product models [25]. VR has been used to improve the user involvement in product development and user experience in product performance. VR provides an enable technology of rich functions for the user involvement in product development through different sensing channels to give the user real feeling of product operations. The product function examination and verification can be conducted in an immersive interactive environment that common CAD systems cannot provide. VR systems provide detailed visual effects of objects to assist users to identify details of product functions for a close-real use experience in a cost-effective way. Through the visual feedback and product interaction, users can evaluate products based on their requirements [34].

The user involvement in the design process is a key for products to meet user requirements. It was found that the failure rate of new product development can be reduced between 20 and 96% by integrating users in the product development [3]. The existing methods of Internet-based interactive platforms and direct market user surveys cannot provide users full experience of product features [34]. VR can help increasing the degree of user involvements in the product development, such as an assembly navigation approach to support human interactions in VEs for effective virtual assembly path planning [9], a mixed VR disassembly environment to evaluate disassembly sequences in the product development [8], and a virtual design review system to guide users in assembly operations and visual fidelity [28].

The user involvement is also significant for the effective adaptable product (AP) design and improvement, especially, for APs to meet users' needs. An AP can be designed to better match user needs if the user needs can be better identified. One of the effective ways to collect user needs is to involve the user in the product design and evaluation.

The development of APs requires data from users in the evaluation of the product function and performance. VR is an ideal tool to meet these requirements in the AP design and evaluation, particularly to identify user needs by involving the user in the design process and solution evaluation.

Therefore, APs should be developed with an early integration of users to satisfy end-user needs. VR provides an efficient way to achieve this need by directly integrating users in the design process to perform design tasks. A VR-based system can be designed and constructed to enable user experience-oriented prototyping to improve user experience in AP design. Using the usability test of products in a VE, it is possible to observe user behaviors in a virtual showroom and conduct real-time surveys [22].

This section introduces a user interactive design system based on VR technologies to provide users a close-real experience in the development of APs. The system provides users an interface built on the VE. Users can review the process of product design by virtual operating and evaluating the product. The system records user operations and feedbacks for designers to improve the product.

The interactive VR system has functions of the product-user interaction, product model processing, function simulation, data recording, and analysis for details of user involvement in product design. The system not only supports user participation in the product development, but also provides data for product developers to improve the product. It can also be used in the operation simulation of upgrading product function modules.

5.4.2 A VR-Based User Interactive System

A VR system is developed to improve the user involvement in the product development [25]. VR interactions are conducted between product models and users for the product evaluation and improvement. Figure 5.34 shows elements of the VR-user interactive system.

The system consists of a user interactive interface, a product model processing module, a function simulation module, and a data recording and analyzing module. Users operate product models in the VE to experience the product, which includes the review and operation of product models. Details of the system elements are as follows:

- User interactive interface: users interact with product models through VR devices. The interface follows user commands to display product operations using VR functions to accomplish different user requirements to experience the product functions.
- Product model processing: product models are imported into the VR system for product operations. The product models are stored in a product model base.

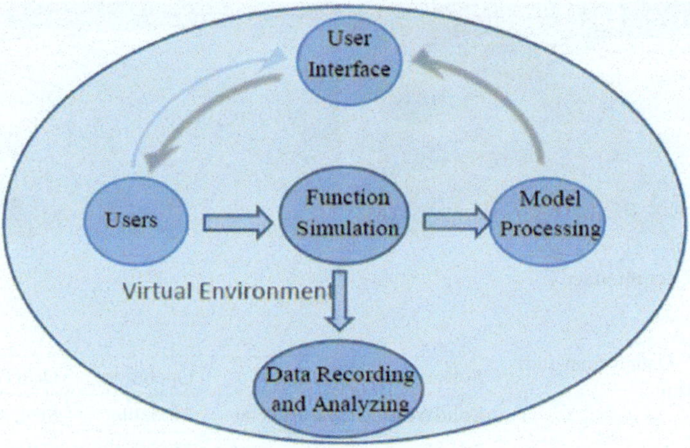

Fig. 5.34 VR-based user interactive system

- Function simulation: it performs interactive activities between users and product models, recognizes user requirements and instructs VR system operations to accomplish different interactions.
- Data recording and analysis: it records the process of the user activities and provides the feedback to designers for the product improvement. User needs and suggestions for the product improvement are concluded through the data analysis.

5.4.2.1 User Interactive Interface

WorldViz VR devices shown in Fig. 5.35 are used in the system development for user interactions with product models. The VE consists of two projection screens for users to conduct interactive actions of product operations and evaluations. The user interface and operation menu are formed using the Python program language as shown in Fig. 5.36. Product models are operated by a user using motion trackers and wand in the VE. The user can choose any part of a product for examinations in the process of interactions.

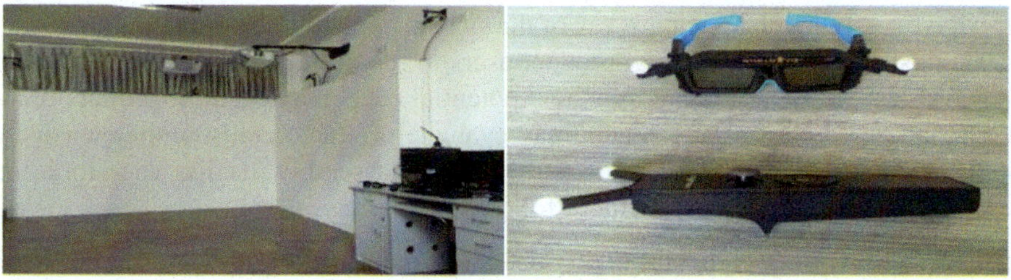

Fig. 5.35 VR devices

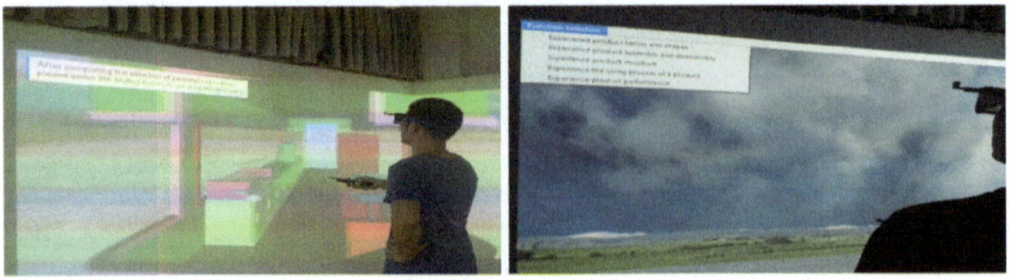

Fig. 5.36 User interface

Table 5.11 Data format and transformation

Software	Format	Operation	Data content
SolidWorks	.part → .sat	Modeling	Size, shape
3ds Max	.sat → .osgb	Modularity	Color, texture
Vizard	.osgb	Simulation	Position, action

5.4.2.2 Product Model Processing

Product model processing converts product models into function simulation models. The models are stored in a model base for simulation operations. The product models are converted into a data format used in the VR system. Details are built for the simulation, such as module operations, configurations, and replacements. Product models in the system are formed to meet user selections in the simulation.

After different product models are formed for the interactive need, the system is ready for the user operation. Along with applications of the system, the number of models in the model base will continue to increase. It is therefore a continuous work for product model processing when each time new models are added into the system. Different product models are dynamically updated by the model processing for the simulation need in the product evaluation and improvement. The model format and data contents are listed in Table 5.11.

5.4.2.3 Design for User Involvement

Design for user involvement is a key of the system application. Users can be product designers, current and potential product users to require different types of the user involvement. Current product users and potential users may examine product functions and performances. Designers may evaluate product details comprehensively including rationality of the product structure and assembly details. Therefore, the design for user involvement includes following contents.

(1) Different user preferences in product colors and shapes for product appearance.
(2) Product assembly and disassembly for user operations and for designers to develop the product.

(3) Rationality of the product structure and structure characteristics by observing the product for upgrading product function modules.
(4) Operation details of the product, such as the driver view of a car, convenience of car operations, and driver comfort in the car.
(5) Product special performance such as the collision effect.

5.4.2.4 Function Simulation

Function simulation processes user interactions with product models. The simulation identifies the user operation to convert user requirements into instructions for system operations. One of its most important parts is the motion capture system. The motion capture system has a tracking mode with six degrees of freedom, combined with functions of optical tracking and inertia positioning to obtain *XYZ* coordinates of the motion object and rotation angles of *XYZ* three directions. The optical motion capture system includes Precision Position Tracking (PPT) Wand, PPT Processor, PPT Cameras, and PPT Eyes. PPT Wand is a wireless handheld input device with six degrees of freedom to browse, navigate, select targets, and other interactive operations in the VE combined with PPT trackers. PPT Eyes is a small, portable, wireless motion tracking device that can be installed by users. When it is used with WorldViz PPT action system, users can browse the product model in the VR system. PPT Processor is mainly used to process the tracking data, and transmit data in the VR system.

The response speed and precision of the function simulation affect the simulation reality directly. The Vizard VR package has the infrastructure built on Open GL high-performance image processing software to realize the rapid and efficient VE rendering. Users wear 3D glasses and PPT Eyes, holding the PPT Wand in the simulation environment to operate the product model. In the operation, the simulation processes interactions in real-time between the user and product models according to the simulation requirement. The simulation is processed by a rendering system before the visualization is projected onto screens. Different interactions can be conducted between the user and VR system in the simulation process. The interactive process is recorded in the computer for the further analysis.

5.4.2.5 Data Recording and Analysis

Data recording and analysis record the process of user participations and operation details, and analyze data for suggestions of user needs and the product improvement. The process collects product data in the simulation process. The data recorder uses the VR motion capture system. In the operation process, recorded data can be real-time displayed on the screen. It can also generate a list of the data and show locations of the operating part. The recorded data are classified into different groups for the analysis. Data analysis identifies the user personal need for the product improvement.

During the user operation, the system automatically generates two types of documents. One is for the level of user satisfactions and the time used in system operations.

The other document is the process of product operations. Accurate locations of the tracker are recorded in the simulation. These data are used to analyze user interactions for designers to identify potential problems in the design improvement.

5.4.2.6 The System Implementation

The system is implemented using the Python language under the WorldViz VR system. Figure 5.37 shows the flowchart. When the program starts, the configuration runs to connect VR devices such as the PPT wand and glasses. A VR interface shows to the user for operations. The system then loads in models for the user to operate the product in the VE. The data flow contains series–parallel data with the unidirectional and bidirectional flows in the interactive system. The data interact with various product models. For example, the function simulation passes data to the system according to user operations, and the system then controls the user interface to generate operations. In the process, the data mainly flow between product models and components in the system, at the same time, the system will pass results to the user through the user interface. After that, the data flow is mainly from the product model processing to the simulation operation. The simulation handles data to show results through the user interface. Some data flow from the simulation to the process of data recording and analyzing.

5.4.3 Case Study—A VR System for Food Truck Design

5.4.3.1 VR System Design and Operation for Food Trucks

Food trucks have a variety of structures to offer different functions based on the food for sales. There are over five thousand food trucks providing 1.2 million different kinds of meals a day from halal food to hot dogs in streets of New York in 2015. There is also an increasing demand for food trucks in the Chinese market to meet different needs. For requirements of consumers and regulators, the design and production of food trucks using the AD concept provides an ideal solution for the fast food industry.

The food truck is proposed as a typical adaptable product using different product function modules as shown in Fig. 5.38. Parts formed by chassis, motor, and steering systems are common platform modules applied for all different food trucks. They will not be changed in the product use phase. Other function modules are parts of the truck body such as doors and windows that can be customized by users with manufacturers. The truck body can have different shapes and colors. Function modules such as cabinets, grills, and gas tanks can be produced by different manufacturers or suppliers. The combination of these modules forms an adaptable food truck based on user preference to meet the user need. A user can examine a variety of product performances in the VR system through interactive operations as follows.

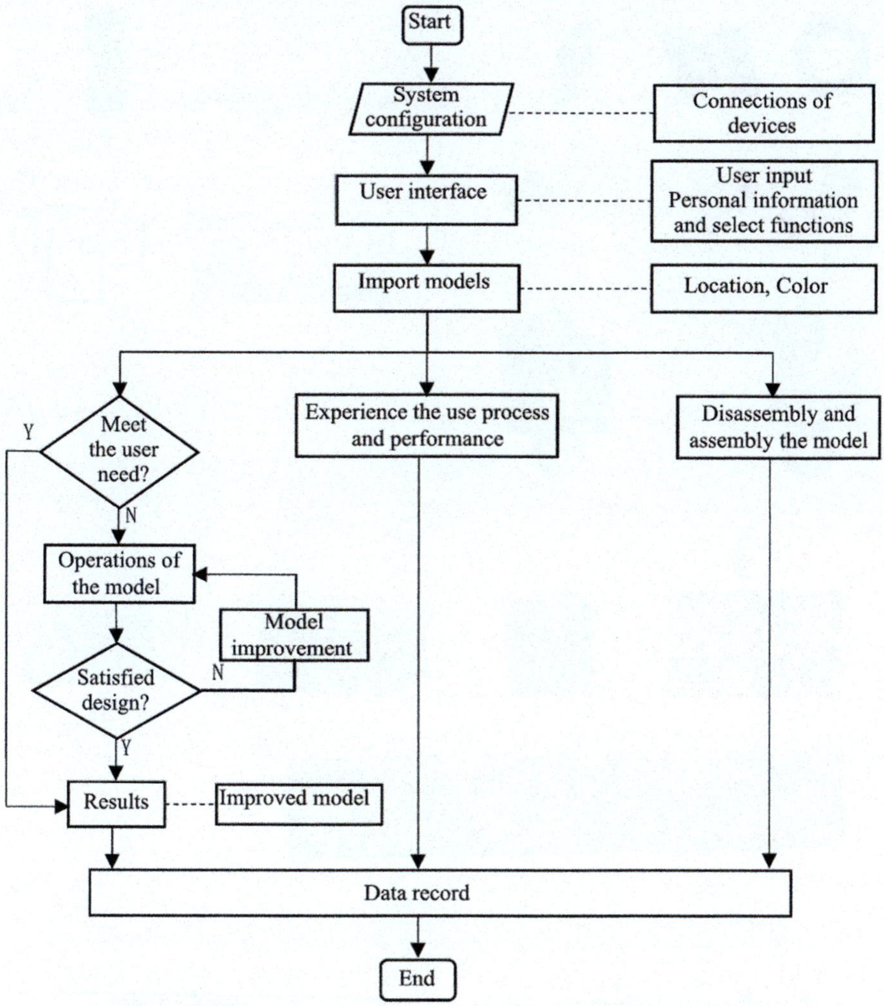

Fig. 5.37 Flowchart of the system

- *Selections of product colors and shapes for product's appearance based on the user preference*

A user can select different shapes and colors of the food truck according to the individual need. The user observes the shape and color of the truck from different distances and perspectives in the VE. The color and shape can be changed using the VR device. There are six colors designed for the truck. The color of different parts can be changed by the user as shown in Fig. 5.39.

On the client side, 3D Viewers are embedded in a HTML file to display a simulation model in Web browsers. The Web-based system starts with selecting IP address

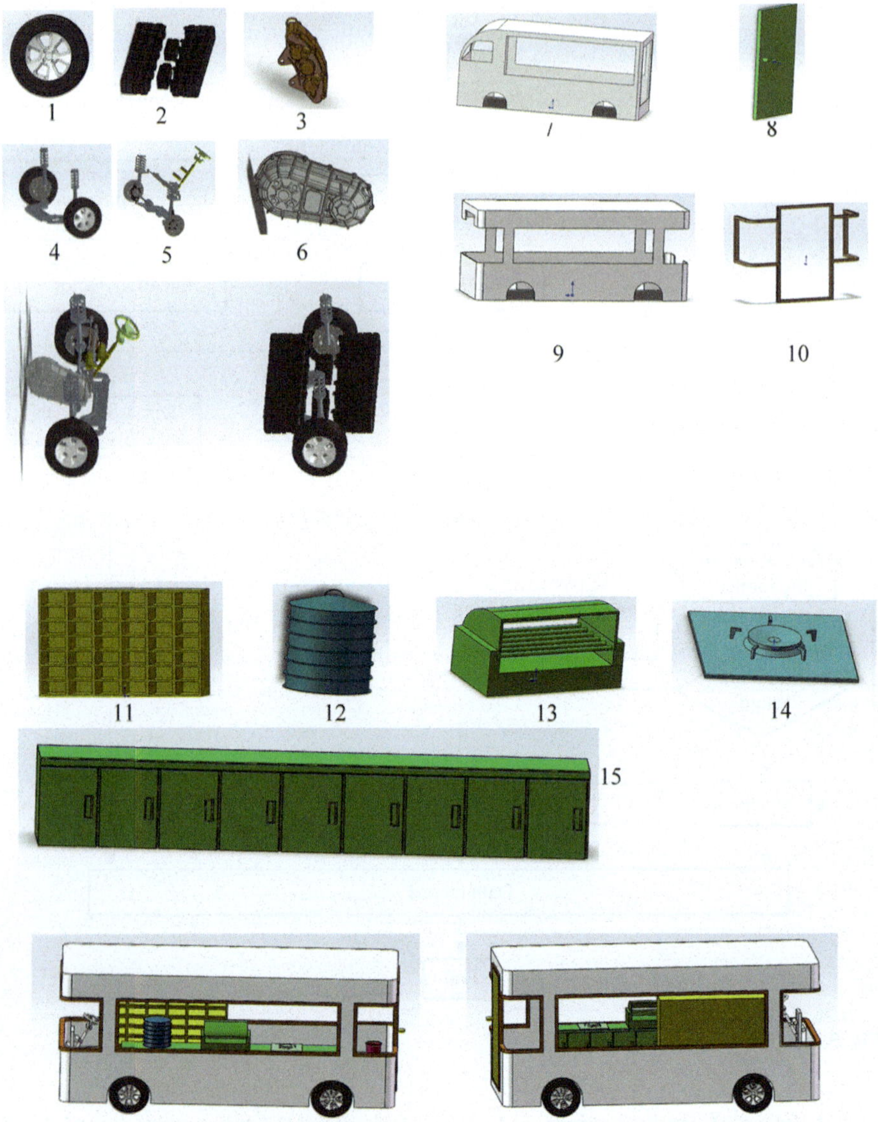

Fig. 5.38 Food trucks and function modules

for retrieving platforms, functions, and modules for configurations of a product. Once all the needed modules data are retrieved from DBs responding to the user selection, the module configuration are automatically constructed by the system. Users can then perform the module configuration using the interactive selection or heuristic algorithms on a Web browser. Using the fully interactive functions, users can examine details of the product configuration.

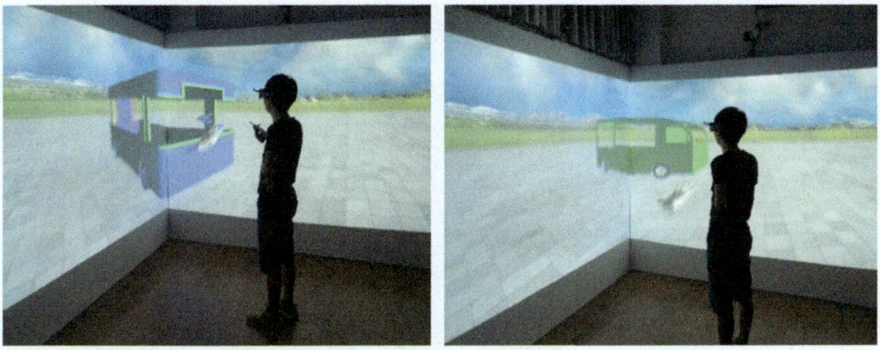

Fig. 5.39 Selection of shapes and colors

- *Operations of product assembly and disassembly*

The food truck and its modules can be simulated for the assembly and disassembly to check feasibility of upgrading modules or train operators as shown in Fig. 5.40.

- *Test of the structure for product upgrading*

This checks the product structure for upgrading product functional modules. Users can select a variety of function modules according to their requirements. The PPT wand is used to operate various kinds of kitchen appliances. An object becomes the light green color when a virtual hand is close to the object. The selected object can be moved with the virtual hand when the user presses a trigger of the wand. The user releases the trigger key when the object is reached the specified location. Figure 5.41 shows the selection and placement of function modules.

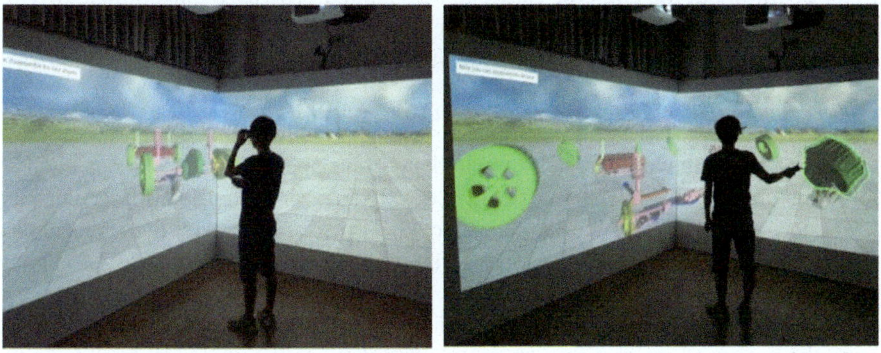

Fig. 5.40 Simulation of the assembly and disassembly

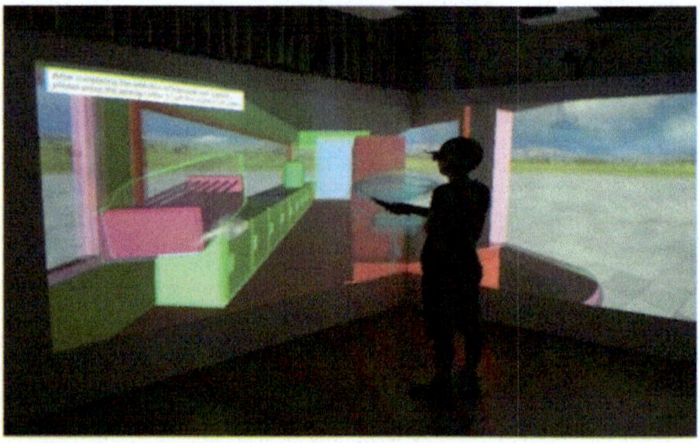

Fig. 5.41 Selection and placement of personalized function modules

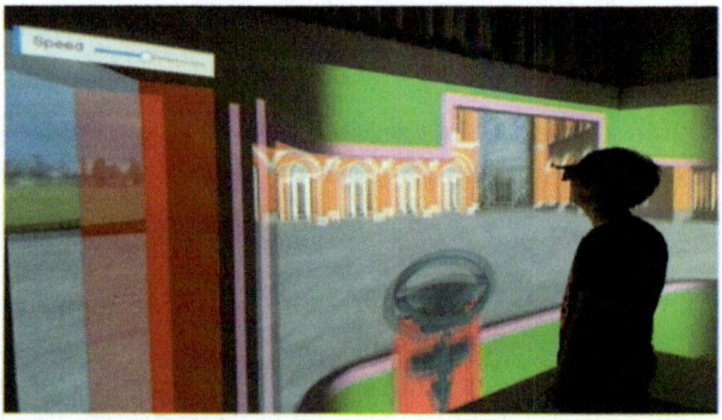

Fig. 5.42 Experience in different driving speeds

- *Examination of the product operation*

Users can simulate the food truck in the operation. User perspective is fixed in the driver position but the angle of view can be changed. The user can simulate driving the food truck by controlling moving directions. At the same time, the driving speed can be adjusted. Figure 5.42 shows the simulation of different driving speeds.

- *Experience in product performances in some special needs*

The truck performance can be simulated in some special circumstances such as colliding objects. Based on the quality, density, friction coefficient, acceleration of gravity, and other physical quantities of the truck and objects, the user can see different collision effects by setting different physical quantities. Figure 5.43 shows a screenshot of the simulation.

Therefore, users can have an in-depth understanding of the food truck performance using the proposed interactive system, which enhances the user participation in the product design and evaluation. When a user experiences the product, the interactive system records a series of data, such as user selections of function modules and

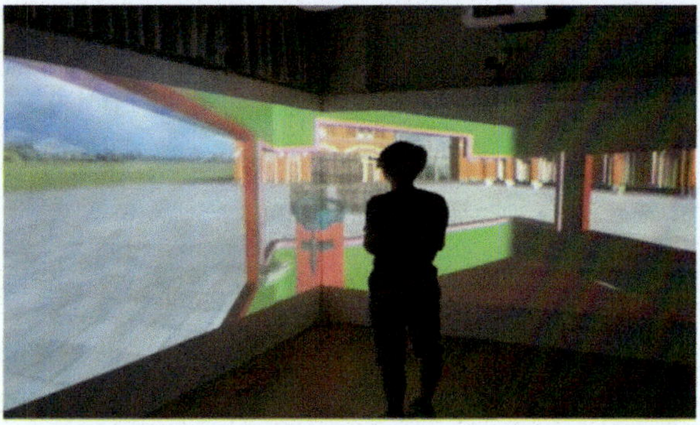

Fig. 5.43 Simulation of the collision

corresponding colors, locations of the modules in the system. Designers can clearly understand the user needs based on these data and then design a food truck to meet the user requirement.

5.4.3.2 System Evaluation

A questionnaire survey is conducted to evaluate the effectiveness of the proposed system by comparing user experience before and after using the system. Prior to the experiment, fifteen participants with varied ages and genders were given a brief verbal introduction to the experiment. Due to the lack of knowledge on the VR-based interactive system, the participants were provided a training program of the system to learn the PPT wand and system operations. They were then asked to operate product models in the system and then to complete a questionnaire about their experience of the food truck. In the questionnaires, the subjects were asked to give scores (ranging from 0 for "unknown" to 100 for "well known") about the experience. Data collected from the questionnaires were analyzed to evaluate the system performance. In addition, the users were also asked to comment advantages and disadvantages of the system.

Using the statistical analysis in the SPSS software, the Cronbach alpha is calculated to test reliability and internal consistency for rating questions in the questionnaires. The *t*-test is used to find the statistically significant difference of mean scores between before and after using the system. Through the data analysis, both of questionnaires get high Cronbach alpha values (0.733) and (0.8233), respectively, which means that the scale of the questions has a high level of the reliability and internal consistency in the both questionnaires. Figure 5.44 shows one of the results.

Figure 5.45 shows the mean scores, difference of two mean scores, and quantity of subjects of Question 1 in the two questionnaires. It is to test users' experience in colors and shapes for the food truck. A statistically significant difference was found before and after the experience of the interactive system, $p\text{-value} < 0.05$. Therefore, it indicates that user experience for the food truck has been significantly improved after using the system.

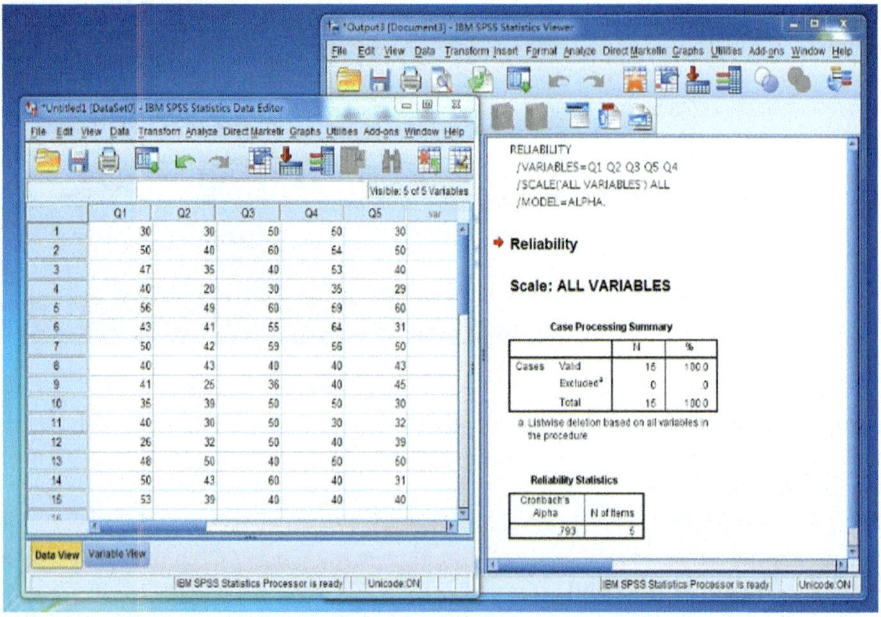

Fig. 5.44 Cronbach alpha of question 1

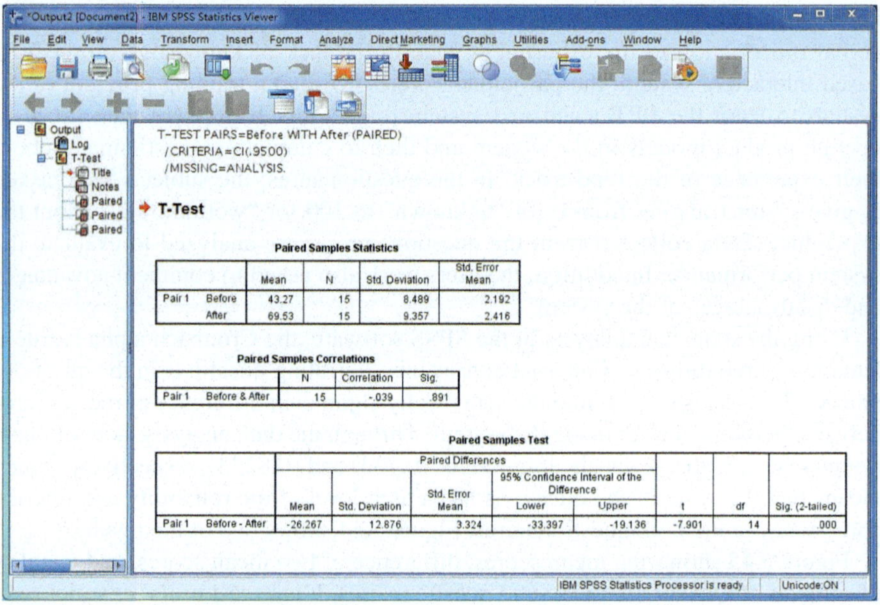

Fig. 5.45 *T*-test results of questionnaire 1

Table 5.12 *T*-test results of questions 2, 3, 4, 5

	Mean (before)	Mean (after)	Mean (different)	*N*	*P*
Question 2	40.27	71.47	31.2	15	< 0.05
Question 3	48.00	64.60	16.6	15	< 0.05
Question 4	46.73	70.60	23.867	15	< 0.05
Question 5	40.00	57.00	17.0	15	< 0.05

Table 5.12 shows scores of Question 2 for the user experience of assembly and disassembly of the food truck and scores of Question 3 about the experience degree of the structure for the food truck. There is a statistically significant effect of the interactive system on both understanding of the food truck configuration, $p < 0.05$, and learning of the assembly and disassembly, $p < 0.05$. Therefore, the results suggest that users acquire not only understanding of the food truck structure but also experience of assembly and disassembly of the food truck.

For Questions 4 and 5 in Table 5.12, the former is for the cognition degree of operating food trucks and the latter is to evaluate the user cognition performance in some special needs for food trucks. It reveals user understanding of the operation, $p < 0.05$ and performance, $p \leq 0.05$ through the interactive system.

For the open questions, 80.0% of participants comment that the interactive system creates a friendly and realistic-learning environment and an almost truly experiencing environment. Using the wand to operate the virtual food truck largely increases the learning and experience interest. Some of them also feel that it would be more confident to operate the food truck after training by the interactive system. However, 55.3% of participants mentioned that it is not easy to catch certain virtual parts accurately and quickly in the operation using the VR device.

In conclusion, the statistical analysis reveals that the VR system creates an effective learning and experiencing environment, which significantly improves user involvements in the product design and evaluation. Most participants agree that the VR system provides a user-friendly, cost-effective, safe, and highly interactive learning and experiencing environment.

5.4.4 Case Study—A VR System for Caravans Design

Caravans are customized vehicles that require users to involve in the design process from the conceptual design to detail finish. The user involvement is a key for the product success in the market. Different methods have been applied by industries to receive user suggestions and demands during the caravan development to meet the user requirements and preferences. Traditional methods are difficult to provide users real feeling of the product in using and testing solutions before the product is made. This section introduces a VR system that provides users a VE to examine design solutions of the caravan.

Fig. 5.46 A platform module of caravans

Caravans are designed and modeled based on the adaptable design concept [34]. A caravan consists of a platform module and different function modules to meet different needs of users. The platform module is shown in Fig. 5.46. It is assumed that interfaces are adaptable to be able to connect the platform and different functional modules.

The interior design of the caravan is decided by users using different function modules. Figure 5.47 shows some of these modules of caravans, such as bed, desk, chair, TV, refrigerator, bathtub, and shower that are modeled and saved in the VR model database. Models in the VR system also include different interfaces to be applied in connections of function modules of the caravan.

A group of potential users operates the proposed VR system. After the users complete the design, the result can be saved, reviewed, and modified. Some design processes are shown in Fig. 5.48.

In this VR-based design system, the module selection and placement are accomplished by the user interaction through a hand wand with tracking sensors in the VR system. All modules are stored in the VR database. The users can view a design and suggest the design changes until they are satisfied.

When users involve the design process in a VR system, the design process and solution are visualized and verified for the feasibility of operations. The design evaluation of each step provides the user a chance to make changes to meet expectations. Therefore, the VR-based design system provides users following functions.

- Visualization of the design elements.
- Improving understanding of product design in product modules and interfaces.
- Interactive with product to test the product performance.

5.4.5 Comments and Analysis

It is essential for designers to understand user requirements and changeable demands in the development of adaptable products. An effective user interface is important for users to experience product in the product design and improvement. Current

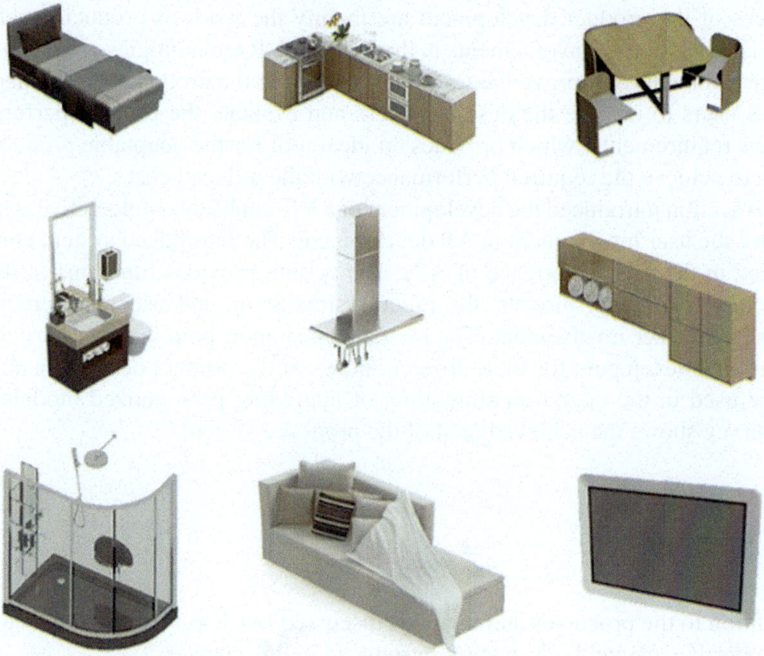

Fig. 5.47 Function modules of caravans

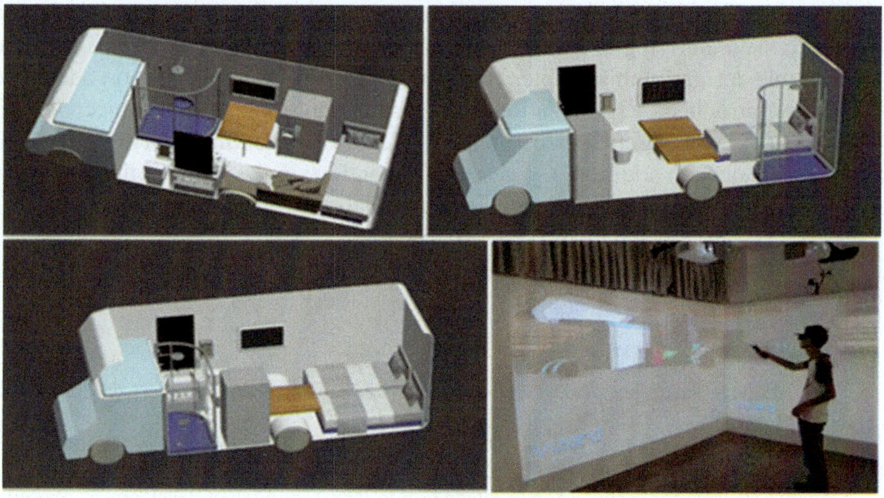

Fig. 5.48 Design of caravans using the VR system

practices of the product development are mainly the work of product developers. There is a lack of user involvements in the process. VR technologies provide a close-real environment to improve the user participation in the product development. VR enables users to involve the design process and evaluate the product performance for their requirements, which provides an ideal tool for the adaptable product (AP) design to achieve the required performance with the reduced cost.

This section introduced the development of a VR-enable user interactive system to improve the user involvement in AP development. The introduced system considers the need in the user experience of APs. The system provides functions of the user interaction to product models, the process simulation, and user data process for details of the user involvement. The system can support both the user participation and product developers for their different needs in the product development. It can also be used in the operation simulation of upgrading personalized modules. The user survey shows the achieved goal of the proposed system.

5.5 Summary

In addition to the processes and methods discussed in Chap. 4, tools and techniques for supporting adaptable design are introduced in this chapter.

As discussed in Chap. 1, prior to the introduction of adaptable design, many design methods, including the modular design and design optimization, have been developed for creating mechanical products to better satisfy customer requirements. Although the underlying objectives are different, these design methods are effective in adaptable design. In this chapter, modular design tools that can support adaptable design are introduced. Optimization tools such as multi-level, multi-objective, and global optimization algorithms are then provided for the identification of the optimal adaptable design. In addition, the case studies and examples showed that the Web-based and virtual reality tools can be used for adaptable design processes.

References

1. K.K. Ali, M.S. Saed, *Product Design for Modularity* (Kluwer Academic, Norwell, MA, USA, 2002)
2. J.S. Arora, *Introduction to Optimum Design* (McGraw-Hill, New York, NY, USA, 1989)
3. K. Backhaus, J. Jasper, K. Westhoff, J. Gausemeier, M. Grafe, J. Stöcklein, Virtual reality based conjoint analysis for early customer integration in industrial product development. Procedia CIRP **25**, 61–68 (2014)
4. J.C. Bezdek, *Pattern Recognition with Fuzzy Objective Function Algorithms* (Plenum, New York, NY, USA, 1981)
5. Y. Chen, Q. Peng, P. Gu, Methods and tools for the optimal adaptable design of open-architecture products. Int. J. Adv. Manuf. Technol. **94**, 991–1008 (2018). https://doi.org/10.1007/s00170-017-0925-6

6. C. Chung, Q. Peng, The selection of tools and machines on web-based manufacturing environments. Int. J. Mach. Tools Manuf **44**(2–3), 315–324 (2004)
7. C. Chung, Q. Peng, A hybrid approach to selective-disassembly sequence planning for de-manufacturing and its implementation on the Internet. Int. J. Adv. Manuf. Technol. **30**, 521–529 (2006). https://doi.org/10.1007/s00170-005-0038-5
8. C. Chung, Q. Peng, Enabled dynamic tasks planning in web-based virtual manufacturing environments. Comput. Industry **59**, 82–85 (2008)
9. C. Chung, Q. Peng, Tool selection-embedded optimal assembly planning in a dynamic manufacturing environment. Comput. Aided Des. **41**(7), 501–512 (2009)
10. H. Fazeli, Q. Peng, Efficient extraction of information from correlation matrix for product design using an integrated QFD-DEMATEL method. Comput.-Aided Des. Appl. **18**(5), 1131–1145. https://doi.org/10.14733/cadaps.2021.1131-1145.
11. J.K. Gershenson, G.J. Prasad, S. Allamneni, Modular product design: a life-cycle view. J. Integr. Des. Process. Sci. **3**(4), 13–26 (1999)
12. D.E. Goldberg, *Genetic Algorithms in Search, Optimization, and Machine Learning* (Addison-Wesley, Reading, MA, USA, 1989)
13. G. Hong, L. Hu, D. Xue, Y.L. Tu, Y.L. Xiong, Identification of the optimal product configuration and parameters based on individual customer requirements on performance and costs in one-of-a-kind production. Int. J. Prod. Res. **46**(12), 3297–3326 (2008)
14. X. Kang, Q. Peng, Tool feasibility analysis for fixture assembly planning. J. Manuf. Sci. Eng. Trans. ASME **130**(4), 041010-1–041010-9 (2008)
15. X. Kang, Q. Peng, Integration of CAD models with product assembly planning in a Web-based 3D visualized environment. Int. J. Interact. Des. Manuf. **8**, 121–131 (2014). https://doi.org/10.1007/s12008-014-0220-9
16. J.R. Koza, *Genetic Programming: On the Programming of Computers by Means of Natural Selection* (The MIT Press, Cambridge, MA, USA, 1992)
17. M. Martinez, D. Xue, A modular design approach for modeling and optimization of adaptable products considering the whole product life-cycle spans. J. Mech. Eng. Sci. **232**(7), 1146–1164 (2018)
18. P.J. Newcomb, B. Bras, D.W. Rosen, Implications of modularity on product design for the life cycle. J. Mech. Des. **120**(3), 483–490 (1998)
19. Q. Peng, Virtual manufacturing for dynamic product development: problems and solutions, in *Proceedings of the Canadian Society for Mechanical Engineering Forum 2010*. June 7–9, 2010, Victoria, BC, Canada (2010)
20. Q. Peng, C. Chung, C. Yu, T. Luan, A networked virtual manufacturing system for SMEs. Int. J. Comput. Integr. Manuf. **20**(1), 71–79 (2007)
21. Q. Peng, F.R. Hall, P.M. Lister, Application and evaluation of VR-based CAPP system. J. Mater. Process. Technol. **107**(1–3), 153–159 (2000)
22. Q. Peng, X. Kang, T. Zhao, Effective virtual reality based building navigation using dynamic loading and path optimization. Int. J. Autom. Comput. **6**(4), 335–343 (2009)
23. Q. Peng, C. Yu, A visualised manufacturing information system for mass customisation. Int. J. Manuf. Technol. Manag. **11**(3/4), 278–295 (2007)
24. Y. Shi, Q. Peng, A spectral clustering method to improve importance rating accuracy of customer requirements in QFD. Int. J. Adv. Manuf. Technol. **107**, 2579–2596 (2020). https://doi.org/10.1007/s00170-020-05204-1
25. H. Song, F. Chen, Q. Peng, Improvement of user experience using virtual reality in open-architecture product design. Proc. Inst. Mech. Eng. Part B J. Eng. Manuf. **232**(13), 1–12 (2017). https://doi.org/10.1177/0954405417711736
26. N. Suh, *Axiomatic Design: Advances and Applications* (Oxford University Press, New York, 2001)
27. H. Wang, D. Xue, An intelligent zone-based delivery scheduling approach. Comput. Industr. **48**(2), 109–125 (2002)
28. Y. Xiao, Q. Peng, Body gesture-based user interaction in design review, in *Proceedings of ASME IDETC/CIE*, August 06–09, DETC2017-67415 (2017)

29. D. Xue, A multi-level optimization approach considering product realization process alternatives and parameters for improving manufacturability. J. Manuf. Syst. **16**(5), 337–351 (1997)
30. D. Xue, J.H. Rousseau, Z. Dong, Joint optimization of performance and cost in integrated concurrent design: the tolerance synthesis part. J. Eng. Des. Autom. **2**(1), 73–89 (1996)
31. D. Xue, H. Wang, D.H. Norrie, A fuzzy mathematics-based optimal delivery scheduling approach. Comput. Industr. **45**(3), 245–259 (2001)
32. H. Yang, D. Xue, Y.L. Tu, Modeling of the non-linear relations among different design and manufacturing evaluation measures for multi-objective optimal concurrent design. Concurr. Eng. Res. Appl. **14**(1), 43–53 (2006)
33. J. Zhang, Y. Chen, D. Xue, P. Gu, Robust design of configurations and parameters of adaptable products. Front. Mech. Eng. **9**(1), 1–14 (2014)
34. Y. Zhang, J.K. Gershenson, An initial study of direct relationships between life-cycle modularity and life-cycle cost. Concurr. Eng. Res. Appl. **11**(2), 121–128 (2003)
35. C. Zhao, Q. Peng, P. Gu, Development of a paper-bag-folding machine using open architecture for adaptability. Proc. Inst. Mech. Eng. Part B J. Eng. Manuf. **229**(S1), 155–169 (2015). https://doi.org/10.1177/0954405414559281
36. L. Zhao, Q. Peng, Development of a CMM training system in virtual environments, in *Proceedings of ASME IDETC/CIE, DETC2010-28274* (2010)

Chapter 6
Applications of Adaptable Design

Since introduction of the adaptable design (AD) concept by Gu and Hashemian [3], significant progress has been made in development of methodologies and computer tools for adaptable design. Many applications have also been implemented based on the research results. In this chapter, some major applications are selected to show how the methods of adaptable product modeling, adaptability evaluation, adaptable design process, and adaptable design tools explained in Chaps. 2–5 were applied to solve engineering problems.

6.1 Adaptable Design of CNC Machine Tools Structure

The application provided in this section is based on the research results developed by Xu et al. [12].

The YH60 series spiral-bevel-gear cutting machines with Siemens controls run multiple spindles in multiple axes all in precise synchronization. These machines are used to cut bevel gears for light trucks. The structure of YH60 machines follows the principle of modular design of CNC machines. Figure 6.1 shows the column, bed, saddle, and workpiece suitcase modules of a CNC spiral-bevel-gear cutting machine.

Based on requirements to improve performance of gear machining and quality of gears, the original machine structures of YH603 were improved to enhance both the static and the dynamic stiffness of the machine, while reducing the weight of the machine. As the modulus range of original YH603 machine is only under 4 mm, this machine cannot be used to cut bevel gears for medium trucks. A new generation of gear cutting machine with modulus range up to 12 mm must be developed. According to the high static and dynamic rigidity requirements for the new machine structure, a dynamic design approach was used together with adaptable design analysis to determine the optimal design of the modular structure of the machine.

P. Gu et al., *Adaptable Design*, Research on Intelligent Manufacturing,
https://doi.org/10.1007/978-981-99-5869-6_6

Fig. 6.1 YH603 CNC spiral-bevel-gear cutting machine

(a) Picture of the initial product

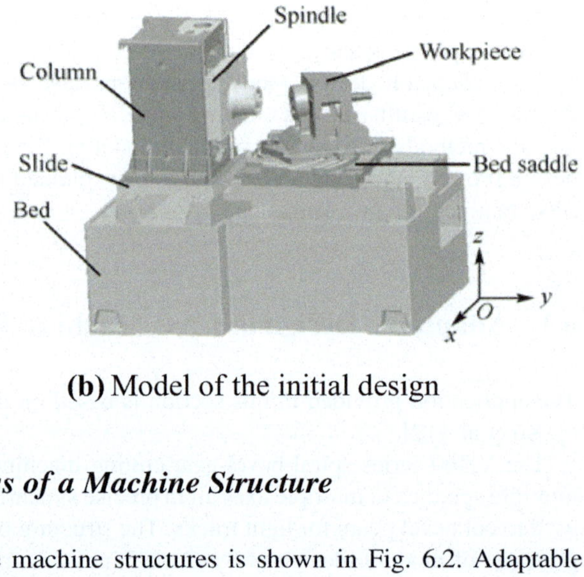

(b) Model of the initial design

6.1.1 Redesign Process of a Machine Structure

The redesign process of the machine structures is shown in Fig. 6.2. Adaptable modular design, static analysis, and dynamic analysis were considered in this process in the following steps.

(1) Modularize the original machine structure (P_0^1) based on design requirements.
(2) Conduct the static and dynamic analysis for all the modules (M_iS_0).
(3) Obtain a series of improved module structures (M_iS_i) by modifying the structures of the modules based on the analysis results.
(4) Obtain the improved overall machines by selecting and combining the optimal modules. The decision to select the best modules was based on evaluation criteria of performance improvement. The interfaces between modules were kept unchanging.
(5) Conduct static and dynamic analysis for the whole machines ($P_1{}^1$ and $P_2{}^1$).

Fig. 6.2 Redesign process of the machine structures

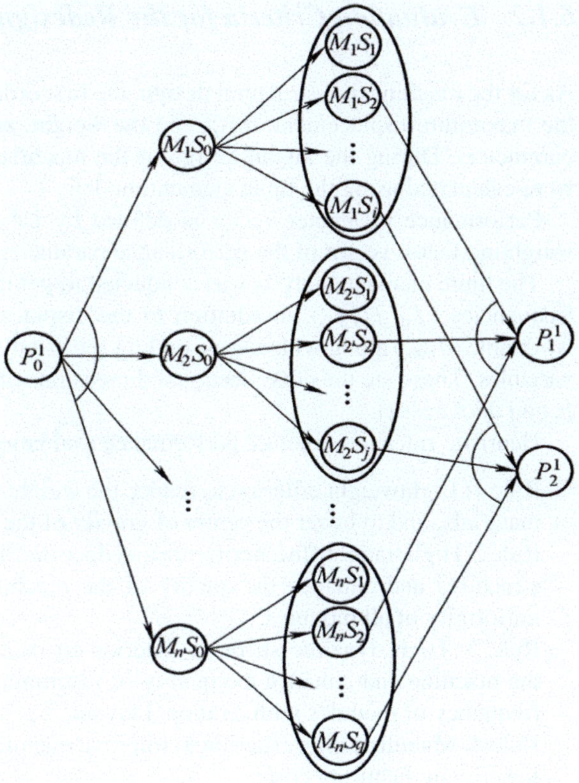

(6) Based on the analysis result, select the best design according to adaptability evaluation criteria.
(7) Optimize the structure of the machine to obtain the final design.

In the above steps, many tools, such as the tool for sensitivity analysis of the structural parameters of the machine and its modules, were used to support the design process.

As discussed previously, a variant of product or module can be redefined by changing the functional requirements and/or geometry of the model. The machine structures were modeled and evaluated using the ANSYS finite element analysis software. Static load cases (cutting force on the frame) were considered. It was assumed that these load cases provided the real world behaviors of machine such that the designer could evaluate the design and improved the design through modification.

6.1.2 Evaluation Criteria for the Redesign Process

As for the machine tool structural design, the first order of the natural frequency, f_0, the maximum displacement, d_{max}, and the weight, m, were used as the functional parameters. During the layout design of the machine, these functional parameters were calculated using the finite element models.

Performance parameter vector is defined by $\mathbf{U}^{\mathbf{P}} = (m^p, d_{max}{}^p, f_0{}^p)^T$, and the weighting factor vector of the performance parameter vector is $\mathbf{W}_j^u = (w_1, w_2, w_3)$.

The finite element analysis was conducted to get mass (m), the first three natural frequencies (f_0, f_1, f_2), in addition to the displacements to static load cases of worktable (d_{max}) along with corresponding sensitivity information for all the design variables. These are the cross-sectional dimensions of the rib plates (width b, height h, and thickness t).

Heuristic rules for structure performance evaluation are as follows.

Rule 1: Lightweight criterion to reduce the weight of overall machine and cost of materials, and to lower the center of gravity of the machine.
Rule 2: High static stiffness criterion to reduce the displacement of overall machine structures and enhance the quality of the machining process, thus improving uniformity of all modules.
Rule 3: High dynamic stiffness criterion to increase the natural frequency of the machine and enhance the quality of machining process, thus improving the frequency of modules with a rational layout.
Rule 4: Manufacturing criterion to improve manufacturability of the structures to lower manufacturing costs.

Criteria for adaptable design of structures:

Criterion 1 (Performance improvement criterion): making performance improvement E_i as large as possible for structure selection.
Criterion 2 (Structure similarity criterion): making the machine structure similarity s^{p0} to original as high as possible for structure modification.
Criterion 3 (Adaptability criterion): making adaptability of redesign structure AF(S_i) or product $A(P)$ as high as possible for structure selection.

The finite element analysis, quantitative analysis, and evaluation were carried out on the machine bed, bed saddle, workpiece suitcase, and column modules. The redesigns of these modules are shown in Table 6.1. In Table 6.1, S_0 represents the original design, and S_i represents the ith redesigned candidate.

The column redesign is used in this section to explain the process. The column supports the spindle system; and, the stiffness of column is an important factor that affects the accuracy of the machining processes. In the cutting process, the column may vibrate along x and y directions, thereby the accuracy is reduced. One of the redesign objectives is, therefore, to increase the static and dynamic stiffness of the column.

Table 6.1 Structures of redesigned modules

Machine structure	Column M_1	Bed M_2	Workpiece suitcase M_3	Slide board M_4	Bed saddle M_5
Original structure S_0					
Proposed structure S_1					
Proposed structure S_2					
Proposed structure S_3					
Proposed structure S_4					
Proposed structure S_5					
Proposed structure S_6					

The first step of the redesign process was to conduct a sensitivity analysis of the critical structural parameters. The data obtained from the analysis are shown in Fig. 6.3, where t_4 represents the thickness of the side and back rib plates of the column and the top and bottom plates; t_1 represents the thickness of the front and exterior rib plates; and, t_3 is thickness of the interior rib plate. Based on the results of sensitivity analysis, the thickness measures of the exterior, top and bottom plates had major effects on dynamic performance. The effects of other parameters were not significant. Hence, the thickness of the rib plates was selected for the redesign of the column, in order to improve its dynamic performance. Figure 6.3 shows the sensitivity analyses of stiffness and natural frequency.

The second step in the redesign process was to conduct a similarity analysis of the critical structural parameters. The weights of the design parameters were defined as follows:

$$w_j^x = (1/6, 1/6, 1/6, 1/6, 1/6, 1/6)$$

According to Xu et al. [12], the column design parameters and similarity between the redesign and the original design are given in Tables 6.2, 6.3 and 6.4.

Fig. 6.3 Sensitivity analysis for the original column

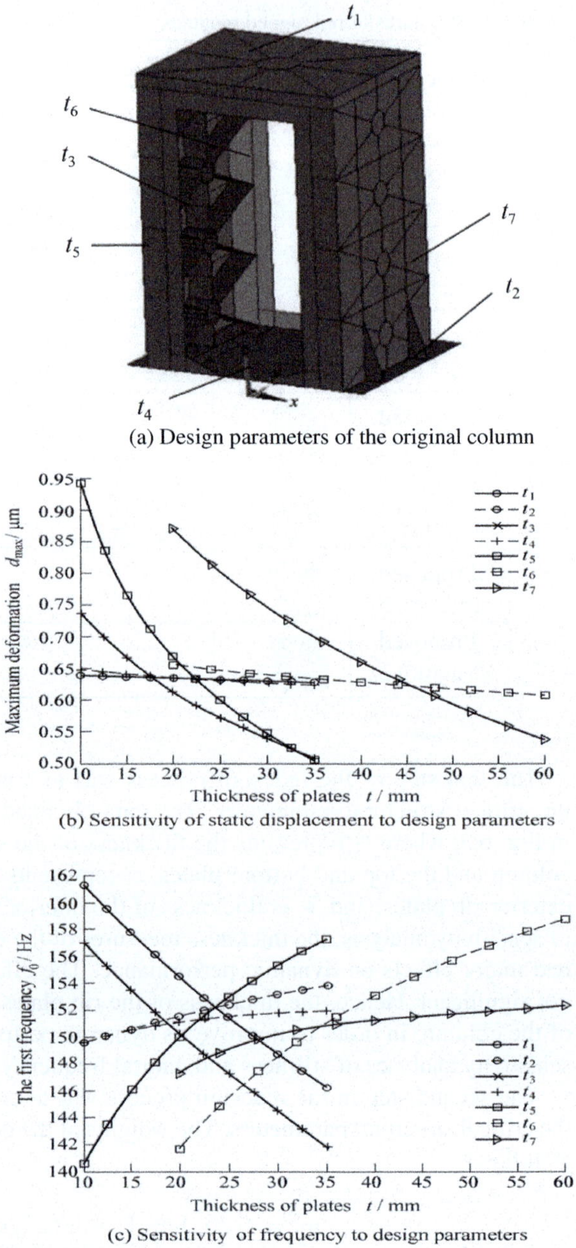

(a) Design parameters of the original column

(b) Sensitivity of static displacement to design parameters

(c) Sensitivity of frequency to design parameters

The third step of the redesign process was to conduct the performance improvement and adaptability analysis of the functional parameters. The weights of the functional parameters were defined as follows:

$$\mathbf{w}^u = (0.5, 0.25, 0.25)$$

Table 6.2 Structural design parameters of the YH603 column in the original and redesigns

Design candidate	Exterior wall thickness t_1/mm	Interior wall thickness t_3/mm	Bottom Plates thickness t_4/mm	Height of rib plates h_1/mm	Layout of interior ribs	Large round of rib corners
Original S_0	25	18	45	40	"*"form	No
Redesign S_1	20	12	45	95	"#"form	No
Redesign S_2	20	12	45	95	"#"form	No
Redesign S_3	20	12	45	125	"#"form	No
Redesign S_4	20	12	45	125	"#"form	Yes
Redesign S_5	15	12	45	125	"#"form	Yes
Redesign S_6	15	12	45	125	"#"form	Yes

Table 6.3 Structural similarity of the column between the original and redesigns

Similarity of each structural parameters $s^{p0}(x_j)$						
Design candidate	Exterior wall	Interior wall	Bottom plate	Rib plate	Layout of interior ribs	Large round of rib corners
Redesign S_1	0.8	0.667	0	0.421	0	1
Redesign S_2	0.8	0.667	0	0.421	0	1
Redesign S_3	0.8	0.667	0	0.320	0	1
Redesign S_4	0.8	0.667	0	0.320	0	0
Redesign S_5	0.6	0.667	0	0.320	0	0
Redesign S_6	0.6	0.667	0	0.320	0	0

According to Xu et al. [12], the structure performance improvement was calculated as shown in Tables 6.5 and 6.6.

The adaptability was calculated as shown in Table 6.7. The comparison of the performance improvement between the original and redesigned structures is also shown in Table 6.6. According to criterion 1 of AD, redesign S_6 was the best choice.

Table 6.4 Adaptability of the column

Design candidate	Structural similarity s^{p0}	Performance improvement E_i	Adaptability of the redesign $AF(S_i)$
Redesign S_1	0.481	1.448	2.790
Redesign S_2	0.481	1.428	2.751
Redesign S_3	0.465	1.444	2.699
Redesign S_4	0.298	1.476	2.103
Redesign S_5	0.266	1.437	1.958
Redesign S_6	0.266	1.494	2.035

Table 6.5 Structural performance of the column for the original and redesigns

Design candidate	The 1st-order frequency f/Hz	Displacement s/μm	Weight w/kg
Original design S_0	119	14.60	308
Redesign S_1	194	7.64	291
Redesign S_2	191	7.64	300
Redesign S_3	195	7.01	315
Redesign S_4	202	6.72	318
Redesign S_5	193	8.74	276
Redesign S_6	205	6.80	309

Table 6.6 Performance improvement of the column between the original and redesigns

Design candidate	Performance improvement of each performance parameter r_{ij}		
	Change of the 1st-order frequency	Displacement change	Weight change
Redesign S_1	– 0.630	– 0.477	– 0.055
Redesign S_2	– 0.605	– 0.477	– 0.026
Redesign S_3	– 0.639	– 0.520	0.023
Redesign S_4	– 0.697	– 0.540	0.032
Redesign S_5	– 0.622	– 0.401	– 0.104
Redesign S_6	– 0.723	– 0.534	0.003

Table 6.7 Adaptability of the bed

Design candidate	Structural similarity s^{p0}	Performance improvement E_i	Adaptability of the redesign $AF(S_i)$
Redesign S_1	0.754	1.446	5.878
Redesign S_2	0.754	1.339	5.443
Redesign S_3	0.733	1.420	5.318
Redesign S_4	0.733	1.292	4.839

Following the same steps as those used in the column redesign process, the redesign and adaptability analysis of other modules were also conducted with results shown in Tables 6.8, 6.9, 6.10 and 6.11. The best choice of each module is shown in Table 6.12.

Table 6.8 Adaptability of the workpiece suitcase

Design candidate	Structural similarity S^{p0}	Performance improvement E_i	Adaptability of the redesign $AF(S_i)$
Redesign S_1	0.838	0.873	5.389
Redesign S_2	0.669	1.039	3.139
Redesign S_3	0.669	1.056	3.191

Table 6.9 Adaptability of the slide board

Design candidate	Structural similarity s^{p0}	Performance improvement E_i	Adaptability of the redesign $AF(S_i)$
Redesign S_1	0.854	0.784	5.370
Redesign S_2	0.854	0.809	5.541

Table 6.10 Adaptability of the bed saddle

Design candidate	Structural similarity s^{p0}	Performance improvement E_i	Adaptability of the redesign $AF(S_i)$
Redesign S_1	0.780	1.220	5.545
Redesign S_2	0.664	1.243	3.699

Table 6.11 Modules of the YH603 machines in the original and redesigns

Design candidate	Column	Bed	Workpiece suitcase	Slide board	Bed saddle
Original design $P_0{}^1$	M_1S_0	M_2S_0	M_3S_0	M_4S_0	M_5S_0
Redesign $P_1{}^1$	M_1S_6	M_2S_3	M_3S_3	M_4S_3	M_5S_2
Redesign $P_2{}^1$	M_1S_7	M_2S_5	M_2S_4	M_2S_3	M_2S_3

Table 6.12 Structural similarity of the modules of the YH603 machines between the original and redesigns

Design candidate	Structural similarity of the modules s^{p0}				
	Column	Bed	Workpiece suitcase	Slide board	Bed saddle
Redesign $P_1{}^1$	0.266	0.733	0.669	0.854	0.664
Redesign $P_2{}^1$	0.850	0.800	1.000	1.000	1.000

6.1.3 Improvement of the Entire Machine

The first step of the redesign process for the entire machine was to assemble the modules to obtain entire machines. According to Table 6.11, the improved overall machines was obtained by combining the optimal modules.

Figure 6.4 shows the overall machine structure and improvement candidates. Based on the improved modules of the original structure of machine YH603 ($P_0{}^1$), a completely new structure ($P_1{}^1$) design was determined. Considering the costs of manufacturing of the newly redesigned machine, the modified version of redesign ($P_2{}^1$) was adopted, in order to avoid reproduction of casting molds.

The second step of the redesign process for the entire machine was to conduct similarity analysis. The similarity of each module between the original and redesigned one was calculated as shown in Table 6.12.

The third step of the redesign process was to conduct the performance improvement and adaptability analysis of the functional parameters. The static and dynamic analyses results of the machine structures are shown in Table 6.13. The performance improvement of the entire machine compared to the original machine was calculated as shown in Table 6.14. According to Xu et al. [12], the adaptability of the redesign was calculated as shown in Table 6.15.

According to criterion 3 of AD, redesign $P_2{}^1$ is the best choice.

As we do not have more concrete information on the adaptation, therefore, the probability of the adaptation tasks is defined as follows

$$\Pr(S_i) = (0.5, 0.5)$$

The adaptability of the original product design is

$$A(P) = 12.255$$

Without substantially increasing the production costs, actual adopted redesign $P_2{}^1$ had enhanced the functionality. Based on the existing casting models and mode analysis, ribs were added to strengthen the weak structure area. The sharp corners were changed to round corners. Other changes, such as changing the thickness of casting walls, were not adopted, as these changes would be costly.

6.1.4 Development of a New Generation of YH605 Gear Cutting Machine

To fulfill the need of gear cutting machine with modulus range up to 12 mm, a new generation of machines, YH605, was developed by the company based on the redesigned YH603 machine ($P_1{}^1$) with significant performance improvement.

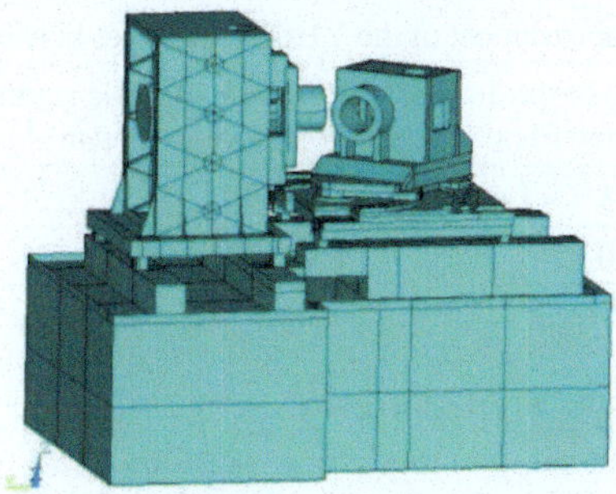

(a) YH603 initial solution P_0^1

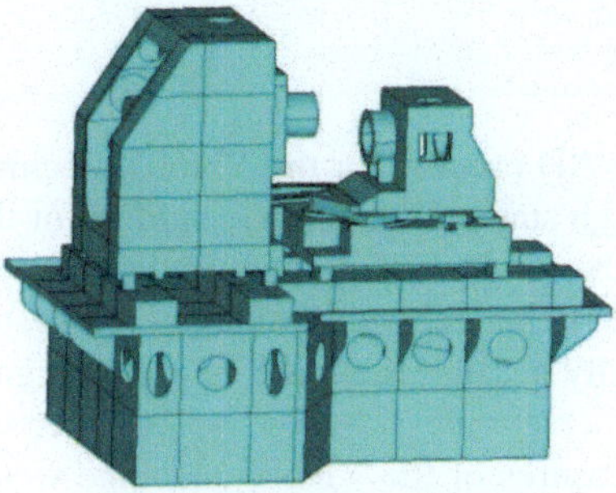

(b) YH603 new solution P_1^1

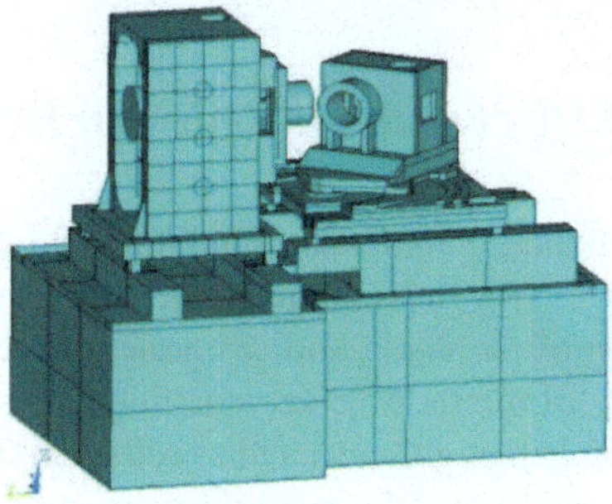

(c) YH603 final solution P_2^1

Fig. 6.4 YH603 machine redesigns

Table 6.13 Performance of the YH603 machines for the original and redesigns

Design candidate	The 1st-order frequency f/Hz	Static rigidity of y axis k/ (N·μm^{-1})	Mass m/kg
Original design P_0^1	69.33	57.13	2890
Redesign P_1^1	99.03	145.82	2472
Redesign P_2^1	78.31	106.43	3037

Table 6.14 Performance improvement of the YH603 machines between the original and redesigns

Design candidate	Change of the 1st-order frequency f/Hz	Change of static rigidity of y axis k/(N·μm^{-1})	Change of Mass m/kg
Redesign $P_1{}^1$	– 1.798	– 1.552	– 0.145
Redesign $P_2{}^1$	– 0.130	– 0.863	0.051

Table 6.15 Adaptability of the YH603 machines

Design candidate	Structural similarity s^{p0}	Performance improvement E_i	Adaptability of the redesign $AF(S_i)$
Redesign $P_1{}^1$	0.637	2.323	6.400
Redesign $P_2{}^1$	0.930	1.268	18.110

Figure 6.5 shows the AD process for the YH605 machine. Special attention and focus were placed on high static and dynamic rigidity for this design. This can also be viewed easily from the shapes and geometric dimensions of the extremely rigid structure and elements of this machine. Particular high system rigidity was required for this machine, primarily because the machine was designed to cut the bevel gears with large sizes.

The performance measures of the YH605 machine were obtained through both computer simulation and physical experiment as shown in Table 6.16.

6.1.5 Analysis and Discussion of Redesign Results

Based on the static and dynamic analyses of the machine structures, weak elements and locations were identified for the machine bed, saddle and workpiece suitcase. In addition to the evaluation of the static and dynamic performance of the redesigned structure, the adaptability of the redesigned machine was also achieved. The results showed that when adaptable design was used to redesign the machine structure, the static and dynamic characteristics of the machine were improved, while the machine's weight was reduced. For the bed saddle, its dynamic performance was enhanced, while its mass was basically unchanged. The static rigidity of the workpiece suitcase was kept unchanged, but its weight was reduced by 20%. Through the redesign, the static and dynamic performance of the whole machine was increased by over 30%, while the weight of the whole machine was reduced by 17%.

During the redesign of the machine structure, efforts were made to keep the thickness of the ribs and plates unchanged, so that the redesigned structures could be manufactured with the same manufacturing platform to minimize costs. Adaptability was obtained to evaluate the level of difficulty for making the required changes. The adaptable design approach provided effective guidance in the redesign process.

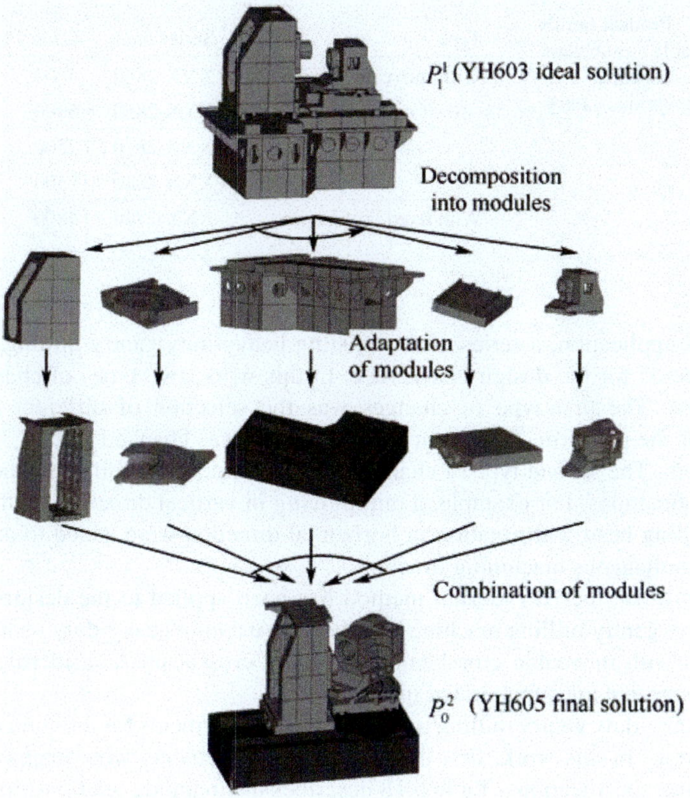

Fig. 6.5 YH603 machine design

Table 6.16 Performance of the YH605 machines

Item	The 1st-order frequency f/Hz	Static rigidity of y axis k/ (N·μm^{-1})	Mass m/kg
Design result	53.1	136.02	9692
Experiment result	50.9	116.13	12,000

6.2 Design of an Adaptable Product Platform for Heavy-Duty Gantry Milling Machines

6.2.1 Design of the Adaptable Product Platform (APP)

The application provided in this section is based on the research results developed by Cheng et al. [2].

Table 6.17 Product family with moveable crossbeams and fixed crossbeams considering various gantry spans

Type	Series name	Gantry span (mm)
With moveable crossbeam	XXX-2840	4800
	XXX-2850	5800
	XXX-2870	7500
	XXX-2880	8500
With fixed crossbeam	XXX-2740	4800
	XXX-2750	5800

In this application, a series of the existing heavy-duty gantry milling machines were selected for the design of the APP. In this work, two types of changes were considered. The first type of changes was the selection of different parameter values for the platform. For example, gantry span was changed from 2,350 mm to 10,500 mm. The second type of changes was the selection of different modules for different functions. For example, a ram moving in vertical direction on the column and a milling head with rotation in horizontal direction were added to achieve the 5-axis simultaneous machining function.

In this work, the APP design method has been applied to the design of a new heavy-duty gantry milling machine based on the existing heavy-duty gantry milling machines with moveable crossbeams and fixed crossbeams considering different gantry spans that are summarized in Table 6.17.

For heavy-duty gantry milling machines, the requirements for different customers are different. In this work, first the customer requirements were mapped into the engineering specifications. Table 6.18 describes the mapping relationships between customer requirements and engineering specifications. Table 6.19 shows the requirements of 19 customers described by engineering specifications. In order to identify the modules of APP, a clustering algorithm was used to classify the customer requirements into common requirements, customization requirements, and personalized requirements. Basic common modules were identified to satisfy common requirements, and add-on modules were identified to satisfy personalized requirements. These basic common modules and add-on modules were identified using the method introduced by Cheng et al. [2]. Customization requirements were achieved by adaptable modules with adaptable parameters. Details for identification of the adaptable parameters and the non-adaptable parameters will be provided in this section. Since these engineering specifications were modeled by different values with different units, normalization was conducted to convert these specifications into dimensionless indices between 0 and 1.

In this research, normalization is conducted by mapping the minimum measure to 0 and the maximum measure to 1 for each of the specifications. In this way, the original value of A, x, is mapped to a value, x', between 0 and 1 using:

$$x' = \frac{x - \min A}{\max A - \min A} \tag{6.1}$$

Table 6.18 Mapping relationships between customer requirements and engineering specifications

Customer requirement	Engineering specification					
	Gantry span	Maximum spindle speed	Spindle motor power	Maximum worktable load	Crossbeam type	Number of axes
High precision				×	×	×
Low cost			×			×
High reliability			×		×	
Stable work	×	×		×		
Convenience for human–computer interaction					×	×
Low noise		×	×	×		

Table 6.20 shows the normalized engineering specifications.

Based on the normalized customer requirements, the 19 customers were clustered into different groups using K-means clustering method with genetic algorithm (GA) as the optimization tool. From the experience in the heavy-duty gantry milling machines, the number of clusters, K, was set as 6 in this work. The clustering results are shown in Table 6.21.

Through the comparison among the different values of customer requirements in the six customer groups, it was found that the spindle motor power and the maximum worktable load were common requirements. The crossbeam type and the number of axes were personalized requirements. The gantry span, the maximum spindle speed, and the travels of gantry frame, slide carriage and ram were customization requirements.

The functional requirements (FRs) and their corresponding design parameters (DPs) were decomposed by the zigzag mapping method based on adaptable design (AD) for the heavy-duty gantry milling machine. These FRs and DRs are shown in Tables 6.22 and 6.23, respectively.

The basic common modules in APP of the heavy gantry milling machine product family without considering parameters were obtained from Table 6.23. In this case study, the power module and protection module to prevent machining chips and debris were identified as the basic common modules. Although the worktable load was a common requirement, since the sizes of different machines for different customers were different, the worktable load was selected as adaptable parameter. Based on different requirements considering the crossbeam type and the number of machining axes, different crossbeam modules, milling head modules, signal control devices and the user interfaces were selected as add-on modules. Since this research focuses on identification of adaptable parameters for adaptable modules and non-adaptable parameters for non-adaptable modules for the platform, details on selection of basic common modules for the platform and add-on modules to be added to the platform will not be provided.

Table 6.19 Customer requirements in engineering specifications

Customer	Gantry span (mm)	Travel of gantry frame (mm)	Travel of slide carriage (mm)	Travel of ram (mm)	Spindle motor power (kw)	Maximum spindle speed (r/min)	Maximum worktable load (T/m^2)	Crossbeam type (0: fixed, 1: movable)	Number of axes (0: 3-axis, 1: 5-axis)
X1	8500	16,500	9000	3000	100	1500	25	0	1
X2	4800	13,500	6000	2000	100	1240	25	1	0
X3	10,500	17,500	11,500	3000	100	1500	25	1	0
X4	9500	16,500	10,000	2500	100	1500	25	1	1
X5	3800	11,500	5000	1500	100	2000	25	0	0
X6	8000	17,500	8000	1500	100	1500	25	0	1
X7	9500	21,500	10,000	2500	100	2000	25	1	1
X8	8000	16,500	8500	2000	100	3000	25	0	1
X9	5800	16,500	7000	1500	100	1500	25	1	0
X10	8500	17,500	10,000	2000	100	2000	25	0	1
X11	6800	17,500	8000	3000	100	1500	25	1	1
X12	10,500	16,500	11,500	2000	100	2000	25	1	1
X13	4800	11,500	6000	1500	100	1240	25	0	0
X14	6800	16,500	8000	2500	100	1240	25	0	1
X15	3800	11,500	5000	1500	100	2000	25	1	1
X16	9500	16,500	10,000	3000	100	1500	25	1	0
X17	5800	17,500	7000	2000	100	1240	25	0	1
X18	4800	17,500	6000	2000	100	1240	25	1	0
X19	7500	16,500	8000	2000	100	1240	25	1	1

Table 6.20 Normalized customer requirements in engineering specifications

Customers	Gantry span	Travel of gantry frame	Travel of slide carriage	Travel of ram	Spindle motor power	Maximum spindle speed	Maximum worktable load	Crossbeam type	Number of axes
X1	0.7015	0.5000	0.6154	1.0000	1.0000	0.1477	1.0000	0.0000	1.0000
X2	0.1493	0.2000	0.1538	0.3333	1.0000	0.0000	1.0000	1.0000	0.0000
X3	1.0000	0.6000	1.0000	1.0000	1.0000	0.1477	1.0000	1.0000	0.0000
X4	0.8507	0.5000	0.7692	0.6667	1.0000	0.1477	1.0000	1.0000	1.0000
X5	0.0000	0.0000	0.0000	0.0000	1.0000	0.4318	1.0000	0.0000	0.0000
X6	0.6269	0.6000	0.4615	0.0000	1.0000	0.1477	1.0000	0.0000	1.0000
X7	0.8507	1.0000	0.7692	0.6667	1.0000	0.4318	1.0000	1.0000	1.0000
X8	0.6269	0.5000	0.5385	0.3333	1.0000	1.0000	1.0000	0.0000	1.0000
X9	0.2985	0.5000	0.3077	0.0000	1.0000	0.1477	1.0000	1.0000	0.0000
X10	0.7015	0.6000	0.7692	0.3333	1.0000	0.4318	1.0000	0.0000	1.0000
X11	0.4478	0.6000	0.4615	1.0000	1.0000	0.1477	1.0000	1.0000	1.0000
X12	1.0000	0.5000	1.0000	0.3333	1.0000	0.4318	1.0000	1.0000	1.0000
X13	0.1493	0.0000	0.1538	0.0000	1.0000	0.0000	1.0000	0.0000	0.0000
X14	0.4478	0.5000	0.4615	0.6667	1.0000	0.0000	1.0000	0.0000	1.0000
X15	0.0000	0.0000	0.0000	0.0000	1.0000	0.4318	1.0000	1.0000	1.0000
X16	0.8507	0.5000	0.7692	1.0000	1.0000	0.1477	1.0000	1.0000	0.0000
X17	0.2985	0.6000	0.3077	0.3333	1.0000	0.0000	1.0000	0.0000	1.0000
X18	0.1493	0.6000	0.1538	0.3333	1.0000	0.0000	1.0000	1.0000	0.0000
X19	0.5522	0.5000	0.4615	0.3333	1.0000	0.0000	1.0000	1.0000	1.0000

Table 6.21 Results of clustering based on K-means clustering method with GA

Cluster	Number of customers	Customers
1	6	X1, X6, X8, X10, X14, X17
2	3	X2, X9, X18
3	2	X3, X16
4	5	X4, X7, X11, X12, X19
5	2	X5, X13
6	1	X15

Based on the requirements including the customization requirements such as the gantry span, the stroke of gantry frame, the stroke of sliding carriage, the stroke of ram, the spindle motor power, and the maximum spindle speed, their corresponding parameters including the adaptable parameters were determined based on AD and sensitivity design structure matrix (SDSM). First all the key design parameters (DPs) for the heavy-duty gantry milling machines were associate with 21 ID numbers as shown in Table 6.24. The SDSM considering these 21 DPs was obtained as shown in Fig. 6.6. The different requirements to select the stroke of gantry frame and the stroke of sliding carriage were represented by the letters "*W*" and "*L*" respectively. It should be noted that because design parameters were modeled by components and modules in this case study, only qualitative sensitivity measures of 1 (i.e., sensitive) and 0 (i.e., not sensitive) were used.

The platform design parameters were then separated into non-adaptable common platform parameters and adaptable parameters by using the segmentation algorithm introduced by Cheng et al. [2]. The segmented SDSM is shown in Fig. 6.7. Based on the experience in design of the heavy-duty gantry milling machines, the modules and components with different types of parameters for the platform were identified as follows:

1. Non-adaptable common modules with non-adaptable common parameters: DP21: foundation, DP22: foundation baseplate, DP511: motor, DP512: reducer gear box, DP232, DP32, and DP42: bevel gear-pair, DP234, DP3321, and DP4321: feed accessories, DP3322 and DP4322: synchronous belt pulley, and so on.
2. Adaptable modules with adaptable parameters: DP31: Slide carriage, DP41: ram, DP233, DP331, DP431: ball screw, and so on.

The results of the APP design show that the power module and protection module without consideration of parameters were always the same, thus selected as the basic common modules. The modules such as the worktable were selected as the adaptable modules due to the different requirements such as different sizes of tables for different machines. In addition, the modules such as rotational milling head were selected as add-on modules to be added to the platform.

Table 6.22 Hierarchical structure of FRs for heavy-duty gantry milling machine (partial)

FR1: Milling functions	FR11: Provide mill		
	FR12: Hold mill		
	FR13: Rotate mill	FR131: Provide power	
		FR132: Support power unit	
		FR133: Change power	
		FR134 Synchronize feed	
FR2: Support and feed in *X*-axis	FR21: Support carrier		
	FR22: Foundation support		
	FR23: Feed in *X*-axis	FR231: Feed support	
		FR232: Transfer power	
		FR233: Feed motion power	
		FR234: Feed motion control	
		FR235: Feed motion path guide	
FR3: Feed in *Y*-axis	FR31: Feed support		
	FR32: Transfer power		
	FR33: Feed motion	FR331: Feed motion power	
		FR332: Feed control	FR3321: Motion control
			FR3322: Synchronize feed
		FR333: Feed motion path guide	
FR4 Feed in *Z*-axis	FR41: Feed support		
	FR42: Transfer power		
	FR43: Feed motion	FR431: Feed motion power	
		FR432: Feed control	FR4321: Motion control
			FR4322: Synchronize feed
		FR433: Feed motion path guide	
FR5: Additional functions	FR51: Feed power and transfer	FR511: Provide power	
		FR512: Transfer power	
	FR52: Locking function		
	FR53: Connection function		
	FR54: Lubrication function		
	FR55: Protection function		

Table 6.23 Hierarchical structure of DPs for heavy-duty gantry milling machine (partial)

DP1: Milling devices	DP11: Milling cutter		
	DP12: Tool holder		
	DP13: Tool rotational motion devices	DP131: Spindle motor	
		DP132: Spindle motor rack	
		DP133: Spindle gear box	
		DP134 Synchronous pulley-belt mechanism	
DP2 Support and feed devices in *X*-axis	DP21: Foundation		
	DP22: Foundation baseplate		
	DP23: Feed devices in *X*-axis	DP231: Sliding block	
		DP232: Bevel gear-pair	
		DP233: Ball screw in *X*-axis	
		DP234: Feed accessories in *X*-axis	
		DP235: Rail for sliding block	
DP3 Feed devices in *Y*-axis	DP31: Sliding carriage		
	DP32: Bevel gear-pair		
	DP33: Feed function parts	DP331: Ball screw in *Y*-axis	
		DP332: Feed control devices	DP3321: Feed accessories in *Y*-axis
			DP3322: Synchronous pulley-belt mechanism in *Y*-axis
		DP333: Rail for sliding block	
DP4 Feed devices in *Z*-axis	DP41: Ram		
	DP42: Bevel gear-pair		
	DP43: Feed function parts	DP431: Ball screw in Z-axis	
		DP432: Feed control devices	DP4321: Feed accessories in *Z*-axis
			DP4322: Synchronous pulley-belt in *Z*-axis
		DP433: Rail for ram	
DP5: Additional devices	DP51 Power devices	DP511: Motor	
		DP512: Reducer gear box	
	DP52: Pressure plate		
	DP53: Shaft coupling		
	DP54: Oil separator		
	DP55: Protective cover		

Table 6.24 ID numbers for design parameters (DPs) in SDSM

ID	DP	ID	DP	ID	DP	ID	DP
1	DP 21	7	DP 235	13	DP 333	19	DP 433
2	DP 22	8	DP 31	14	DP 41	20	DP 511
3	DP 231	9	DP 32	15	DP 42	21	DP 512
4	DP 232	10	DP 331	16	DP 431		
5	DP 233	11	DP 3321	17	DP 4321		
6	DP 234	12	DP 3322	18	DP 4322		

		FRs		DPs																				
		W	L	1	2	3	4	5	6	7	8	9	10	11	12	13	14	15	16	17	18	19	20	21
FRs	W																							
	L																							
DPs	1	0	0	1	0	0	0	0	0	0	0	0	0	0	0	0	0	0	0	0	0	0	0	0
	2	0	0	0	1	0	0	1	1	0	0	0	0	0	0	0	0	0	0	0	0	0	0	0
	3	0	0	0	0	1	1	0	0	1	0	0	0	0	0	0	0	0	0	0	0	0	0	0
	4	0	0	0	0	0	1	0	0	1	0	0	0	0	0	0	0	0	0	0	0	0	0	0
	5	0	0	0	1	0	0	1	1	0	0	1	0	0	0	0	0	0	0	0	0	0	0	0
	6	0	0	0	1	0	0	0	1	0	0	1	0	0	0	0	0	0	0	0	0	0	0	0
	7	0	0	0	0	0	0	0	0	1	1	0	0	0	0	0	0	0	0	0	0	0	0	0
	8	0	0	0	0	0	0	0	0	0	1	0	0	0	0	0	0	0	0	0	0	0	0	0
	9	0	0	0	0	0	0	1	1	0	0	1	0	0	0	0	0	0	0	0	0	0	0	0
	10	0	0	0	0	0	0	0	0	0	0	0	1	0	0	1	0	0	0	0	0	0	0	0
	11	0	0	0	0	0	0	0	0	0	0	0	0	1	1	0	0	1	0	0	0	0	0	0
	12	1	0	0	0	0	0	0	0	0	0	0	0	0	1	0	0	0	0	0	0	0	0	0
	13	1	1	0	0	0	0	0	0	0	0	0	1	0	0	1	0	1	0	0	0	0	0	0
	14	1	1	0	0	0	0	0	0	0	0	0	0	0	0	1	1	0	0	0	0	0	0	0
	15	1	0	0	0	0	0	0	0	0	0	0	0	0	0	1	1	1	0	1	1	0	0	0
	16	1	0	0	0	0	0	0	0	0	0	0	0	0	0	0	0	1	1	0	0	0	0	0
	17	1	0	0	0	0	0	0	0	0	0	0	0	0	0	0	0	1	0	1	1	0	0	0
	18	1	0	0	0	0	0	0	0	0	0	0	0	0	0	0	0	0	1	1	1	0	0	0
	19	1	0	0	0	0	0	0	0	0	0	0	0	0	0	0	0	0	1	0	0	1	0	0
	20	1	0	0	0	0	0	0	0	0	0	0	0	0	0	0	0	0	0	0	0	0	1	0
	21	1	0	0	0	0	0	0	0	0	0	0	0	0	0	0	0	0	0	0	0	0	0	1

Fig. 6.6 SDSM of heavy-duty gantry milling machine (before segmentation)

		FRs		Non-adaptable common platform parameters												Adaptable parameters								
		W	L	1	2	5	6	9	11	12	15	17	18	20	21	3	4	7	8	0	3	14	16	19
FRs	W																							
	L																							
Non-adaptable common platform parameters	1	0	0	1	0	0	0	0	0	0	0	0	0	0	0	0	0	0	0	0	0	0	0	0
	2	0	0	0	1	1	1	0	0	0	0	0	0	0	0	0	0	0	0	0	0	0	0	0
	5	0	0	0	1	1	1	1	0	0	0	0	0	0	0	0	0	0	0	0	0	0	0	0
	6	0	0	0	1	0	1	1	0	0	0	0	0	0	0	0	0	0	0	0	0	0	0	0
	9	0	0	0	1	0	1	1	0	0	0	0	0	0	0	0	0	0	0	0	0	0	0	0
	11	0	0	0	0	0	0	0	1	1	1	0	0	0	0	0	0	0	0	0	0	0	0	0
	12	0	0	0	0	0	0	0	1	1	0	0	0	0	0	0	0	0	0	0	0	0	0	0
	15	0	0	0	0	0	0	0	1	0	1	1	1	0	0	0	0	0	0	0	0	0	0	0
	17	0	0	0	0	0	0	0	0	0	1	1	1	0	0	0	0	0	0	0	0	0	0	0
	18	0	0	0	0	0	0	0	0	0	1	1	1	0	0	0	0	0	0	0	0	0	0	0
	20	0	0	0	0	0	0	0	0	0	0	0	0	1	0	0	0	0	0	0	0	0	0	0
	21	0	0	0	0	0	0	0	0	0	0	0	0	0	1	0	0	0	0	0	0	0	0	0
Adaptable parameters	3	1	1	0	0	0	0	0	0	0	0	0	0	0	0	1	1	1	0	0	0	0	0	0
	4	1	1	0	0	0	0	0	0	0	0	0	0	0	0	1	1	1	0	0	0	0	0	0
	7	1	0	0	0	0	0	0	0	0	0	0	0	0	0	1	1	1	1	0	0	0	0	0
	8	1	0	0	0	0	0	0	0	0	0	0	0	0	0	0	0	1	1	0	0	0	0	0
	10	1	0	0	0	0	0	0	0	0	0	0	0	0	0	0	0	0	0	1	1	0	0	0
	13	1	0	0	0	0	0	0	0	0	0	0	0	0	0	0	0	0	0	1	1	0	0	0
	14	1	0	0	0	0	0	0	0	0	0	0	0	0	0	0	0	0	0	0	0	1	0	0
	16	1	0	0	0	0	0	0	0	0	0	0	0	0	0	0	0	0	0	0	0	0	1	1
	19	1	0	0	0	0	0	0	0	0	0	0	0	0	0	0	0	0	0	0	0	0	1	1

Fig. 6.7 SDSM of heavy-duty gantry milling machine (after segmentation)

6.2.2 *Implementation of a Heavy-Duty Gantry Milling Machine with APP*

According to the designed APP for heavy-duty gantry milling machines, a CNC bridge-type double-gantry boring/milling machine XXX-2890 has been implemented as shown in Fig. 6.8. Because this machine was designed to mill and bore large structures with sizes up to 8 m and two rams were required to machine the structures from two directions simultaneously to improve the efficiency, a double-column gantry with a span of 9 m, a moving crossbeam, 2 rams with vertical motion, 2 rotational milling heads, and many other modules were designed and implemented as shown in Fig. 6.8.

Based on the APP, by adjusting the adaptable parameters such as the stroke of sliding carriage, the gantry span of the customized gantry machine was identified as

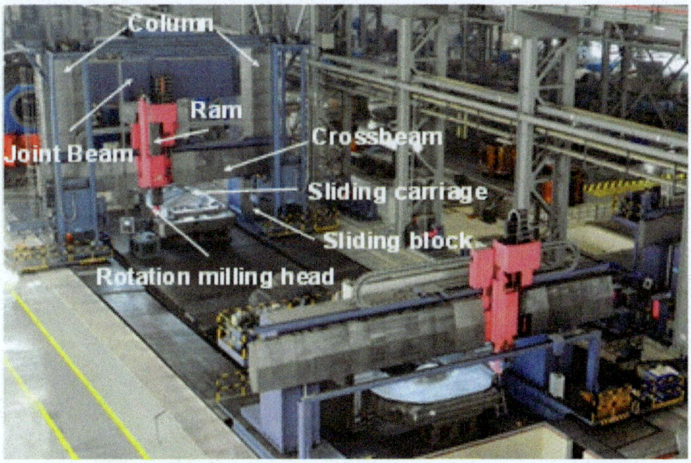

Fig. 6.8 CNC bridge-type double-gantry boring and milling machine XXX-2890

9 m. Because only adaptable modules in APP were designed based on new requirements, the lead time to design the XXX-2890 boring and milling machine was reduced from 12 months to 8 months. Since the new values of adaptable parameters and add-on modules (e.g., rotational milling head) were designed and implemented, the dynamic performance and machining accuracy of this new machine were tested. In this work, finite element analysis (FEA)-based simulation and physical experiments were conducted for tests.

The simplified model for FEA is shown in Fig. 6.9. QT600 was selected as the main material for this milling/boring machine. Properties of this material are shown in Table 6.25. In this analysis, the element type of solid 45 in ANSYS was selected, and the fixed grid element of tetrahedrons with size of 100 mm were used. A total of 811,196 grid nodes and 1,894,571 elements were created for FEA.

The modal frequencies of the milling machine under 120 Hz were obtained by the FEAs. During the physical experiments, in order to fully identify the dynamic characteristic parameters of the moving beam gantry, the excitations to this whole gantry in both horizontal direction and vertical direction were applied. The signal acquisition and analysis instrument by LMS Company (Belgium) was used to collect and process the data. The inherent frequencies of the gantry in horizontal and vertical directions under 100 Hz were obtained as: 20.38, 35.00 and 84.81 Hz in horizontal direction and 19.89, 49.10 and 97.56 Hz in vertical direction. The inherent frequencies and corresponding vibration modes from both the FEA-based simulation and physical experiments were compared as shown in Fig. 6.10 and Table 6.26.

Since the maximum relative error was obtained as 13.88% from Table 6.26, the simulation results were considered satisfactory for dynamic analysis of the designed CNC machine. When the rotational speed of the spindle was selected with a value

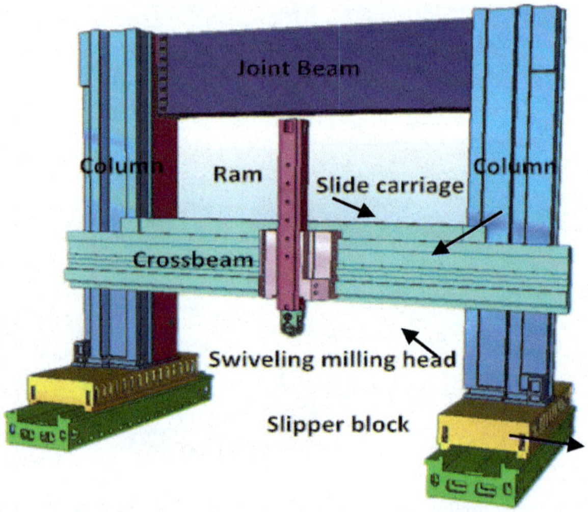

Fig. 6.9 Structure model of XXX-2890 boring/milling machine

Table 6.25 Material properties of QT600 for the milling/boring machine

Material	Density (kg/m^3)	Elasticity modulus (Gpa)	Poisson's ratio
QT600	7200	174	0.275

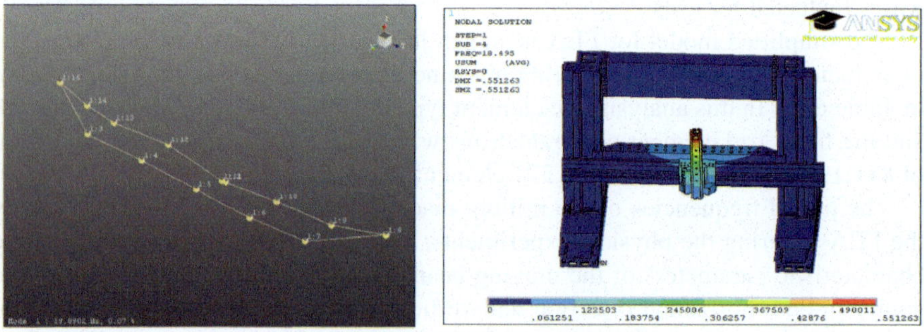

(a) Results from physical experiments (b) Results from FEA-based simulation

Fig. 6.10 Vibration modes from both the physical experiments and the FEA-based simulation at 19.89 Hz

different from the inherent frequencies, the machining errors due to vibrations can be reduced.

Table 6.26 Comparison between results from simulations and experiments

Order	FEA-based modal analysis (Hz)	Physical experiment (Hz)	Difference (Hz)	Relative error (%)
1	18.495	19.89	1.395	7.01
2	22.932	20.38	2.552	12.52
3	34.277	35.00	0.723	2.07
4	47.233	49.10	1.867	3.80
5	81.306	84.84	3.534	4.17
6	111.099	97.56	13.539	13.88

Experiments for verification of machining accuracy of the implemented machine tool were also conducted by machining of a standard cone with cast iron material as shown in Fig. 6.11. Figure 6.12 shows the machining process of the standard cone with the implemented machine. Table 6.27 lists the machining errors considering different types of tolerance requirements.

Fig. 6.11 A standard cone

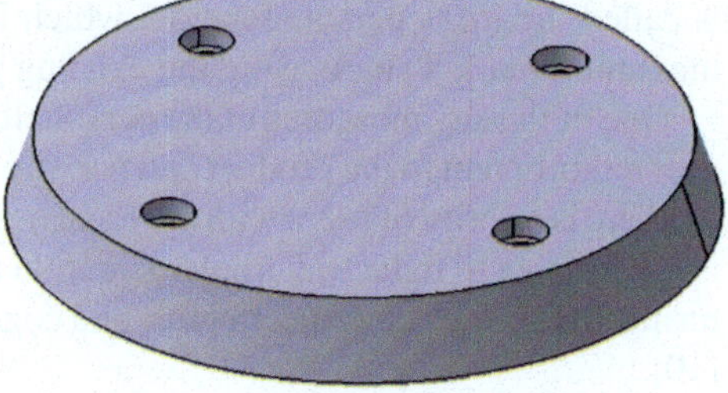

Fig. 6.12 Machining of the standard cone

Table 6.27 Machining errors

No.	Tolerance type	Tolerance requirement	Machined error
T1	Circularity (mm)	0.10	0.0084
T2	Concentricity (mm)	Φ 0.10	Φ 0.0105

6.3 Adaptable Design for Mass Personalization

The application provided in this section is based on the research results developed by Peng et al. [10].

6.3.1 Introduction

A product that meets the maximum user requirement will have the highly competitive ability in the market. A product to meet the individual user need is called the personalized product. The personalized product that can be made using mass production is called the mass personalization, which is highly pursued by industry to gain the maximum market share using the existing production capacity.

One of the key measures of the personalized product is the interaction between the user requirement to the product and the way of the product to respond to the user need. As there is a verity of users with different needs for a product, the mass personalization is a challenge to industrial implementation using the traditional design and manufacturing method. It needs an effective and cognitive method for product personalization [10].

Product architecture is an interactive pattern of functional modules in the product to form a technical system, which allows components of a product to interact and correlate with each other. Product architecture therefore normally dominates the manufacturing process to form the product detail. The architecture approach provides a very important perspective for mass personalization of product. Using an open concept in the product architecture, called the open-architecture product (OAP), a bridge can be formed to link product developers and users for the mass personalization of product.

OAP was proposed to provide a special manufacturing mode for flexible product structures using functional modules and adaptable interfaces [6]. OAP can support the massive personalization by facilitating the involvement of users and third-party vendors in the product development. The superiority of an OAP is to construct a personalized product to support particular functional needs with the ease of replacement of personalized modules in the original product [5]. This requires not only the modularization of products, but also the compatibility of product interfaces and structures.

Adaptable design (AD) uses the modular design to build product architecture for interactions of functional modules to meet user requirements. AD can plan function

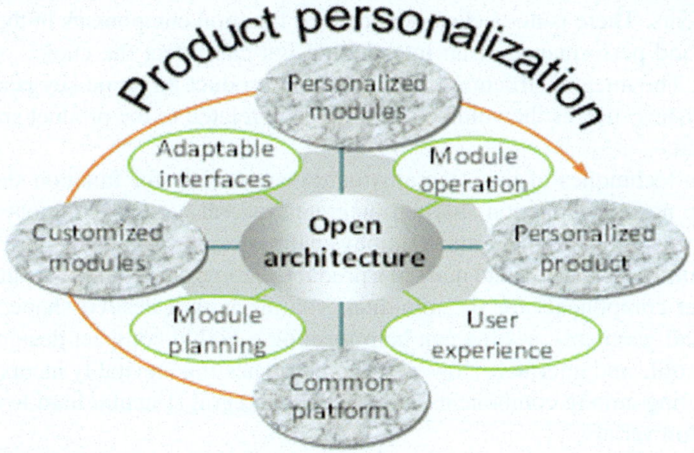

Fig. 6.13 PP elements for mass personalization

requirements using product lifecycle parameters. Product configurations, parameters and functions can be changed in the product lifespan to satisfy user changeable requirements. A product design using AD is featured by the extendibility of functions, upgradeability of modules, and customizability of components.

Therefore, OAPs can be developed using the AD method to implement the mass personalization. An OAP is proposed using three types of function modules: common platform modules made by manufacturers using the mass production, customized modules provided by manufacturers for user choices, and personalized modules designed and supplied by users or third-party suppliers [9] AD enables users to select different configurations in the OAP design stage, and to change product configurations in operations with various options based on the need.

The mass personalization can be initiated by introducing the open concept to form the product. This section introduces development of personalized product (PP) using OAP for a promise direction of the mass personalization [10]. Following parts will first review the related research. Methods of the PP development are then proposed as shown in Fig. 6.13. Function module planning, design, and interfaces are discussed in detail. Industrial PPs are developed using the proposed method.

6.3.2 General Background

6.3.2.1 Product Structure

Product structure is the interactive pattern of physical components to implement product functions. Common product structures include integral structure and modular structure. The integral structure is a single product block to implement function requirements. There is not a clear boundary of function components in the product. The demand performance of an intended design can affect the choice of product structure. The integral structure is often used in product for some key performance characteristics such as the volume and efficiency related to the product size, shape, and weight.

Design techniques of the integral structure are mainly for function sharing and geometric nesting. Function sharing integrates several parts or components into a single physical unit so that redundant physical parts and components can be eliminated to minimize the size and mass of a product. Geometric nesting arranges profiles of product components for the minimum volume with a desired shape. Function sharing and geometric nesting can help forming a highly compact design with the desired profile and interface. However, geometric nesting inevitably incurs the interface coupling among components. It makes the integral structure hard to adjust for the function variety.

The modular structure breaks a product structure into discrete modules so that the product can fully function with interchangeable modules through well-defined interfaces. The modular structure can increase product feasibility, variety, and durability by replacing or upgrading add-on modules without affecting other parts of the product.

Therefore, the architecture, module, and interface are three basic characteristics of a modular product. The architecture is the combination pattern of modules. The module is an independent physical block corresponding to its function elements based on the physical and functional similarity. Modules are connected using a set of interfaces. The difference between a simple block and a module can be identified based on their realization of a specific function. The interface is physical connections between modules to achieve functional interactions, or a port that logically or physically integrates boundaries between modules or boundaries between modules and their environments.

Comparing with the integral structure, the modular structure reduces function sharing, incurring the redundant physical expense such as the interface, more parts and suboptimal use of the space, mass, and energy. The performance-based standardization also causes the excessive capability for each individual application. In short, the integral architecture gains the technical advantage, and the modular architecture favors the business performance.

Most existing products use a somewhat hybrid structure. Research shows that restricted technical constraints lead to a high degree of the integral structure, otherwise products tend to the modularization. The cost and possible benefits can be analyzed from the integration of function modules for product performance. Solutions can be concluded as follows. (1) there is no single structure that is optimal in all cases, whereas there are always sub-optimal architectures, (2) for the adaptable performance, the modular structure is an optimal choice, (3) for the compact performance, the integral architecture is an optimal option. Open architecture has not been fully discussed to take advantages of the existing product structure in the development of mechanical products.

6.3.2.2 AD for OAP

Adaptable design (AD) and axiomatic design have the similar consideration to meet flexible requirements of different functions with corresponding design characteristics, where only a subset of functions with their physical characteristics is required at any time or circumstance so that the system has adaptability [11]. The existing research shows that the product adaptability comes from the adaptable architecture, modules and interfaces.

There are mainly four factors that influence product architecture, including the market variance, usage variance, technology change, and design method. The architecture may relate to the task definition of a firm and industry structure. The interface of the product also affects the architecture. AD considers these factors when forming an adaptable product.

Since the AD is based on the modular structure, the adaptable product architecture follows a manner of the modular combination. Two typical combination modes of modules were suggested based on the module spatial sequence and interface conditions. One is the fully modular structure where each module only connects to its direct neighbors, the other is the bus-module where all modules are connected to a common platform. Two combination modes can grow and blend to form more complex modular combinations.

Specific to the AD, an important requirement is that when a module is adjusted, it only affects its downward sub-functions. With the architecture, any downward functional elements only interact with its upward functional elements without affecting other parts of a product. OAP uses public interfaces to expand the product variety. It is formed by the common platform, customized modules and personalized components. Therefore, the OAP is essentially the modular structure. It emphasizes the product personalization.

Interfaces play a key role in the connection, transformation, and interaction of product modules. Especially, interfaces support the module replacement in the lifespan of OAPs. Using effective interfaces for the product personalization is also an effective way if the content of personalization is not enough to provide personalized offers. An adaptable interface supports modules replaceable, upgradeable, and functionally variable abilities in OAPs.

Modules are connected through interfaces to form a modular product. The independence axiom of axiomatic design can guide the modularization of formed modules to be functional and physical uncoupled. In this aspect, a concept of openness can be used to evaluate the interface for sharing levels of resources. The research has shown that the standardization of interfaces can benefit product improvement. Interfaces are analyzed in terms of physical interaction factors to evaluate the general product adaptability. Interface efficacy has been introduced to evaluate the interface efficiency by integrating the interface graph representation, criteria matrix, and House of Quality [4].

6.3.2.3 Product Personalization

Product personalization can be measured based on different factors. One factor considers the product adaptability for personalized needs as specific adaptability and general adaptability. The specific adaptability is the function predicted for the product lifespan. It can be evaluated by comparison of the effort of product adaption with respect to the effort of new product creation. The specific adaptability is evaluated through the extendibility of functions, upgradeability of modules, and customizability of components. The extendibility of functions considers the potential function extension, upgradeability of modules for the product improvement, and customizability of components to improve the personal convenience of product adaption. Measurements of these three aspects can be normalized and combined as the specific adaptability. The general adaptability is the product function that is not considered initially. It is evaluated by comparing the actual or full architecture of a product with its ideal form of the segregated architecture and quantitatively measured based on characteristic parameters of interfaces and interactions.

The product adaptability for personalized needs can also be defined using the essential adaptability and behavior adaptability. The essential adaptability reflects the effort such as time, resource, and energy to modify a current product for new function requirements. An easy modification of functions will bring a good adaptability of the product. The behavior adaptability reflects customers' satisfactions of adapted products. It shows the cost-effective level of the adaptation activity. An adaptation should be implemented when both essential adaptability and behavior adaptability are high. At three adaptable levels of the product, modules, and parts for personalized needs, the design efficiency can be used to evaluate the implementation of adaptability considering three major factors of product architecture, complexity of interface, and operation ability [8]. At each level, methods can model design candidates and calculate the robustness so that an optimal design candidate can be identified with the best solution. These research solutions have made a significant progress in the product personalization.

In summary, a product can be designed using integral or modular structures. The integral structure uses function sharing and geometric nesting to form highly compact products. Modular structure uses modules and interfaces to improve the product flexibility. AD is based on the modular structure for product adaptability using the adaptable architecture, modules, and interfaces. OAPs can effectively support the implementation of product personalization. Following parts of this section introduce methods and applications of design for personalized products.

6.3.3 Methods of Design for Mass Personalization

6.3.3.1 Module Planning for Personalized Products Based on OAPs

Module types of personalized products can be planned by the integration of extended quality function deployment (QFD) and axiomatic design methods using the modular method introduced in Sect. 5.1.1 [15]. The degree of variety (DV) is expanded as a quantitative measure for various components to decide the difference of product function modules. A weighting coefficient of the variance index is used as a quantitative description of the variance measure to decide the module type.

A design structure matrix (DSM) can be formed to cluster product components to indicate degrees of the component DV. According to the DV, components are divided into three types of modules based on two thresholds of the variant. The thresholds are set based on variant technical requirements (TRs) over constant TRs. From the assumption of need changes, different values of thresholds can be applied based on manufacturer data and user requirement changes. Following relationships of components and their DV values, components can be clustered into different modules of an OAP.

6.3.3.2 Detail Design of Personalized Products

An electric vehicle is driven by electrical power through energy/force operations as shown in Fig. 4.26. The process is conducted to transform the design concept into detailed mechanical elements through the energy/force transformation as shown in Fig. 4.27. A flowchart of the search process is shown in Fig. 4.28.

6.3.4 Case Study

A multiple-purpose electrical car (MEC) is developed for personalization. MEC is used by small businesses or families for commuting or shopping in a short distance. The user investigation found the need in different steering wheel options and trunks for different loads. MEC is designed as a personalized product to meet a variety of user requirements in the product lifespan [10]. Steps of MEC development using the introduced method are as follows.

(1) MEC modules planning based on user requirements,
(2) Detailed modules and interfaces formation,
(3) Design solutions for personalization.

Table 6.28 Customer requirements (CRs) and function needs (FNs)

CRs	FNs
Carry different loads	Container for bulk loads; platform for packed boxes
Easy control	Handle bar steering; steering wheel; Foot acceleration and brake
Use electrical power	Driving system; house and public electrical charge
Run on different roads	Spring support to reduce road impact
Use in different weather conditions	Prevent wind and rain
Safety	Car body; chassis; frame
Driver comfort	Car seat

6.3.4.1 Module Planning

Customer requirements for MEC are collected from the user survey. The mapping from customer requirements (CRs) to product function needs (FNs) is summarized in Table 6.28.

An extended QFD is built to plan MEC modules. The variant of CRs is identified based on MEC lifespan applications. Possible changes of requirements are called expected changes (ECs) that are in the last column of the matrix shown in Table 6.29 for estimated values of user requirement changes. The extended QFD of MEC contains the current function needs (FNs) and future expected changes (ECs). Using Eqs. (5.1)–(5.3), the variance coefficient of technical requirements (VC), technical importance weight (TIW), relative importance of technical importance weight (rTIW), and weighted coefficient of variances (wCOV) are calculated as listed in Table 6.29 to plan function modules.

DV is decided using Eqs. (5.4)–(5.6) to identify the variation of FNs. The greater the DV, the more likely the FN to perform as a personalized module. The common platform is decided for FNs with less than 60% variation. Personalized modules are decided to conduct FNs with more than 70% variation. Customized modules are for FNs variations in between 60 and 70%. Therefore, seven modules are decided to build the MEC for different function requirements.

6.3.4.2 Detailed Modules Formation

Using methods proposed in Sect. 6.3.3.2, the product conceptual modules are transferred into physical components. Thirty-five mechanical components are decided to form the physical structure of seven function modules based on search of mechanical components from the part database and common practice of the vehicle design, as shown in Table 6.30 and Fig. 6.14. The relation of the components and modules is listed in Table 6.31.

Following discussion takes the common platform module as an example for the detail design. The platform module performs functions of the steering, driving,

Table 6.29 Extended QFD for MEC modules planning

CRs \ FNs	Container for bulk loads	Platform for packed boxes	Handle bar steering	steering wheel	Foot acceleration	Foot brake	House electric charge	Public electric charge	Spring support to reduce road impact	Prevent wind	Prevent rain	Car body	Car chassis	Car frame	Car seat	CRs weight	Expected change
Carry different loads	6	6	1	1	0	0	0	0	1	0	0	3	3	1	0	5	3
Easy control	0	0	6	6	6	6	1	1	0	1	1	0	0	0	0	5	1
Use electric power	0	0	0	0	0	0	6	6	0	0	0	0	0	0	0	5	1
Run on different roads	3	3	1	1	1	1	1	1	6	1	1	1	1	1	3	3	1
Use in different weather conditions	3	3	1	1	1	1	1	1	1	6	6	1	1	1	1	3	3
Safety	1	1	3	3	1	1	3	3	1	3	3	3	3	6	3	5	1
Driver comfort	0	0	3	3	1	1	0	0	3	6	6	3	3	3	6	3	2
VC	31	31	22	22	13	13	14	14	19	35	35	22	22	19	21		
TIW	53	53	65	65	44	44	56	56	40	59	59	45	45	50	45		
rTIW	6.8%	6.8%	8.3%	8.3%	5.6%	5.6%	7.2%	7.2%	5.1%	7.6%	7.6%	5.8%	5.8%	6.4%	5.8%		
wCOV	2.11	2.11	1.84	1.84	0.73	0.73	1.01	1.01	0.98	2.65	2.65	1.27	1.27	1.22	1.21		
DV	79.59%	79.59%	69.27%	69.27%	27.71%	27.71%	37.98%	37.98%	36.82%	100%	100%	47.96%	47.96%	46.02%	45.78%		
Module type	Vp	Vp	Vc	Vc	C	C	C	C	C	Vp	Vp	C	C	C	C		

Correlation coefficient (Strong -6, Medium- 3, Weak -1, None-0 . Expected changes of customer need (high (H = 3), medium (M = 2), low (L = 1)).

Table 6.30 Physical components

Part	Part name	Part	Part name
1	Mainframe	19	Brake panel and cylinder
2	Central steering arm	20	Speed controller
3	Spindle (left, right)	21	Safety frame
4	Hub	22	Batteries
5	Tie rod and ball joint	23	Windshield frame
6	Front spring and mounting brackets	24	Windshield
7	Front suspension arm	25	Handlebar
8	Rear spring and mounting bracket	26	Steering shaft
9	Suspension link	27	Support member
10	Electric motor and differential unit	28	Universal joint set
11	Driving shaft and universal set	29	Steering wheel
12	Rear suspension frame	30	Steering shaft and reducer
13	Rear brake drum & driving axles	31	Support member
14	Front body	32	Universal joint set
15	Hood	33	Container for bulk loads
16	Bumper and lights	34	Platform for packing boxes
17	Wheel	35	Roof
18	Driver seat		

chassis, body, safety frame, and wheels. It is the most complicated structure in the seven modules. The design develops detailed structures of the module as follows.

The steering system is an energy-flow system that can use a spline steering shaft for the input rotation. Wheel deflections are for steering. Other mechanical elements include ball joints and spindles. For the two motion linkages and a rotation distribution, tie rods and a central steering arm are applied to form the Ackerman steering geometry as illustrated in Fig. 6.15a.

The driving device is an energy-flow system that consists of batteries, a speed controller for the foot acceleration, a brake panel unit for the foot brake, and two wheels to drive and brake as shown in Fig. 6.15b. An electric motor/differential unit is used for the electric-motion conversion and motion distribution.

The chassis is a force-path subsystem. By comparing different classic automotive structures for technical feasibility and cost, a chassis frame is designed using swing arm and leaf spring suspensions as shown in Fig. 6.15c. The combination of the steering subsystem with the chassis is shown as Figs. 6.15d, e. The chassis links the steering system using interface E. The rotation is performed by the steering shaft and bearings.

Combination of the driving system and chassis is shown in Fig. 6.15f, g. To avoid the transmission confliction between the fixed electric motor/differential unit and two vibrating wheels, straight rotation transmissions are adjusted as two-universal-joint

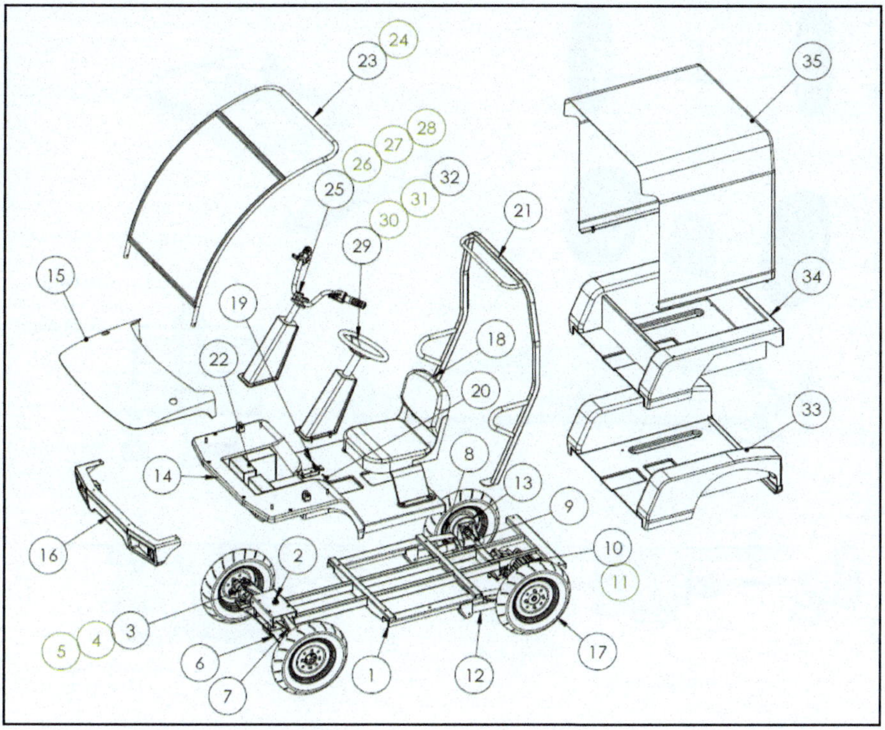

Fig. 6.14 Components of the MEC

Table 6.31 Components and modules

Part #	Module	Type
1–22	1—MEC platform	Common
23–24	6—Windshield	Personalized
25–28	2—Handlebar steering	Customized
29–32	3—Steering wheel steering	Customized
33	4—Container	Personalized
34	5—Loading platform	Personalized
35	7—Roof	Personalized

transmissions. Two mounting brackets are used. The two-universal-joint transmissions are implemented using two driving shafts and universal joints. The electrical motor and differential unit are fixed on the chassis to transmit power to wheels.

These mechanical components are combined into a complete module, the chassis, to support steering and driving systems as shown in Fig. 6.15h. Figure 6.15j is the final formed platform module with added front body and seat units shown in Fig. 6.15i.

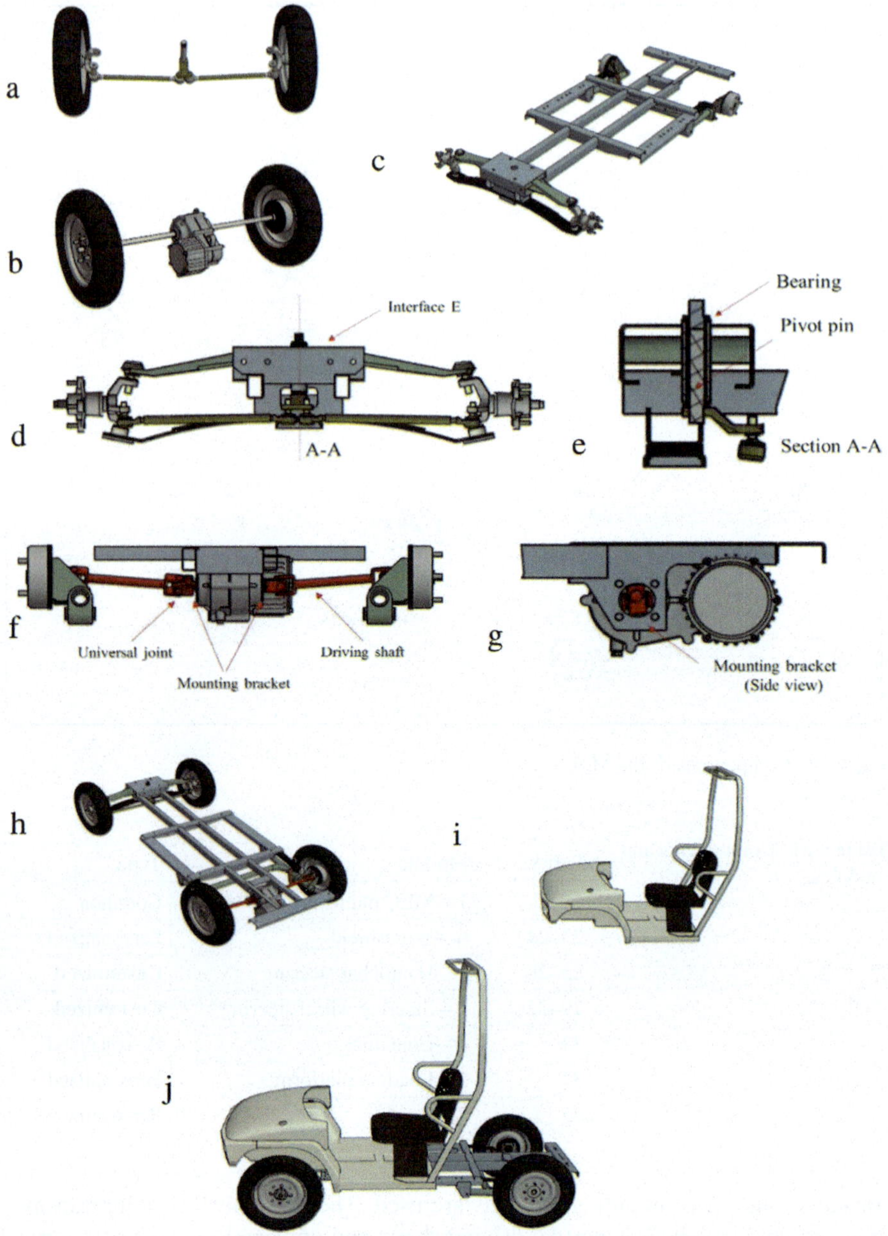

Fig. 6.15 Formation of the common platform module

Finally, seven modules are formed for different MEC models to meet different user requirements. The common platform module is required by all configurations of MEC models. Interfaces are required to connect the common platform module and other function modules. The attention is paid to the link between the common platform module and personalized modules that can be easily replaced by users in the MEC application.

6.3.4.3 Interface Design

For bulk loads and packed boxes, interface A is designed considering the spatial need and loading weight as shown in Fig. 6.16. Where Case A is for bulk loads, B is for packed boxes. For windshield and roof modules to prevent the airflow and rain drops, interfaces B, C, and D are designed to connect them to the common platform module as shown in Fig. 6.17. Interface E is used to connect two modules of the steering option.

Since module functions are considered independently to meet requirements of the open architecture, these modules can be added or replaced by users in the product

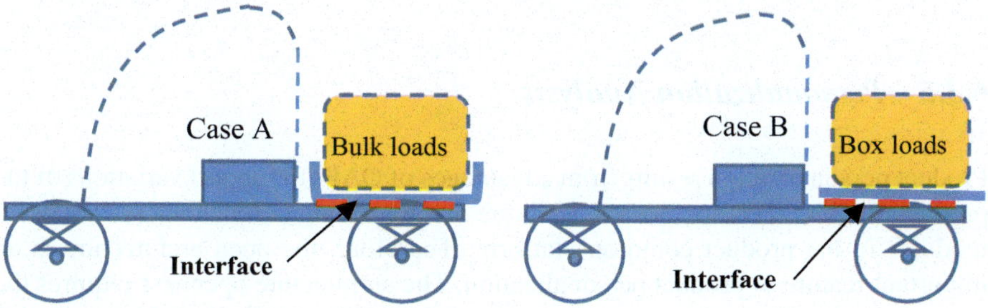

Fig. 6.16 Interface A

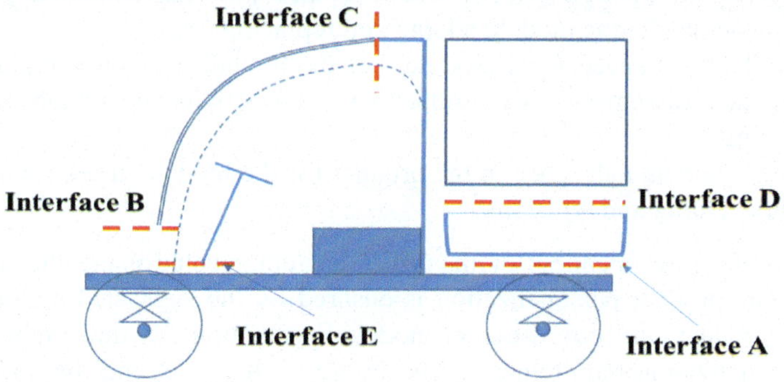

Fig. 6.17 Interfaces A, B, C, D, E

lifespan. Considering the interface alignment, fastening, interaction, and assembly/disassembly, mechanical structures of each interface are designed as follows.

(1) Interface A supports the loading weight from the container module for bulk loads, or the platform module for packed boxes. The preferred form is a set of flat planes with side profile constraints fastened to limit the six-degree freedom. This can also make the module a flat bottom to support loads.
(2) Interfaces B and C are used to connect the windshield module. Each of them adopts two fastening points so that total four fastening points to limit the six-degree freedom of the module.
(3) Interface D connects modules 1 and 7. The preferred form is a set of flat planes with side profile constraints fastened to limit the six-degree freedom.
(4) Interface E is used to connect modules 2 or 3 to module 1 for two steering options. The preferred form for connection/disconnection is a set of flat planes.
(5) Screws and nuts used in the interfaces are same type and size so that all assembly/disassembly can be handled by users using a screwdriver and a wrench.

Therefore, the seven modules can be combined through interfaces to form more than ten different MEC models. Figure 6.18 shows these modules, and two models formed by the platform and selected customized and personalized modules.

6.3.5 Personalization Analysis

Product personalization results from advantages of OAPs to support variations of the product function. Personalization needs are met by adding or replacing personalized modules in the product common platform. Therefore, the open architecture is an important feature of product personalization. The architecture openness requires the allowance of a product to accept personalized function modules. Levels to meet this requirement are classified as follows.

(1) A product has enough space to add or replace a module without any obstacle. Users can handle the module adding and replacing easily.
(2) A product has space for a module, but the module adding or replacing will cause the disassembly of the product. Users can still handle module adding and replacing.
(3) There is not enough space in the product for the module, users cannot handle module adding and replacing.

The architecture openness of the MEC is implemented using the introduced method. The product personalization is ensured by the suggested modular structure. The product can form different models by function module combinations to meet personalization requirements. The future need can be met by adding more personalized modules either provided by users or developed by the third-party industries.

MEC modules		
 1. Common platform	 2. Handlebar (customized)	 3. Steering wheel (customized)
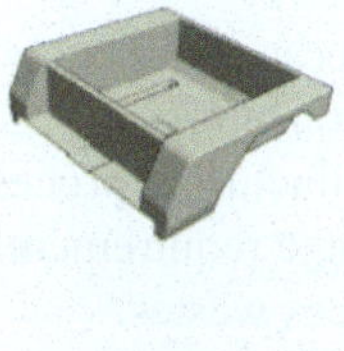 4. Container (Personalized)	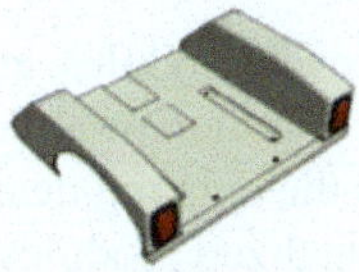 5. Rear Platform (Personalized)	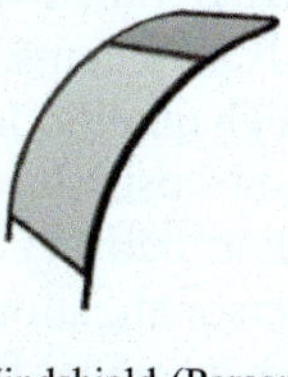 6. Windshield (Personalized)
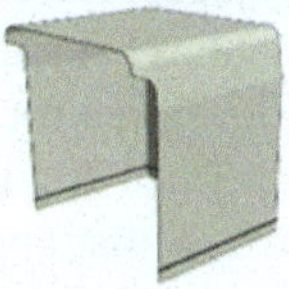 7. Roof (Personalized)		
 **Model A** (Modules 1+3+4+6+7)	 **Model B** (Modules 1+2+5)	*Different models can be formed based on customer needs, these are two examples*

Fig. 6.18 MEC modules and models

6.3.6 Concluding Remarks

Open architecture has the potential to support mass personalization in following two aspects.

(1) As different users and industries can involve in product development, new designs can be developed to satisfy various market segments, leading to mass personalization in product developments.
(2) Product modules are developed with technologies from different sources that can be integrated through adaptable design to lead mass personalization in the technology integration.

Therefore, the open architecture can address the challenge of mass personalization in following solutions.

- Better functionality: It allows product design to incorporate new technological advancements,
- High quality: It utilizes the proven modules, associated processes and knowledge,
- Low costs: It reuses designs, common modules, and components,
- Short delivery time: It reuses the existing design and manufacturing processes,
- Personalization: It uses personalized modules to meet change requirements,
- Environment friendly: It extends the product life and reduces wastes.

However, the mass personalization is still in its early research stage. As a concept, it is clear with advantages. As a method, although is incomplete, it has been used for various product development. Significant efforts are required to develop it for a cost-effective method of the mass personalization. In order to apply the open architecture in the industrial mass personalization, a tool should be developed so that engineers can use it in their daily design process.

6.4 Adaptable Design of Coating Machines

The application provided in this section is based on the research results developed by Chen et al. [1].

Industrial coating machines are the typical printing equipment for coating certain liquid on a substrate, such as paper, to improve surface properties. Due to the variety of customer needs in the market, the manufacturer needs to develop products whose functions are adaptable to changes of customer requirements. It is a challenge for a company to meet different product requirements using the traditional coating machines. Therefore, an adaptable coating machine was proposed to meet the product change. It uses the open-architecture structure to build an adaptable coating machine with independent functional modules. The coating machine with the open architecture is designed by applying the adaptable design method [1].

The layout of the coating machine includes six functional modules: M_1-unwinding device, M_2-infeed module, M_3-coating module, M_4-drier module, M_5-outfeed device, and M_6-rewinding device as shown in Fig. 6.19.

Figure 6.20 shows connections of the adaptable coating machine made from common platforms, interfaces and different modules. Modules M_1, M_2, M_5, and M_6 are used to build the platform of the machine, while modules M_3 and M_4 are

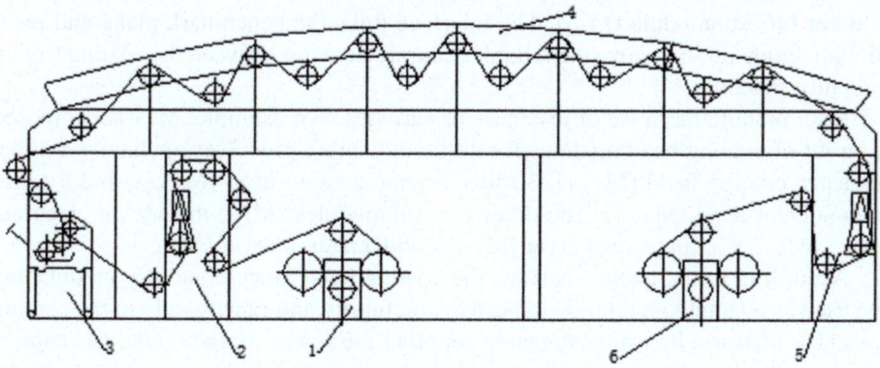

Fig. 6.19 Layout of modules for the coating machine. 1—Unwinding device, 2—Infeed device, 3—Coating unit, 4—Outfeed device, 6—Rewinding device

customized modules or customer defined modules designed to satisfy the individual requirement. The platform modules are mass production modules. M_2 consists of infeed column ($M_{2.1}$) and infeed tension sub-module ($M_{2.2}$). M_5 consists of outfeed column sub-module ($M_{5.1}$) and out feed tension sub-module ($M_{5.2}$). The platform module provides the support for the machine control, positioning, and energy supply for other modules. Different modules are used to meet different product needs.

M_3 is a coating unit module, consisting of the coating column submodule ($M_{3.1}$) with an open interface, coating trolley submodule ($M_{3.3}$) and coating head submodule ($M_{3.2}$). Coating head ($M_{3.2}$) connects with the coating column through an open interface with the support of coating trolley, as shown in Fig. 6.21. M_4 is a dryer unit module, consisting of a dryer frame submodule ($M_{4.1}$) with an open interface, and

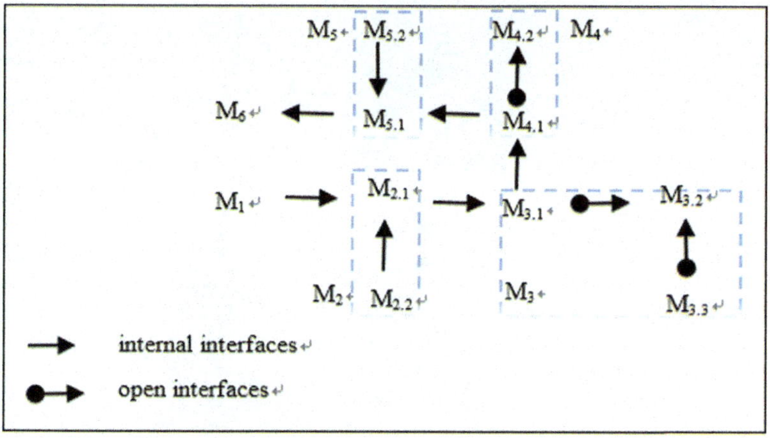

Fig. 6.20 Connections in the adaptable coating machine

a dryer box submodule ($M_{4.2}$). The interface links the benchmark plane and positioning guide rail to realize the reliable accurate docking between the coating trolley and host machine.

Each module has a set of instances or varieties. For example, $M_{3.2}$ is composed of a set of coating head modules for different coating processes, including a direct concave coating head ($M_{3.2\text{-}1}$), 3-roller reverse coating head ($M_{3.2\text{-}2}$), and 5-roller non-solvent head ($M_{3.2\text{-}3}$). The dryer box submodules ($M_{4.2}$) include an electrical dryer ($M_{4.2\text{-}1}$), a microwave dryer ($M_{4.2\text{-}2}$), and a natural dryer ($M_{4.2\text{-}3}$).

According to the above analysis, the coating head module and dry module can be used as adaptable modules connected structurally and functionally to the coating machine platform by using adaptable coupling interfaces. Based on the mechanical structure of this specific coating machine, through physically connecting with the column module, the coating module is connected to the basic platform.

Fig. 6.21 Coating unit with open interfaces

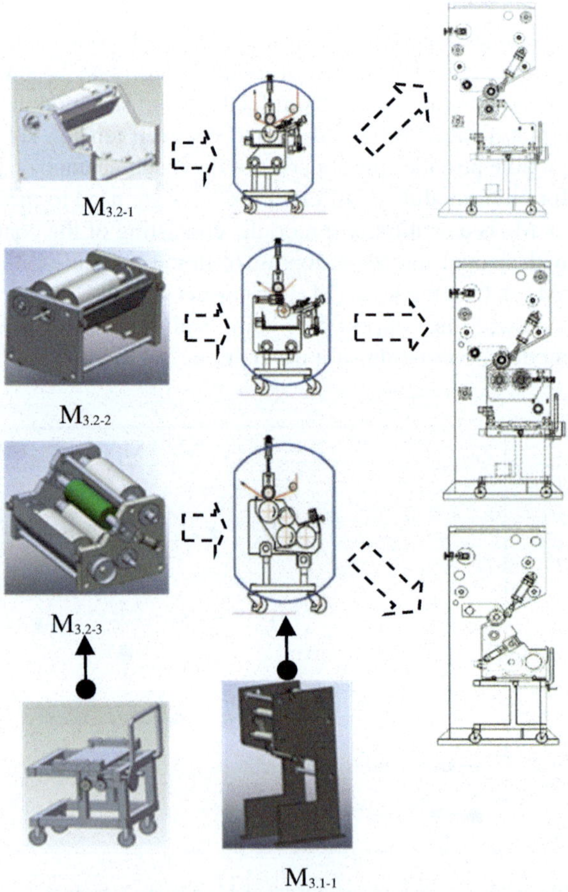

Since the adaptable coating machine consists of many modules, various types of coating trolleys and driers can be installed and modified for different types of products changes. To meet an individual need from solvent coating with much more emissions of harmful volatile to solvent free coating with almost no emissions of the harmful volatile, the manufacturer has developed three different types of coating machines based on above architecture to satisfy functional requirements at three different phases, as shown in Fig. 6.22.

The cost is considered for product adaptability c_a and openness c_I, as shown in Eqs. (6.2) and (6.3), respectively.

$$c_a(P_t) = \frac{c(P_t)}{A(P_t)} \tag{6.2}$$

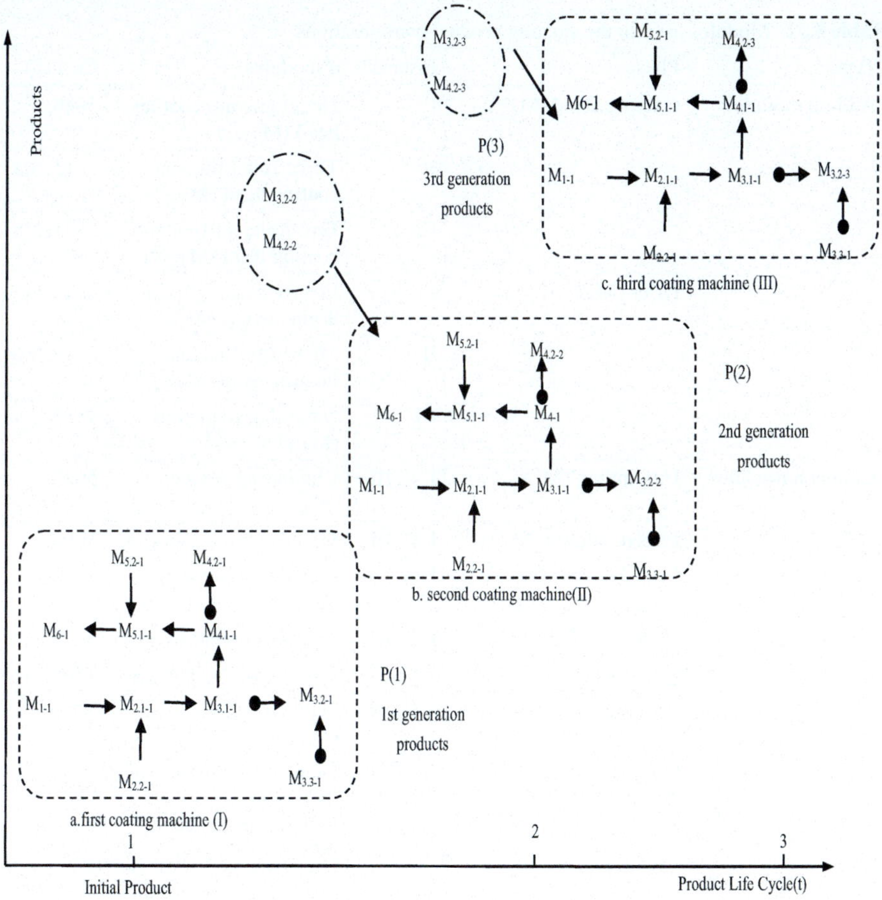

Fig. 6.22 Configurations of three coating machines

$$c_I(P_t) = \frac{c(P_t)}{A(P_t)}(1 + I(P_t)) \quad (6.3)$$

where $c(P_t)$ is the sum of the cost of all the modules in the product configuration P_t. $A(P_t)$ is the adaptability of the product configuration P_t. $I(P_t)$ is openness of the product configuration P_t.

Modules and product configurations in different phases are summarized in Tables 6.32 and 6.33, as well as costs of these modules and configurations. As shown in Table 6.33, the exiting products consist of product configurations in phases I, II, and III.

To meet the new customer need, some new modules with different costs have also been developed for the future use as shown in Table 6.34. The cost, adaptability, and openness for the existing product configurations can also be measured as shown in Table 6.35.

Table 6.32 Modules used in the existing product configurations

Type	Phase	Instances of modules		Cost ($)
Add-on modules	Coating Head ($M_{3.2}$)	I	Direct gravure coating head ($M_{3.2\text{-}1}$)	5046
		II	Three-roller transfer coating head ($M_{3.2\text{-}2}$)	13,308
		III	Five-roller solvent-free coating head ($M_{3.2\text{-}3}$)	31,128
	Drier (M_4)	I	Electric heating roller drying oven ($M_{4\text{-}1}$)	17,347
		II	Microwave heating floating oven ($M_{4\text{-}2}$)	19,386
		III	Natural drying oven ($M_{4\text{-}3}$)	9885
Common modules	Unwinding (M_1)	I, II, III	Unwinding device ($M_{1\text{-}1}$)	6623
	Infeed column ($M_{2.1}$)	I, II, III	Infeed column ($M_{2.1\text{-}1}$)	5000
	Infeed tension ($M_{2.2}$)	I, II, III	Infeed tension device ($M_{2.2\text{-}1}$)	10,723
	Coating column ($M_{3.1}$)	I, II, III	Coating column ($M_{3.1\text{-}1}$)	4000
	Coating trolley ($M_{3.3}$)	I, II, III	Coating trolley ($M_{3.3\text{-}1}$)	1500
	Outfeed column ($M_{5.1}$)	I, II, III	Outfeed column ($M_{5.1\text{-}1}$)	3000
	Outfeed tension ($M_{5.2}$)	I, II, III	Outfeed tension device ($M_{5.2\text{-}1}$)	6308
	Rewinding (M_6)	I, II, III	Rewinding device ($M_{6\text{-}1}$)	7885

Table 6.33 Existing product configurations

Phases	Common modules	Add-on modules	Total cost ($)
Phase I	$M_{1\text{-}1}$,$M_{2.1\text{-}1}$,$M_{2.2\text{-}1}$,$M_{3.1\text{-}1}$,$M_{3.3\text{-}1}$,$M_{5.1\text{-}1}$,$M_{5.2\text{-}1}$,$M_{6\text{-}1}$	$M_{3.2\text{-}1}$,$M_{4\text{-}1}$	67,432
Phase II	$M_{1\text{-}1}$,$M_{2.1\text{-}1}$,$M_{2.2\text{-}1}$,$M_{3.1\text{-}1}$,$M_{3.3\text{-}1}$,$M_{5.1\text{-}1}$,$M_{5.2\text{-}1}$,$M_{6\text{-}1}$	$M_{3.2\text{-}2}$,$M_{4\text{-}2}$	77,733
Phase III	$M_{1\text{-}1}$,$M_{2.1\text{-}1}$,$M_{2.2\text{-}1}$,$M_{3.1\text{-}1}$,$M_{3.3\text{-}1}$,$M_{5.1\text{-}1}$,$M_{5.2\text{-}1}$,$M_{6\text{-}1}$	$M_{3.2\text{-}3}$,$M_{4\text{-}3}$	86,052

Table 6.34 New modules for future product configurations

Type of module	Instances of the modules	Cost ($)
Coating Head ($M_{3.2}$)	Direct gravure coating head ($M_{3.2\text{-}5}$)	4500
	Five-roller solvent-free coating head ($M_{3.2\text{-}4}$)	30,020
Drier (M_4)		
Unwinding (M_1)	Unwinding device ($M_{1\text{-}2}$)	6520
	Unwinding device ($M_{1\text{-}3}$)	6800
Infeed Column ($M_{2.1}$)	Infeed Column ($M_{2.1\text{-}2}$)	4500
	Infeed Column ($M_{2.1\text{-}3}$)	5500
Infeed Tension ($M_{2.2}$)	Infeed Tension device ($M_{2.2\text{-}1}$)	9725
	Infeed Tension device ($M_{2.2\text{-}3}$)	11,720
Coating Column ($M_{3.1}$)	Coating Column ($M_{3.1\text{-}2}$)	3500
	Coating Column ($M_{3.1\text{-}3}$)	4200
Coating trolley ($M_{3.3}$)	Coating trolley ($M_{3.3\text{-}2}$)	1400
	Coating trolley ($M_{3.3\text{-}3}$)	1600
Outfeed column ($M_{5.1}$)	Outfeed column ($M_{5.1\text{-}2}$)	2900
	Outfeed column ($M_{5.1\text{-}3}$)	3150
Outfeed Tension ($M_{5.2}$)	Outfeed tension device ($M_{5.2\text{-}2}$) Outfeed tension device ($M_{5.2\text{-}3}$)	6100 6500
Rewinding (M_6)	Rewinding device ($M_{6\text{-}2}$)	7500
	Rewinding device ($M_{6\text{-}3}$)	8000

Table 6.35 Cost, adaptability, and openness of the existing configurations

Configuration	$C(P_t)$/($)	$A(P_t)$	$I(P_t)$	$c_a(P_t)$/($)	$c_I(P_t)$/($)
I	67,432	1	2	67,432	202,296
II	77,733	1.579	1.421	49,229	119,184
III	86,052	2.047	1.477	42,036	104,128

6.5 Design of an Open Architecture Paper-Bag Folding Machine

The application provided in this section is based on the research results developed by Zhang et al. [14].

The paper-bag folding machine is designed to fold packing papers into the required shape. The existing paper-bag folding machines were designed to satisfy specific folding requirements of packing papers for customers. Although various functional modules can be changed to satisfy different requirements, all functional modules are specified in the product development stage. During the machine operation, different personalized requirements on the types/shapes of paper bags may be required. These personalized packaging requirements during the machine operation might not be satisfied by the existing paper-bag folding machines. In order to meet new requirements, a new machine needs to be developed.

In this design, an open architecture paper-bag folding machine is developed to satisfy different requirements. Usually, bagging modules in the paper-bag folding machine are developed according to specific packaging requirements in terms of paper bags types and/or shapes. As shown in Fig. 6.23, the bagging modules that need to be changed in the operation stage are considered as add-on modules. The corresponding product open interfaces need to be developed to connect the bagging modules provided by customers/third-party vendors to the machine platform. According to the operational processes of a paper-bag folding machine, six bagging modules and six open interfaces are required for the open architecture machine. Since those modules and open interfaces are similar, only one bagging module and one open interface are considered in this design.

Fig. 6.23 Paper-bag folding machine (Courtesy of Shantou Institute for Light Industrial Equipment Research)

6.5.1 An Initial Design of the Open Interface

(1) Modeling of the open interface

For the paper-bag folding machine, the bagging modules (i.e., add-on modules) are designed according to the specific shapes of bags. An open interface is designed to connect the bagging modules and the platform. An initial design of the open interface for the paper-bag folding machine is shown in Fig. 6.24, and components of the open interface are summarized in Table 6.36.

(2) Quantification of the open interface functional adaptability

Inputs and outputs of the open interface are used for quantification of the open interface functional adaptability. An ideal changing range of inputs and outputs of the bagging module interface is expressed as:

$$R_R = (I_1, I_2, O_1) \tag{6.4}$$

where I_1, I_2, O_1 represent the ideal inputs and outputs of the open interface as shown in Table 6.37.

Due to constraints of parts in the open interface, the feasible changing range of inputs and output of the open interface design can be expressed as:

$$R_{O1} = (I_1^{S1}, I_2^{S1}, O_1^{S1}) \tag{6.5}$$

where ${I_1}^{S1}, {I_2}^{S1}, {O_1}^{S1}$ represent the inputs and outputs of the open interface as shown in Table 6.38.

According to Tables 6.37 and 6.38, and Eqs. (6.4) and (6.5), the functional adaptability of this open interface is obtained as [14]:

$$A_F = 0.55 \tag{6.6}$$

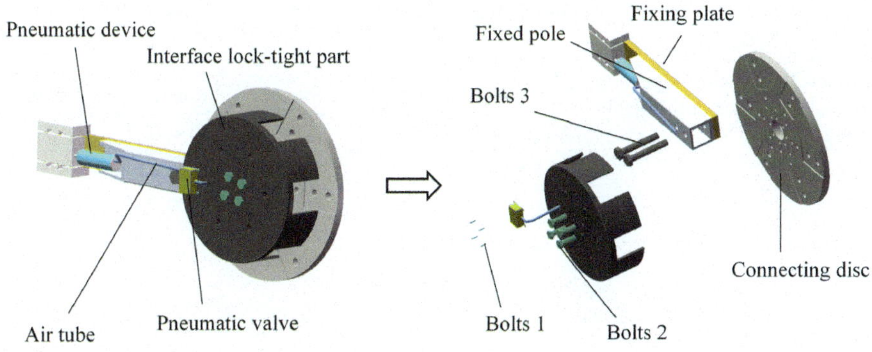

Fig. 6.24 An initially designed bagging module

Table 6.36 Components of the open interface

Platform interface	Parts	$P^I{}_1$	Connecting disc
		$P^I{}_2$	Interface lock-tight part
		$P^I{}_3$	Bolts 1
		$P^I{}_4$	Pneumatic valve
		$P^I{}_5$	Bolts 2
	Assembling relations among the parts	$R^I{}_1$	Assembling relation between $P^I{}_1$ and $P^I{}_2$
		$R^I{}_2$	Assembling relation among $P^I{}_1$, $P^I{}_2$, and $P^I{}_5$
		$R^I{}_3$	Assembling relation between $P^I{}_2$ and $P^I{}_4$
Interface connector	Parts	$P^{IC}{}_1$	Bolts 3
		$P^{IC}{}_2$	Fixing plate
		$P^{IC}{}_3$	Fixed pole
		$P^{IC}{}_4$	Air tube
	Assembling relations among the parts	$R^C{}_1$	Assembling relation between $P^{IC}{}_2$ and $P^{IC}{}_3$
		$R^C{}_2$	Assembling relation among $P^{IC}{}_2$, $P^{IC}{}_3$, and $P^{IC}{}_1$
Assembling relations between interface and connector		$R^{IC}{}_1$	Assembling relation between $P^{IC}{}_2$ and $P^I{}_1$
		$R^{IC}{}_2$	Assembling relation between $P^I{}_2$ and $P^{IC}{}_3$
		$R^{IC}{}_3$	Assembling relation among $P\ ^{IC}{}_1$ and $P^I{}_1$, $P^I{}_2$
		$R^{IC}{}_4$	Assembling relation between $P^I{}_4$ and $P^{IC}{}_4$

Table 6.37 Ideal inputs and output ranges of open interface

Input/output parameter	Symbol	Range	Unit
Air pressure	I_1	[0.4, 0.7]	MPa
Preload	I_2	[2.0, 10.0]	KN
Press force	O_1	[100, 150]	N

Table 6.38 Feasible inputs and output ranges of the open interface

Input/output parameter	Symbol	Range	Unit
Air pressure	$I_1{}^{S1}$	[0.4, 0.7]	MPa
Preload	$I_2{}^{S1}$	[2.0, 7.5]	KN
Press force	$O_1{}^{S1}$	[100, 140]	N

(3) Quantification of open interface structural adaptability

Parts $P^I{}_3$, $P^I{}_4$, $P^I{}_5$, $P^{IC}{}_1$, and $P^{IC}{}_4$ shown in Table 6.36 and Fig. 6.27 are selected based on the international standard, and the corresponding evaluation factors of adaptability are equal to 1. Parts $P^I{}_1$, $P^I{}_2$, $P^{IC}{}_2$, and $P^{IC}{}_3$ are designed based on the enterprise standard, evaluation factors of adaptability are set as 0.2. The open interface structural adaptability can be calculated as [14]:

$$A_S = (5 \times 1 + 4 \times 0.2)/9 = 0.64 \tag{6.7}$$

(4) Quantification of the open interface manufacturing adaptability

As shown in Table 6.36, the platform interface and interface connectors are composed of bolts, fixing plate and pole, pneumatic valve, lock-tight part and connecting disc. Based on data collected in the manufacturing site, the manufacturing cost of parts of the interface connector, $C_{M(P^C)}$, is equal to 200 RMB. The assembling cost of the parts of interface connector, $C_{A(R^C)}$, is 30 RMB. Therefore, the manufacturing cost of the interface connector, $C_M = C_{M(P^C)} + C_{A(R^C)} = 230$ RMB, or 0.23 thousand RMB. The manufacturing adaptability of the open interface can be calculated as [14]:

$$A_M = e^{-0.23} = 0.79 \tag{6.8}$$

(5) Quantification of the open interface operational adaptability

For the open interface, assembling relations between the interface and connector are shown in Table 6.36. Based on data collected in the manufacturing site, the assembling cost of the platform interface and interface connector, C_A, is 90 RMB, and the disassembling cost of the platform interface and interface connector, C_D, is 30 RMB. Therefore, the cost for assembling and disassembling of platform interface and interface connector can be calculated by $C_O = C_A + C_D = 120$ RMB, or 0.12 thousand RMB. The open interface operational adaptability, A_O, can be calculated as [14]:

$$A_O = e^{-0.12} = 0.88 \tag{6.9}$$

(6) Sensitivity analysis

In this design, the overall adaptability of the open interface is calculated considering functional, structural, manufacturing, and operational adaptabilities. A sensitivity analysis is conducted as shown in Table 6.39. From the analysis results in Table 6.39, $P^I{}_2$, $P^{IC}{}_2$, and $P^{IC}{}_3$ of the open interface are identified as parts with the maximal influence on the interface adaptability. Moreover, $R^I{}_1$, $R^I{}_2$, $R^C{}_2$, and $R^{IC}{}_2$ are identified as assembling relations having the maximal influence on the interface adaptability.

Table 6.39 Sensitivity analysis of interface adaptability

			Degree of sensitivity at functional level (DS_F)	Degree of sensitivity at structural level (DS_S)	Degree of sensitivity at manufacturing level (DS_M)	Degree of sensitivity at operational level (DS_O)
Platform interface	Parts	$P^I{}_1$	0.05	0.08	0	0
		$P^I{}_2$	0.05	0.1	0	0
		$P^I{}_3$	0.05	0.05	0	0
		$P^I{}_4$	0.1	0.05	0	0
		$P^I{}_5$	0.1	0.05	0	0
	Assembling relations among the parts	$R^I{}_1$	0.03	0.1	0	0
		$R^I{}_2$	0.1	0.1	0	0
		$R^I{}_3$	0.05	0.05	0	0
Interface connector	Parts	$P^{IC}{}_1$	0.08	0.05	0.1	3
		$P^{IC}{}_2$	0.05	0.08	0.1	0
		$P^{IC}{}_3$	0.08	0.08	0.25	0
		$P^{IC}{}_4$	0.15	0.08	0.25	0
	Assembling relations among the parts	$R^C{}_1$	0.08	0.1	0.15	0
		$R^C{}_2$	0.15	0.1	0.15	0
Assembling relations between interface and connector	$R^{IC}{}_1$		0	0	0	0
	$R^{IC}{}_2$		0	0	0	0
	$R^{IC}{}_3$		0	0	0	0
	$R^{IC}{}_4$		0	0	0	0

6.5.2 *Modified/Improved Design of Open Interface of the Paper-Bag Folding Machine*

(1) Modeling of the modified open interface

According to results of the sensitivity analysis shown in Table 6.39, the open interface is redesigned as shown in Fig. 6.25. Compared to Fig. 6.24, the interface lock-tight part, bolts, fixed pole, and assembling relations among parts $R^I{}_1$ and $R^I{}_2$ in Table 6.36 are modified. The redesigned open interface is shown in Table 6.40.

(2) Quantification of functional adaptability of the modified open interface

For the redesigned open interface, the feasible changing range of inputs and outputs of the modified open interface, $R^O{}_2$, can be expressed as:

$$\mathbf{R}_2^{\mathrm{o}} = (\mathrm{I}_1^{\mathrm{S2}}, \mathrm{I}_2^{\mathrm{S2}}, \mathrm{O}_1^{\mathrm{S2}}) \tag{6.10}$$

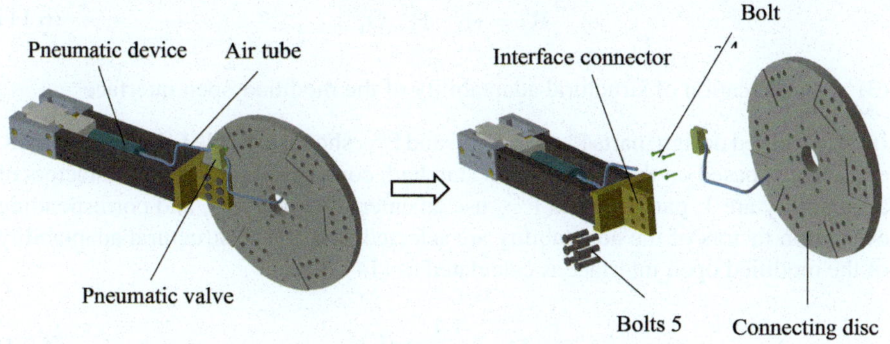

Fig. 6.25 Redesigned open interface of the paper-bag folding machine

Table 6.40 Modeling of the modified open interface

Platform interface	Parts	$P^I{}_1$	Connecting disc
		$P^I{}_2$	Pneumatic valve
		$P^I{}_3$	Bolts 4
	Assembling relations among the parts	$R^I{}_1$	The assembling relation between $P^I{}_1$ and $P^I{}_2$
		$R^I{}_2$	The assembling relation among $P^I{}_1$, $P^I{}_2$, and $P^I{}_3$
Interface connector	Parts	$P^{IC}{}_1$	Bolts 5
		$P^{IC}{}_2$	Interface connector
		$P^{IC}{}_3$	Air tube
	Assembling relations among the parts	$R^C{}_1$	The assembling relation between $P^{IC}{}_1$ and $P^{IC}{}_2$
Assembling relations between interface and connector		$R^{IC}{}_1$	The assembling relation among $P^{IC}{}_1$, $P^{IC}{}_2$, and $P^I{}_1$
		$R^{IC}{}_2$	The assembling relation between $P^I{}_2$ and $P^{IC}{}_3$

where $I_1{}^{S2}$, $I_2{}^{S2}$, $O_1{}^{S2}$ represent the inputs and outputs of the modified open interface as shown in Table 6.41.

According to Tables 6.37 and 6.41, and Eqs. (6.1) and (6.7), the functional adaptability of the open interface can be obtained as [14]:

Table 6.41 Input and output ranges of the modified open interface

Input/output parameter	Symbol	Range	Unit
Air pressure	$I_1{}^{S2}$	[0.4, 0.7]	MPa
Preload	$I_2{}^{S2}$	[2.0, 8.5]	KN
Press force	$O_1{}^{S2}$	[100, 145]	N

$$\mathbf{R}_2^o = (I_1^{S2}, I_2^{S2}, O_1^{S2}) \tag{6.11}$$

(3) Quantification of structural adaptability of the modified open interface

In the modified design, parts $P^I{}_2$, $P^I{}_3$, $P^{IC}{}_1$, and $P^{IC}{}_3$ shown in Table 6.40 and Fig. 6.28, are selected based on the international standard, corresponding evaluation factors of adaptability are 1, parts $P^I{}_1$ and $P^{IC}{}_2$ use an enterprise standard, and corresponding evaluation factors of the adaptability are selected as 0.2. The structural adaptability of the modified open interface is calculated as [14]:

$$A_S = \frac{4.4}{6} = 0.73 \tag{6.12}$$

(4) Quantification of manufacturing adaptability of the modified open interface

As shown in Fig. 6.25, the modified open interface is composed of bolts, interface disk component, and interface connector part. The manufacturing cost of parts of the interface connector, $C_{M(P^C)}$, is 50 RMB. The assembling cost of parts of the interface connector, $C_{A(R^C)}$, is 25 RMB. Therefore, the manufacturing cost of the interface connector $C_M = C_{M(P^C)} + C_{A(R^C)} = 75$ RMB, or 0.075 thousand RMB. The manufacturing adaptability of the modified open interface is calculated as [14]:

$$A_M = e^{-0.075} = 0.93 \tag{6.13}$$

(5) Quantification of operational adaptability of the modified open interface

For the modified design, assembling relations between the open interface and connector are shown in Table 6.40. Based on data collected in the manufacturing site, the assembling cost of the platform interface and interface connector, C_A, is 50 RMB, and the disassembling cost of the platform interface and interface connector, C_D, is to 20 RMB. Therefore, the cost for assembling and disassembling of the platform interface and interface connector is $C_O = C_A + C_D = 70$ RMB, or 0.07 thousand RMB. The operational adaptability of the modified open interface, A_O, is calculated as [14]:

$$A_O = e^{-0.07} = 0.93 \tag{6.14}$$

Comparing with the initial design of the open interface, adaptability of the redesigned open interface has been enhanced. Results of this case study show that the newly developed method can be used to: (1) quantify the four types of adaptabilities of the open interface; (2) identify particular interface components that have the high influence on the interface adaptability; and (3) enhance the adaptability of an open interface effectively.

6.6 Design of an Adaptable Equipment for Testing of Generators

The application provided in this section is based on the research results developed by Xue et al. [13].

6.6.1 *The Design Problem*

The problem of this case study was to design an equipment to test the horizontal wind power generators which were manufactured later on in a company in the next 9 years. Development of the testing equipment is divided into three phases as shown in Table 6.42. In the first phase, four types of wind power generators will be tested. A motor will be used to generate the required power to replace the wind power for the wind power generator. Due to the low power output level, the vibrations caused by wind forces and the wear caused by frictions are not considered. Each type of the wind power generators will be tested for 700–780 h during this time period. The minimum testing time for all four types of generators in each phase is 3000 h. In the second phase, due to the increase of the required power level, a more powerful motor is selected. In addition, vibrations caused by the wind forces need to be considered. In the third phase, in addition to the increases of motor power and vibration forces, the influence of the high temperature caused by frictions should also be considered.

The requirements for the minimum motor powers and vibration forces in the 3 phases are given in Table 6.43. In this work, the total efficiency considering the motor and the transmission mechanism is assumed to be 90%. Therefore, the minimum power of the motor required in Phase I is calculated by 1000 KW/90% = 1111 KW. The minimum power levels for Phase II and Phase III are obtained in the same way.

Table 6.42 Development of the testing equipment

Phase	Phase I	Phase II	Phase III
Years	1–3	4–6	7–9
Wind Power Generator Series	600 KW, 750 KW, 850 KW, 1000 KW	1,200 KW, 1,300 KW, 1,500 KW, 1,800 KW	2,000 KW, 2,500 KW, 3,000 KW, 5,000 KW
Vibrations Caused by Forces	Not Considered	Mild Forces Considered	Severe Forces Considered
Wear Caused by High Temperature	Not Considered	Not Considered	Considered
Testing time for each series	700–780 h	700–780 h	700–780 h
Minimum total testing time	3000 h	3000 h	3000 h

Table 6.43 Two requirements in the three phases

Phase	Phase I	Phase II	Phase III
Years	1–3	4–6	7–9
Minimum Power	1111 KW	2000 KW	5556 KW
Peak Force	0 N	12,000 N	17,500 N

The sinusoidal forces are used to simulate the vibration forces caused by winds. The peak values of the sinusoidal forces in Phase II and Phase III are given in Table 6.43.

6.6.2 The Optimal Adaptable Design

Three configurations are designed for the testing equipment to be used in the three phases, as shown in Fig. 6.26. The configuration of the equipment in Phase I is composed of an electric motor, the wind power generator, a gear box, 2 coupling devices, an adaptable table, and modular fixtures. Interfaces are designed on the table to install the vibration force generators and the temperature control unit in the later phases. In Phase II, two vibration force generators are added to the two ends of the table to create the vibration forces. In Phase III, the temperature control unit is installed on the table to reduce the temperature of the power generator. In addition, the vibration force generators are also upgraded to provide more forces in Phase III.

Among all the modules of this testing equipment, the table, the vibration force generators, and the modular fixture are designed as adaptable modules. Figure 6.27 shows the adaptable table and adaptable vibration force generator designed in this application. Special interfaces are created on the table for attaching the vibration force generators and the temperature control unit in the Phase II and Phase III of this equipment (Fig. 6.27a). Many slots and holes are also designed on the table for the modular fixture to hold different types of gear boxes, electric motors, and power generators in different phases. The modular fixture is composed of many components and modules for positioning and clamping through reconfiguration of these components and modules. The adaptable vibration force generator uses magnetic coils to convert electric energy into magnetic forces (Fig. 6.27b). In a vibration force generator, the magnetic coil module is composed of many magnetic coil units. When higher peak force is expected, more magnetic coil units then need to be added to the vibration force generator.

The adaptable wind power generator testing equipment is evaluated in terms of costs. In this case study, two types of costs are considered: equipment cost and operation cost.

Equipment cost is calculated by adding the costs of all the required modules for the three phases of this equipment. Suppose N phases are considered, and C_{ij} is the j-th module in the i-th phase that is used for the first time in the adaptable product, the cost of the adaptable equipment can then be calculated by:

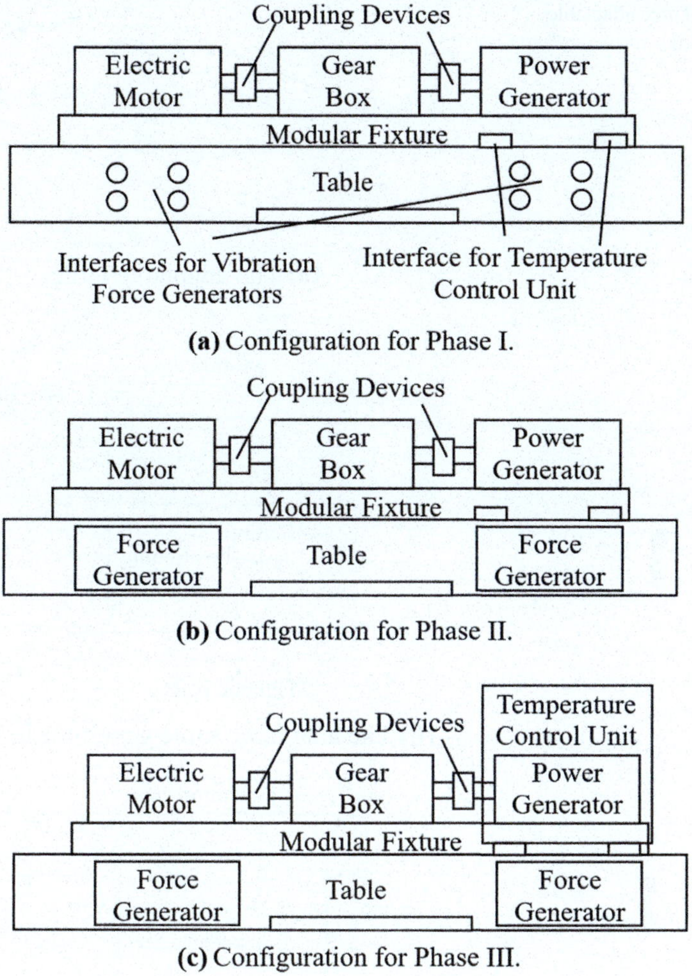

Fig. 6.26 Three configurations of the adaptable equipment in the three phases [13]

$$C_{\text{equipment}} = \sum_{i=1}^{N} \sum_{j=1}^{M_i} C_{ij} \tag{6.15}$$

where M_i is the number of new modules used in the i-th phase. The modules that can be used in different phases and their costs are given in Table 6.44. In this case study, the Electric Motor 1 with the maximum power of 1200 KW can only be used in Phase I. The Electric Motor 2 with the maximum power of 2000 KW can be used in both Phase I and Phase II. The Electric Motor 3 with the maximum power of 5600 KW can be used in all the three phases. Different gear boxes and coupling devices with different functions can also be used in different phases as shown in Table 6.44.

Fig. 6.27 Three adaptable modules [13]

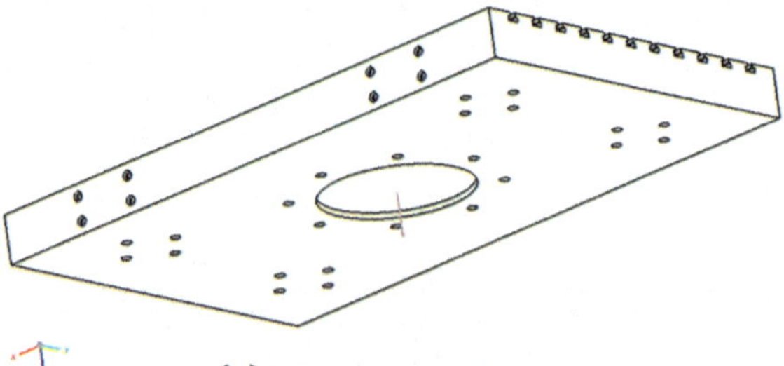

(a) The adaptable table.

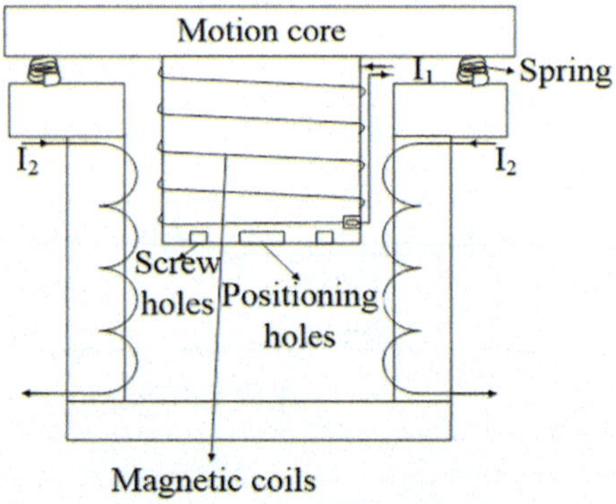

(b) The adaptable force generator in Phase II.

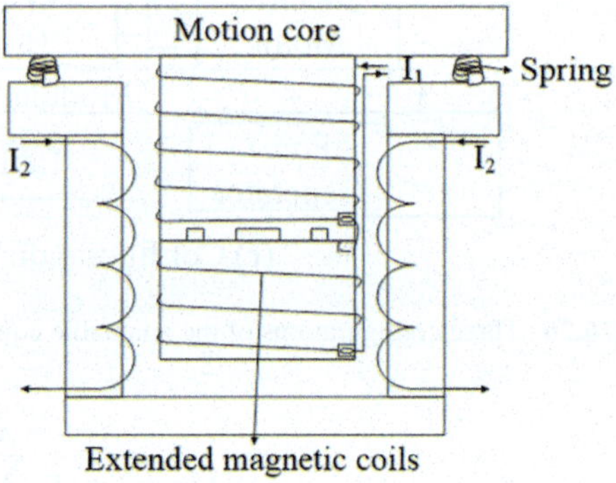

(c) The adaptable force generator in Phase III.

Operation cost is calculated by adding all the costs of electricity to run the electric motors. The cost rate of electricity is determined by the load factor which is the ratio between the required power and the maximum available power. The relations between load factors and electricity cost rates are given in Fig. 6.28. In this case study, the ratio between the required power of the testing wind power generator and the power of the electric motor is used as the load factor. The total operation cost of the adaptable

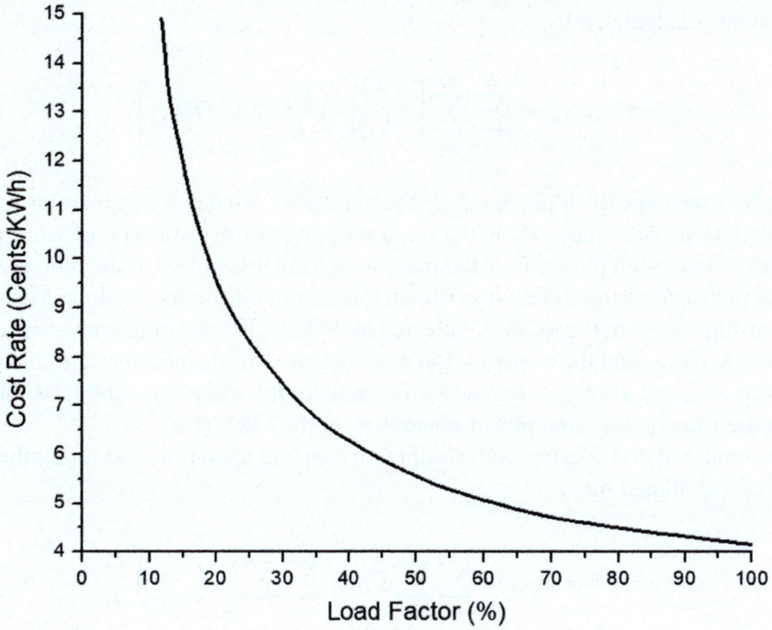

Fig. 6.28 Relationships between load factors and cost rates (Tennessee Valley Authority and Distributors of TVA Electric Power)

Table 6.44 Modules for the adaptable equipment

Category	Phase	Module	Cost ($)
Electric motor	I	Electric motor 1 (1200 KW, 1,500 rpm)	105,000
	I, II	Electric motor 2 (2000 KW, 750 rpm)	180,000
	I, II, III	Electric motor 3 (5600 KW, 600 rpm)	500,000
Gear box	I	Gear box 1	4000
	I, II	Gear box 2	6000
	I, II, III	Gear box 3	8000
Coupling devices	I	Coupling set 1	1000
	I,II	Coupling set 2	1500
	I,II,III	Coupling set 3	2000
Fixture	I, II, III	Modular fixture	15,000
Table	I, II, III	Adaptable table	30,000
Vibration force generators	II, III	Adaptable vibration force generator set	30,000
	III	Additional magnetic coil units	10,000
Temperature control unit	III	Temperature control unit	8000

equipment is calculated by:

$$C_{\text{operation}} = \sum_{i=1}^{N} \sum_{j=1}^{L_i} \left[f\left(\frac{P_{ij}}{P_i} \right) \frac{P_{ij}}{E} (1 + \alpha T_i) t_{ij} \right] \tag{6.16}$$

where N is the number of phases, L_i is the number of wind power generator types to be tested in the i-th phase, P_{ij} is the required power of the j-th type of wind power generators in the i-th phase, P_i is the maximum available power of the motor used in the i-th phase, *f()* is the function to obtain the cost rate from the load ratio using the relationships given in Fig. 6.28, E, selected as 90%, is the total efficiency considering the electric motor and the transmission mechanism, α is the inflation rate selected as 2%/year, T_i is the average life-cycle time value in the i-th phase, and t_{ij} is the time to test the j-th type of wind power generators in the i-th phase.

The total cost considering both equipment cost and operation cost for all the three phases is calculated by:

$$C_{\text{total}} = C_{\text{equipment}} + C_{\text{operation}} = \sum_{i=1}^{N} \sum_{j=1}^{M_i} C_{ij} + \sum_{i=1}^{N} \sum_{j=1}^{L_i} \left[f\left(\frac{P_{ij}}{P_i} \right) \frac{P_{ij}}{E} (1 + \alpha T_i) t_{ij} \right] \tag{6.17}$$

In this case study, many design solution candidates can satisfy the design requirements. For example, all the three motors can be used in Phase I. When Electric Motor 1 is used in Phase I, this motor needs to be changed to either Motor 2 or Motor 3 in Phase II, because the Electric Motor 1 can satisfy the requirements only in Phase I. Although the use of a more powerful motor in an earlier phase can reduce the number of the totally required motors, it can also increase the energy cost rate due to the lower load factor of the powerful motor. In the same way, all the 3 gear boxes and 3 sets of coupling devices can be used in Phase I. The number of configurations for Phase I is calculated as $3 \times 3 \times 3 = 27$ considering the alternatives of 3 electric motors, 3 gear boxes, and 3 sets of coupling devices. For Phase II, the number of configurations is calculated as $2 \times 2 \times 2 = 8$. For Phase III, only one configuration can be selected. Therefore, the total number of design configuration candidates considering the 3 phases is calculated as $27 \times 8 \times 1 = 216$.

For each design configuration candidate, the testing times are selected to test different generators in different phases. In this work, parameter optimization is carried out to identify the optimal testing time parameters to minimize the total cost. For example, when the following design configuration candidate, $\boldsymbol{C}_i(T)$ $(1 \leq i \leq 216)$, is created.

Phase I: Electric Motor 1, Gear Box 1, Coupling Set 1, Modular Fixture, Adaptable Table

Phase II: Electric Motor 2, Gear Box 2, Coupling Set 2, Modular Fixture, Adaptable Table, Adaptable Vibration Generator Set

Phase III: Electric Motor 3, Gear Box 3, Coupling Set 3, Modular Fixture, Adaptable Table, Adaptable Vibration Generator Set, Additional Magnetic Coil Units, Temperature Control Unit

For this design configuration candidate, testing time parameters are selected as variables.

$$\boldsymbol{P}_i(T) = (t_{11}, t_{12}, t_{13}, t_{14}, t_{21}, t_{22}, t_{23}, t_{24}, t_{31}, t_{32}, t_{33}, t_{34}) \tag{6.18}$$

where t_{jk} is the time to test the k-th type of wind power generators in the j-th phase.

Parameter optimization for this design configuration candidate is modeled by:

$$\begin{aligned} &\min_{w.r.t.\boldsymbol{P}_i(T)} C_{\text{total}} \\ &\text{Subject to :} \\ &\qquad 700 \text{ h} \leq t_{jk} \leq 780 \text{ h},\ j = 1, 2, 3;\ k = 1, 2, 3, 4 \\ &\qquad \sum_{k=1}^{4} t_{jk} \geq 3000 \text{ h},\ j = 1, 2, 3 \end{aligned} \tag{6.19}$$

The optimal testing time parameters are obtained as:

Phase I: $t_{11} = 780$ h, $t_{12} = 780$ h, $t_{13} = 740$ h, $t_{14} = 700$ h
Phase II: $t_{21} = 780$ h, $t_{22} = 780$ h, $t_{23} = 740$ h, $t_{24} = 700$ h
Phase III: $t_{31} = 780$ h, $t_{32} = 780$ h, $t_{33} = 740$ h, $t_{34} = 700$ h

The minimum total cost for this design configuration candidate is obtained as:

$$O_i(T) = \$1,897,500$$

Among all the 216 design configuration candidates, the best design configuration candidate, ***C****(T)*, is achieved through configuration optimization.

$$\min_{w.r.t.\boldsymbol{C}_i(T)} \{O_1(T), O_2(T), \ldots, O_{216}(T)\} \tag{6.20}$$

The optimal design, ***C****(T)*, is identified as.

Phase I: Electric Motor 2, Gear Box 3, Coupling Set 3, Modular Fixture, Adaptable Table
Phase II: Electric Motor 2, Gear Box 3, Coupling Set 3, Modular Fixture, Adaptable Table, Adaptable Vibration Generator Set
Phase III: Electric Motor 3, Gear Box 3, Coupling Set 3, Modular Fixture, Adaptable Table, Adaptable Vibration Generator Set, Additional Magnetic Coil Units, Temperature Control Unit

The equipment cost, the operation cost, and the minimum total cost are calculated as

$$C_{\text{total}} = C_{\text{equipment}} + C_{\text{operation}} = \$783{,}000 + \$1{,}035{,}900 = \$1{,}818{,}900 \quad (6.21)$$

It should be noted that instead of Electric Motor 1, Electric Motor 2 is used in Phase I. Although the Electric Motor 2 requires higher operation cost in Phase I due to the higher energy cost rate caused by lower load factor, the electric motor doesn't need to be upgraded in Phase II since the Electric Motor 2 can satisfy the requirements in both Phase I and Phase II.

6.6.3 *Comparative Studies*

The obtained optimal adaptable design solution is compared with design solutions created using other design methods. In this case study, the following 3 solutions created using 3 other design methods are considered.

- Traditional design with three pieces of equipment for three phases
- Traditional design with the best equipment for three phases
- Adaptable design without optimization

(1) *Traditional design with three pieces of equipment for three phases*

In this design, instead of using adaptable modules, different fixtures, tables, and vibration force generators are used for different phases. Therefore, 3 pieces of testing equipment are designed for the 3 phases. The costs of these non-adaptable modules are given in Table 6.45. In addition, the electric motors, gear boxes, and coupling devices with the lowest costs while satisfying the requirements are used in these three phases. The modules of the electric motors, gear boxes, and coupling devices are the same as those given in Table 6.44.

The three pieces of equipment are modeled by three configurations as shown in Fig. 6.29.

Phase I: Electric Motor 1, Gear Box 1, Coupling Set 1, Fixture 1, Table 1

Table 6.45 Modules for the three pieces of equipment

Category	Phase	Module	Cost ($)
Fixture	I	Fixture 1	5000
	II	Fixture 2	8000
	III	Fixture 3	10,000
Table	I	Table 1	15,000
	II	Table 2	20,000
	III	Table 3	25,000
Vibration force generators	II	Vibration force generator set 2	20,000
	III	Vibration force generator set 3	35,000

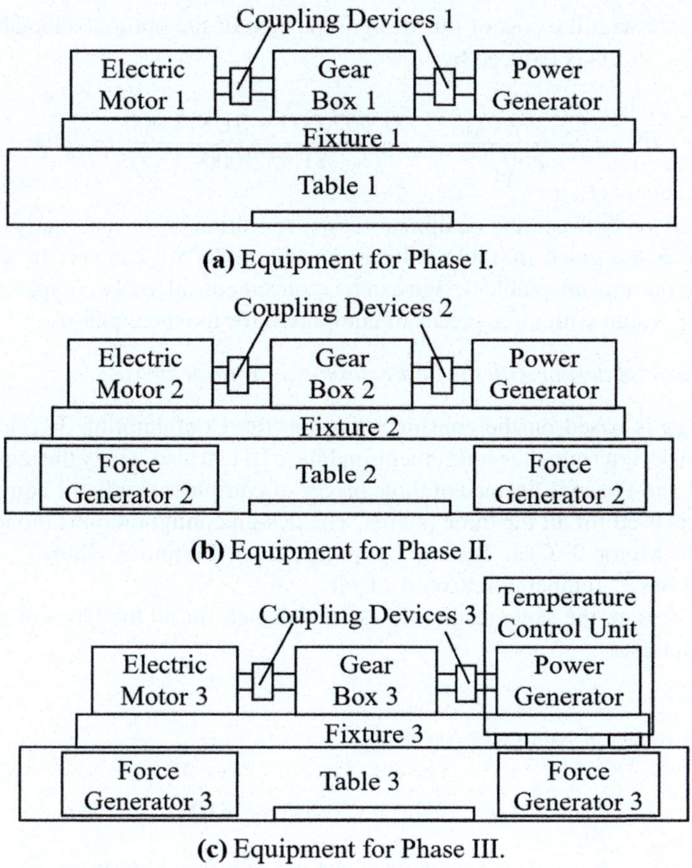

Fig. 6.29 Three pieces of equipment for the three phases [13]

Phase II: Electric Motor 2, Gear Box 2, Coupling Set 2, Fixture 2, Table 2, Vibration Generator Set 2

Phase III: Electric Motor 3, Gear Box 3, Coupling Set 3, Fixture 3, Table 44, Vibration Generator Set 3, Temperature Control Unit

In this design, the same testing times are selected for all the types of generators in the 3 phases.

Phase I: $t_{11} = 750$ h, $t_{12} = 750$ h, $t_{13} = 750$ h, $t_{14} = 750$ h
Phase II: $t_{21} = 750$ h, $t_{22} = 750$ h, $t_{23} = 750$ h, $t_{24} = 750$ h
Phase III: $t_{31} = 750$ h, $t_{32} = 750$ h, $t_{33} = 750$ h, $t_{34} = 750$ h

The equipment cost, the operation cost, and the minimum total cost are calculated as:

$$C^{(1)}_{\text{total}} = C^{(1)}_{\text{equipment}} + C^{(1)}_{\text{operation}} = \$953{,}500 + \$1{,}003{,}500 = \$1{,}957{,}000 \quad (6.22)$$

Compared with the cost of this design, the cost of the optimal adaptable design given in Eq. (6.21) is reduced by

$$\eta^{1)} = \frac{C_{\text{total}}^{(1)} - C_{\text{total}}}{C_{\text{total}}^{(1)}} = \frac{\$1{,}957{,}000 - \$1{,}818{,}900}{\$1{,}957{,}000} = 7.1\% \qquad (6.23)$$

Comparison between the equipment costs, operation costs, and total costs in the three phases are given in Table 6.46. From this table, we can see the equipment cost of the optimal adaptable design can be reduced considerably compared with the traditional design with three pieces of equipment for the three phases.

(2) *Traditional design with the best equipment for three phases*

This design is based on the consideration that the Configuration 3 created in the traditional design from the requirements in Phase III can also satisfy the requirements in Phase I and Phase II. Instead of three pieces of equipment, only the equipment for Phase III is used for all the three phases. The design configuration is modeled by

Electric Motor 3, Gear Box 3, Coupling Set 3, Fixture 3, Table 3, Vibration Generator Set 3, Temperature Control Unit

In this design, the same testing times are selected for all the types of generators in the 3 phases.

Table 6.46 Results of comparative studies

		Phases I	Phase II	Phase III	Total cost	Improvement
Optimal Adaptable design	Equipment cost ($)	235,000	30,000	518,000	783,000	
	Operation cost ($)	170,900	244,100	620,900	1,035,900	
	Total cost ($)	405,900	274,100	1,138,900	**1,818,900**	
Traditional design with 3 equipment for 3 phases	Equipment cost ($)	130,000	235,500	588,000	953,500	7.1%
	Operation cost ($)	132,600	245,100	625,800	1,003,500	
	Total cost ($)	262,600	480,600	1,213800	***1,957,000***	
Traditional design with configuration III for three phases	Equipment cost ($)	588,000	0	0	588,000	8.0%
	Operation cost ($)	336,500	426,100	625,800	1,388,400	
	Total cost ($)	924,500	426,100	625,800	***1,976,400***	
Adaptable design without optimization	Equipment cost ($)	155,000	217,500	528,000	900,500	4.5%
	Operation cost ($)	132,600	245,100	625,800	1,003,500	
	Total cost ($)	287,600	462,600	1,153,800	***1,904,000***	

$$t_{i1} = 750\text{ h}, t_{i2} = 750\text{ h}, t_{i3} = 750\text{ h}, t_{i4} = 750\text{ h } (i = 1, 2, 3)$$

The equipment cost, the operation cost, and the minimum total cost are calculated as

$$C_{\text{total}}^{(2)} = C_{\text{equipment}}^{(2)} + C_{\text{operation}}^{(2)} = \$588{,}000 + \$1{,}388{,}400 = \$1{,}976{,}400 \quad (6.24)$$

Compared with the cost of this design, the cost of the optimal adaptable design given in Eq. (6.21) is reduced by

$$\eta^{(2)} = \frac{C_{\text{total}}^{(2)} - C_{\text{total}}}{C_{\text{total}}^{(2)}} = \frac{\$1{,}976{,}400 - \$1{,}818{,}900}{\$1{,}976{,}400} = 8.1\% \quad (6.25)$$

Comparison between the equipment costs, operation costs, and total costs in the three phases are given in Table 6.46. From the data in this table, we can see that although the equipment cost of the optimal adaptable design is higher than the cost of the single equipment for Phase III, the operation cost of the optimal adaptable equipment can be reduced considerably due to the flexibility of the adaptable design to use less expensive modules and less powerful electric motors in earlier phases.

(3) Adaptable design without optimization

In this design, fixtures, tables, and vibration force generators are designed as adaptable modules, the same as those given in the optimal adaptable design. Instead of using optimization to achieve the optimal configurations and parameters, some heuristics rules are used to identify the configurations and parameters of the adaptable equipment for the three phases. Because the energy cost rates are determined by the load factors of the electric motors, the electric motors that satisfy the minimum requirements are used in these three phases for reducing the operation costs in this design. In addition, the gear boxes and coupling devices with the lowest costs while satisfying the requirements are used in these three phases. The configurations for the three phases are modeled by:

Phase I: Electric Motor 1, Gear Box 1, Coupling Set 1, Modular Fixture, Adaptable Table

Phase II: Electric Motor 2, Gear Box 2, Coupling Set 2, Modular Fixture, Adaptable Table, Adaptable Vibration Generator Set

Phase III: Electric Motor 3, Gear Box 3, Coupling Set 3, Modular Fixture, Adaptable Table, Adaptable Vibration Generator Set, Additional Magnetic Coil Units, Temperature Control Unit.

The same testing times are selected for all the types of generators in the 3 phases.

Phase I: $t_{11} = 750$ h, $t_{12} = 750$ h, $t_{13} = 750$ h, $t_{14} = 750$ h
Phase II: $t_{21} = 750$ h, $t_{22} = 750$ h, $t_{23} = 750$ h, $t_{24} = 750$ h
Phase III: $t_{31} = 750$ h, $t_{32} = 750$ h, $t_{33} = 750$ h, $t_{34} = 750$ h.

Table 6.47 Comparison of equipment costs considering only adaptable modules

Module	Adaptable design ($)	Traditional design ($)
Table	30,000	60,000
Fixture	15,000	23,000
Vibration force Generators	40,000	55,000
Total cost	85,000	138,000
Improvement		38.4%

The equipment cost, the operation cost, and the minimum total cost are calculated as

$$C_{\text{total}}^{(3)} = C_{\text{equipment}}^{(3)} + C_{\text{operation}}^{(3)} = \$900{,}500 + \$1{,}003{,}500 = \$1{,}904{,}000 \quad (6.26)$$

Compared with the cost of this design, the cost of the optimal adaptable design given in Eq. (6.21) is reduced by

$$\eta^{(3)} = \frac{C_{\text{total}}^{(3)} - C_{\text{total}}}{C_{\text{total}}^{(2)}} = \frac{\$1{,}904{,}000 - \$1{,}818{,}900}{\$1{,}904{,}000} = 4.5\% \quad (6.27)$$

In this case study application, among all the modules for the wind power testing equipment, only the table, fixture, and vibration force generators are designed as adaptable modules. A comparative study between the optimal adaptable design and the traditional design with 3 equipments for the 3 phases in terms of equipment costs considering only the 3 adaptable modules is also carried out. The result of this comparative study is given in Table 6.47. Compared with the traditional design, the cost of the adaptable design is reduced by

$$\eta = \frac{C_{\text{traditional}} - C_{\text{adaptable}}}{C_{\text{traditional}}} = \frac{\$138{,}000 - \$85{,}000}{\$138{,}000} = 38.4\% \quad (6.28)$$

From this table, we can see the equipment cost can be reduced considerably by the development of the adaptable equipment which is used to replace the 3 pieces of equipment created using the traditional design method.

6.7 Adaptable Design of a Rotation Table

This example shows assembly planning for a rotation table of the adaptable industrial paper-bag folding machine [7]. Figure 6.30 shows the 3D model of main frame M_1. Figure 6.31 is the 3D model of rotation table M_2.

There are eight bag clamping devices that have the same number of parts and same shape. They are alike-type assemblies and can be divided from M_2. $M_{2\text{-}1}$ is a bag

Fig. 6.30 Main frame M_1

Fig. 6.31 Rotation table M_2

clamping device. Ring gauge $M_{2\text{-}2}$ is independent from the rotation table module. The remainder is represented by $M_{2\text{-}3}$. M_1 is chosen as the support platform module and P_{14} is the base part. The detail of $M_{2\text{-}3}$ is shown in Fig. 6.32.

6.7.1 Assembly Modelling

A BoM of the rotation table is formed as shown in Fig. 6.33. Figure 6.34 is structural diagrams of the rotation table. Table 6.48 is its assembly constraint matrix WS_1.

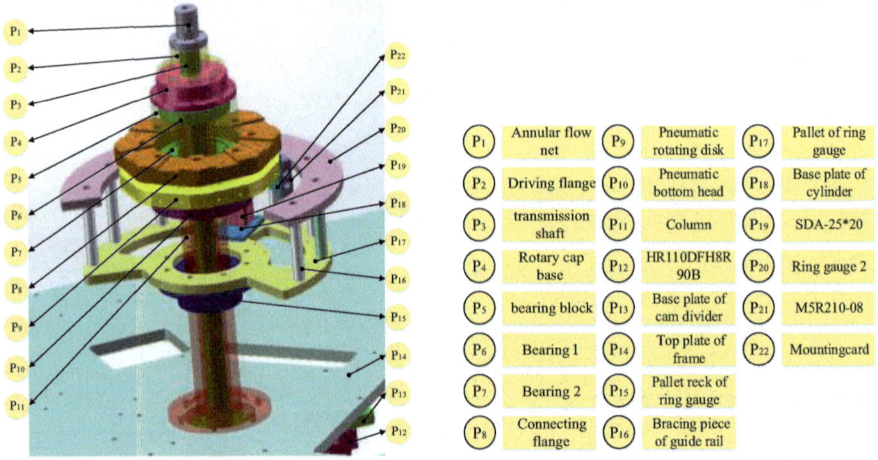

Fig. 6.32 Details of rotation table $M_{2\text{-}3}$

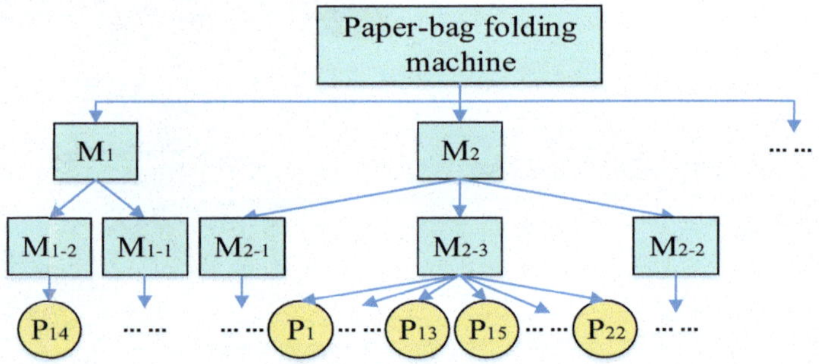

Fig. 6.33 BoM of the paper-bag folding machine

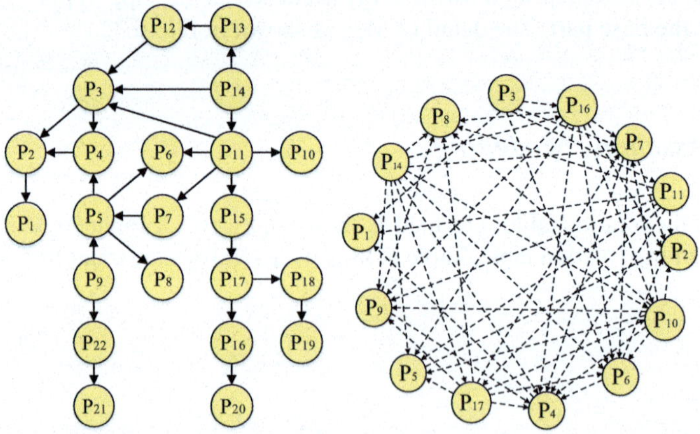

Fig. 6.34 Structural diagrams of the rotation table

Table 6.48 Assembly constraint matrix WS_1

	1	2	3	4	5	6	7	8	9	10	11	12	13	14	15	16	17	18	19	20	21	22
1	0	-1	-9	-9							-9											
2	1	0	-1	-1	-9	-9	-9			-9	-9				-9							
3	9	1	0	1		9	9				9	-1			9							
4	9	1	-1	0	-1	-9	-9		-9	-9	-9			-9	-9		-9					
5		9		1	0	1	-1	1	-1	-9	-9			-9	-9		-9					
6		9	-9	9	-1	0	-9			-9	-1			-9	-9		-9					
7		9	-9	9	1	9	0			-9	-1			-9	-9		-9					
8					-1			0	-9	-9	-9			-9	-9		-9					
9				9	1			9	0	-9	-9			-9	-9		-9					1
10		9		9	9	9	9	9	9	0	-1			-9	-9		-9					
11	9	9	-9	9	9	1	1	9	9	1	0			-1	1		9					
12			1									0	-1									
13												1	0	-1								
14				9	9	9	9	9	9	9	1		1	0	9		9					
15		9	-9	9	9	9	9	9	9	9	-1			-9	0		1					
16																0	-1			1		
17				9	9	9	9	9	9	9	-9			-9	-1	1	0	1				
18																	-1	0	1			
19																		-1	0			
20																-1				0		
21																					0	-1
22									-1												1	0

6.7.2 *Evaluation of Modules Assembled in a Single Subassembly*

A direction matrix of the rotation table is generated as shown in Table 6.49. The element 0 exists in all directions. All these parts cannot be assembled as a module in one direction. The module is divided as follows. Parts including P_{12} and P_{13} are represented by $M_{2\text{-}3\text{-}1}$, and the other is represented by $M_{2\text{-}3\text{-}2}$. But P_3 of $M_{2\text{-}3\text{-}2}$ is surrounded by P_1, P_2, P_4, P_{11}, P_{14} and P_{12}. P_1, P_2, P_4, P_{11} and P_{12} belong to $M_{2\text{-}3}$, while P_{14} belongs to M_1, those parts belong to different modules. Thus $M_{2\text{-}3}$ cannot be assembled as a single subassembly. P_3 and other parts must be assembled after P_3 is separated from module $M_{2\text{-}3}$. This process repeats until all remaining parts can be assembled as a subassembly. The adjusted BoM is shown in Fig. 6.35.

Table 6.49 Direction matrix of the rotation table

0	1	2	3	4	5	6	7	8	9	10	11	12	13	14	15	16	17	18	19	20	21	22
x	1	0	0	0	0	0	0	0	0	0	0	1	0	1	0	1	1	1	1	0	1	1
y	0	0	0	0	0	0	0	0	0	0	0	1	1	0	0	0	0	0	0	0	1	1
z	1	0	0	0	0	0	0	0	0	0	0	1	0	1	0	1	1	1	1	0	1	1
-x	1	0	0	0	0	0	0	0	0	0	0	1	0	1	0	1	1	1	1	1	0	0
-y	1	1	1	1	1	1	1	1	1	1	1	0	0	1	1	1	1	1	1	1	1	1

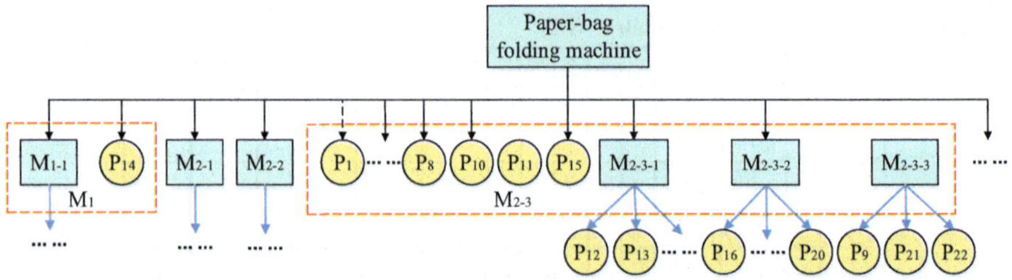

Fig. 6.35 Adjusted bill of materials

6.7.3 *Assembly Sequence Planning for Parts Inside Module of the Rotational Table*

P_{14} is chosen as the base part. Parts that meet the parallel assembly are searched in Step 2. P_{23} is then chosen to be a support part of P_3 in Step 3. After then, an assembly constraint matrix WSk23 is generated as shown in Table 6.50. Set elements in WSk23 as 0, the assembly sequence matrix WSk23-1 can be formed.

Table 6.50 Assembly constraint matrix of parts

23	14	13	12	3	11	15	17	16	20	18	19	10	7	9	22	21	5	8	6	4	2	1
14	0	1			1	9	9					9	9	9			9	9	9	9		
13	-1	0	1																			
12		-1	0	1																		
3			-1	0	9	9							9						9	1	1	9
11	-1			-9	0	1	9					1	1	9			9	9	1	9	9	9
15	-9			-9	-1	0	1					9	9	9			9	9	9	9	9	
17	-9				-9	-1	0	1		1		9	9	9			9	9	9	9		
16							-1	0	1													
20								-1	0													
18							-1			0	1											
19										-1	0											
10	-9				-1	-9	-9					0	9	9			9	9	9	9	9	9
7	-9			-9	-1	-9	-9					-9	0				1		9	9	9	
9	-9				-9	-9	-9					-9		0	1		1	9		9		
22														-1	0	1						
21															-1	0						
5	-9				-9	-9	-9					-9	-1	-1			0	1	1	1	9	
8	-9				-9	-9	-9					-9		-9			-1	0				
6	-9			-9	-1	-9	-9					-9	-9				-1		0	9	9	
4	-9			-1	-9	-9	-9					-9	-9	-9			-1		-9	0	1	-9
2				-1	-9	-9						-9	-9				-9		-9	-1	0	1
1				-9	-9															9	-1	0

6.7.4 Assembly Sequence Planning for Modules

The assemble constraint matrix WT of modules is formed based on module layers in BoM as shown in Table 6.51. The final assembly sequence can be planned as follows.

Only P_{14} is listed in the first level, it connects to $M_{1\text{-}1}$ by analyzing WT0, that is Mh1 = Mb2 = $\{P_{14}\}$. The corresponding row and column elements of P_{14} in WT0 are exchanged with the second row and column elements to form WT1. They are then set to 0 to get $WT_{1\text{-}1}$. All remaining modules that can be assembled are put into the second layer as $M_{2\text{-}3\text{-}1}$ and $M_{2\text{-}2}$, Mh2 = $\{M_{2\text{-}3\text{-}1}, M_{2\text{-}2}\}$. While Mb2 = $\{M_{2\text{-}3\text{-}1}, M_{2\text{-}2}\}$, $M_{2\text{-}3\text{-}1}$ belongs to customized modules and $M_{2\text{-}2}$ is the platform module. Therefore, $M_{2\text{-}2}$ is assembled firstly in this step. WT2 and $WT_{2\text{-}1}$ are formed. This process continues until all modules are assembled according to Fig. 6.35.

After this step, H = {Mh1, ..., Mh(17)}, A = {Ma1, ..., Ma(17)}. The final assembly sequences are formed as follows.

(a) $M_{1\text{-}1}$, P_{14}, $M_{2\text{-}2}$, $M_{2\text{-}3\text{-}1}$, P_3, P_{11}, P_{15}, $M_{2\text{-}3\text{-}2}$, P_{10}, P_7, $M_{2\text{-}3\text{-}3}$, P_5, P_6, P_8, P_4, P_2, P_1.

(b) $M_{1\text{-}1}$, P_{14}, $M_{2\text{-}2}$, $M_{2\text{-}3\text{-}1}$, P_3, P_{11}, P_{15}, $M_{2\text{-}3\text{-}2}$, P_{10}, $M_{2\text{-}3\text{-}3}$, P_7, P_5, P_6, P_8, P_4, P_2, P_1.

Table 6.51 Assembly constraint matrix WT of modules

		1	2	3	4	5	6	7	8	9	10	11	12	13	14	15	16	17	18
		P_1	P_2	P_3	P_4	P_5	P_6	P_7	P_8	$M_{2\text{-}3\text{-}3}$	P_{10}	P_{11}	$M_{2\text{-}3\text{-}1}$	$M_{1\text{-}1}$	P_{14}	P_{15}	$M_{2\text{-}3\text{-}2}$	$M_{2\text{-}1}$	$M_{2\text{-}2}$
1	P_1	0	-1	-9	-9							-9							
2	P_2	1	0	-1	-1	-9	-9	-9			-9	-9				-9			
3	P_3	9	1	0	1		9	9				9	-1			9			
4	P_4	9	1	-1	0	-1	-9	-9		-9	-9	-9			-9	-9	-9		
5	P_5		9		1	0	1	-1	1	-1	-9	-9			-9	-9	-9		
6	P_6		9	-9	9	-1	0	-9	1		-9	-1			-9	-9	-9		
7	P_7		9	-9	9	1	9	0			-9	-1			-9	-9	-9		
8	P_8					-1			0	-9	-9	-9			-9	-9	-9	1	
9	$M_{2\text{-}3\text{-}3}$				9	1			9	0	-9	-9			-9	-9	-9		
10	P_{10}		9		9	9	9	9	9	9	0	-1			-9	-9	-9		
11	P_{11}	9	9	-9	9	9	1	1	9	9	1	0			-1	1	9		
12	$M_{2\text{-}3\text{-}1}$			1									0						
13	$M_{1\text{-}1}$													0	1				
14	P_{14}				9	9	9	9	9	9	9	1		-1	0	9	9		1
15	P_{15}		9	-9	9	9	9	9	9	9	9	-1			-9	0	1		
16	$M_{2\text{-}3\text{-}2}$				9	9	9	9	9	9	9	-9			-9	-1	0		
17	$M_{2\text{-}1}$								-1									0	
18	$M_{2\text{-}2}$														-1				0

6.8 Summary

Adaptable product can be upgraded in the operation stage to satisfy different changes of user requirements. Development of adaptable product can bring both economic and environmental benefits. Adaptable design is the design paradigm which aims to create adaptable products.

Since the introduction of adaptable design, many industrial applications have been implemented to demonstrate the effectiveness of the adaptable design. In this chapter, industrial application examples are provided to illustrate concepts, methods, and tools used for the adaptable design.

References

1. Y. Chen, Q. Peng, P. Gu, Methods and tools for the optimal adaptable design of open-architecture products. Int. J. Adv. Manuf. Technol. **94**, 991–1008 (2018). 0.1007/s00170-017-0925-6
2. Q. Cheng, W. Li, D. Xue, Z. Liu, P. Gu, K. Li, Design of adaptable product platform for heavy-duty gantry milling machines based on sensitivity design structure matrix. J. Mech. Eng. Sci. **231**(24), 4495–4511 (2017)
3. P. Gu, M. Hashemian, Adaptable design. Ann. CIRP **53**(2), 539–557 (2004)
4. C. Hu, Q. Peng, P. Gu, Interface adaptability for an industrial painting machine. Comput.-Aided Des. Appl. **11**(2), 182–192 (2014)
5. C. Hu, Q. Peng, P. Gu, Adaptable interface design for open-architecture products. Comput.-Aided Des. Appl. **12**(2), 1–10 (2015)
6. Y. Koren, S. Hu, P. Gu, Open-architecture products. Ann. CIRP **62**, 719–729 (2013)
7. H. Ma, Q. Peng, J. Zhang, P. Gu, Precedence constraint knowledge-based assembly sequence planning for open-architecture products. Procedia CIRP (2016)
8. Q, Peng, Y. Liu, P. Gu, Improvement of product adaptability by efficient module interactions, in *Proceedings of the ASME Design Engineering Technical Conference*. DETC2014-35183 (2014)
9. Q. Peng, Y. Liu, P. Gu, Z. Fan, Development of an open-architecture electric vehicle using adaptable design, in *Lecture Notes in Mechanical Engineering. Springer International Publishing*, Switzerland (2013)
10. Q. Peng, Y. Liu, J. Zhang, P. Gu, Personalization for massive product innovation using open architecture. Chin. J. Mech. Eng. **31**(1), 34 (2018)
11. N. Suh, *Axiomatic Design: Advances and Applications* (Oxford University Press, New York, 2001)
12. Y. Xu, Y. Chen, G. Zhang, P. Gu, Adaptable design of machine tools structures. Chin. J. Mech. Eng. **21**(3), 7-15 (2008)
13. D. Xue, G. Hua, V. Mehrad, P. Gu, Optimal adaptable design for creating the changeable product based on changeable requirements considering the whole product life-cycle. J. Manuf. Syst. **31**(1), 59–68 (2012)
14. J. Zhang, G. Xue, H. Du, G. Akhil, Q. Peng, P. Gu, Enhancing interface adaptability of open architecture products. Res. Eng. Des. **28**(4), (2017)
15. C. Zhao, Q. Peng, P. Gu, Development of a paper-bag-folding machine using open architecture for adaptability. Proc. Inst. Mech. Eng. Part B J. Eng. Manuf. **229**(S1), 155–169 (2015). https://doi.org/10.1177/0954405414559281

Chapter 7
Concluding Remarks and Development Trends

7.1 Concluding Remarks

Manufacturing challenges resulted from increasing competition for better product functionality, quality, features, customization, environmental friendliness, lower cost, and shorter delivery time cannot be completely addressed by advanced manufacturing technologies alone as some of them are originated from product design. To address these multiple and conflict requirements and objectives, this book proposed a new approach called adaptable design. The book discussed the key issues for adaptable design including adaptable product modeling, adaptability evaluation, adaptable design process, and adaptable design tools. Applications of adaptable design are also provided.

The adaptable design (AD) considers the dynamics of market demands, new product introduction and innovation, and therefore allows the introduction of the design innovation easier than the conventional design. The proposed AD architecture, platforms, modules, and adaptable interfaces as well as functional and physical structure independences focus on product development, manufacturing, and associated business processes that provide the opportunity for maintaining and enhancing the product qualities. Such adaptable architecture allows new features or changes in the existing product for customization that can be localized in certain modules or module variants to keep the cost at the competitive level. Because of the adaptabilities, the time to market for developing a new product can be reduced by adapting the existing design to a new design with most or many of the proven manufacturing and business processes. The environmental aspects can be enhanced by incorporating the lifecycle environmental objectives in the design process. Designs and products are ensured to meet environmental requirements through adapting products to new requirements without reproducing products, by reusing of existing machines, modules, or components.

To make a product design more adaptable, the product architecture is critical. The segregated architectures that can be implemented through platform, modular, and

P. Gu et al., *Adaptable Design*, Research on Intelligent Manufacturing,
https://doi.org/10.1007/978-981-99-5869-6_7

special designed adaptable interfaces provide the foundation for adaptable design. In addition to traditional platform and modular design, the platforms and modules designed in the AD context are adaptable and their interfaces are also adaptable to meet new changes required for customers', environmental, and market needs.

It is understandable to make a product more adaptable, its platform and modules should be more adaptable, and interfaces should provide sufficient flexibility in facilitating the realization of adaptations. In addition to the lack of fundamental theories, methodologies, and comprehensive research in adaptable platform, modulization, and interfaces, there are challenges in realization of the regular modular product design considering design complexities, reliability of data, and efficient and effective clustering techniques because the Unsupervised Learning is the core of clustering techniques. This may be resolved by seeking the domain expertise of design rather than adopting raw data for clustering algorithms.

To achieve the desired adaptability in products and designs, the function independence is critical. The independence would not only facilitate the adaptation of the existing designs for new requirements, but make the modification of the existing products easier. Therefore, it is recommended that the physical function requirements are separately treated from other design requirements to ensure the maximum possibility of the function independence. Suh [11, 12] has provided great details of the axiomatic design with the function independence.

Although concepts, methodologies, and applications of adaptable design are provided in this book, there are a number of issues that require continuing research efforts including adaptable platform, module, adaptable interfaces, functional independence as well as associated techniques for realization of AD. Following conclusions are summarized to understand the importance of AD:

- Benefits of adaptable design are primarily related to economic and environmental aspects. Users can adapt an existing product, rather than buying a new one, to achieve new functional requirements. Comparing with remanufacturing and recycling, an adaptable product can be further extended in its lifespan to reduce the environmental waste.
- Some key issues in adaptable design are discussed, including functional modeling, design modeling, design evaluation, and design process modeling. Many existing design methods such as modular design, product platform design, product family design, and mass customization design are effective tools that can be used in adaptable design. In this context, adaptable design is more general, while modular and platform designs can be considered as special applications of adaptable design.
- Adaptable design is not only effective in developing adaptable mechanical products, but it can also be used in other engineering areas to design adaptable systems and processes.

7.2 Development Trends

Adaptable design is a design paradigm for both economic and environmental benefits. The underlying philosophy of adaptable design is the ability to adapt to new requirements and reuse a product and design when circumstances change such as replacing multiple products by one adaptable product with a set of add-on accessories or attachments. Although basic concepts, methods, and tools of adaptable design are reported in this book, following research and application efforts are needed.

(1) Implications of Adaptable Product Architecture

An adaptable product architecture would allow the addition of new features or change of the existing features for customization that can be localized to certain modules or module variants to keep the cost at the competitive level. In order to support competitive product development, further investigation on implications of adaptable architecture is needed to study the influence of the adaptable architecture on product lifecycle aspects. Potential implications of an adaptable architecture include but not limited to the following aspects (Fig. 7.1):

Responds to market requirements: Modules (i.e., replication, crossable, mutable, and crossmutable modules) can be easily changed through standard and adaptable interfaces to rapid respond to market requirements.

Design adaptability: As maximal reusable modules and interfaces can be achieved by an adaptable product architecture to satisfy changing requirements, enhanced design adaptability can be obtained by using adaptable product architecture.

Flexibility of manufacturing system: Identifications of different types of modules and their interfaces could facilitate flexible manufacturing system development as potential changes of manufacturing requirements of crossable, mutable, and crossmutable modules are different.

Product changes and variety: Product with an adaptable architecture can be easily changed and/or adapted to satisfy different changing requirements through reconfiguring different types of modules.

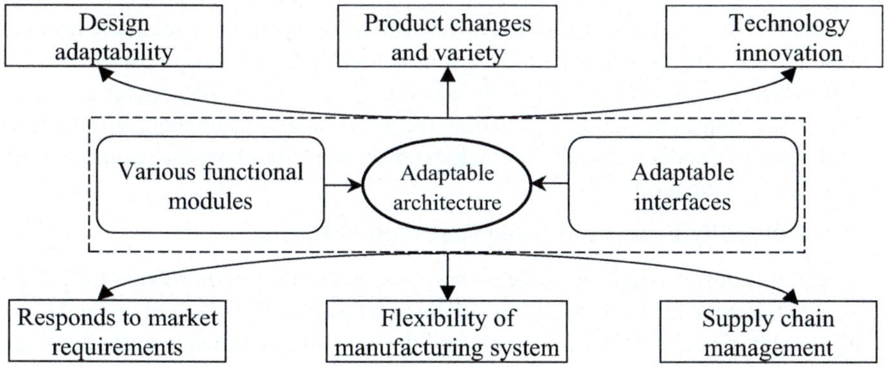

Fig. 7.1 Adaptable architecture and its implications

Supply chain management: Probability of components' changes varies with replication, crossable, mutable, and crossmutable modules. Supply chains of those components and modules should be managed accordingly.

Technology innovation: Mutable and crossmutable modules in an adaptable architecture are with much higher probability to be innovated for competitive product development. Technology innovations for mutable and crossmutable module development are needed.

(2) Product Evolution Mechanism for Adaptable Design

Sophisticated technologies (e.g., network, information, digital twin, and intelligence technologies) bring increased competitions of industries as product features and lifecycle performance information can be easily obtained by customers for comparisons. Consequently, competition results of products in the same category can be reflected by the sales' numbers or market shares. To enhance product competitiveness and preferred market shares, design adaptations need to be carried out by designers, leading to constantly product evolutions.

To support competitive product design and development, investigations of the product evolution mechanism are needed using the game theory by considering customer, functional, and physical domains.

In the customer domain, product specifications and their combinations are compared by customers for supporting product purchase decision-making. Product competitiveness in the market is highly indicated by product sales data and/or market shares. A game model of competing products with different specifications and their combinations can be proposed for modeling of competition relations of those products. Based on the game model of competing products, rules of game analysis of product specifications can be proposed for monitoring customers' selection of product through the specification comparison.

In the functional domain, the game analysis of product specifications (GAPS) can be carried out for enhancing product competitiveness. Mutation, crossover, and reproduction of specifications are modeled as design adaptation operations for the optimization of product competitiveness (i.e., preferred sales numbers and/or market shares).

In the physical domain, with the consideration of dependency relations between specifications and components, product components and their design parameters can be optimized for required specification combinations. By considering both changeable market requirements and potential technology advancements, an adaptable architecture of product can also be achieved through the dynamic game analysis of product specifications.

(3) Adaptability Evaluation Using Information Entropy

Adaptable design is a design process with associated information flows in defining design requirements, generating design solutions through design creativity and information mapping between requirements and design solutions, design evaluation, and design iteration based on the evaluation results. In adaptable design, for each of the above tasks, additional information is required and analyzed so that the adaptation tasks can be anticipated. For example, for current design requirements, potential future changes in design and products should be considered. Another important design task is evaluation. The goal of adaptable design is to enhance the adaptability of designs and products. For different design candidates, evaluation of the adaptability can be used to select the most adaptable one when other criteria are also satisfied. To make the quantitative evaluation of adaptability, a more mathematical rigor approach [13] can develop a comprehensive information entropy-based evaluation method and the associated computing software so that the method can be used in industries for large and complex engineering systems' design.

(4) Artificial Intelligence Applications

Artificial intelligence (AI) techniques have been used in many engineering processes such as engineering design, manufacturing, assembly, inspection, and controls to enhance automation levels of these processes. As discussed in previous chapters, design involves iterative processes from defining design requirements to evaluation of design solutions. The design synthesis and analysis have been repeatedly used to generate ideas and analysis of design selections and decisions. AI techniques can be utilized in both design synthesis and analysis processes. One of the major efforts of AI in design is to learn knowledge embedded in datasets as implicit knowledge. With technologies of the big data, cloud technology, high-performance computing, sensing technology, and Internet of Things (IoT), AI technologies such as Machine Learning (ML) techniques and algorithms have significantly enhanced generating knowledge and reasoning rules, which provide learning methods and approaches to generate solutions to complex engineering design problems. ML is an important component of the growing field of data science. Through the use of statistical methods, algorithms are trained with datasets to conduct classifications or make predictions. It has also been used to uncover key insights in data mining fields. Deep learning (DL) is a special type of ML using available datasets and computing power. According to Nti et al. [9], ML enables machines and systems to learn automatically and can improve the system operation without explicitly instruction. ML can be classified into four main categories: Supervised Learning (SL), Unsupervised Learning (UL), Semi-Supervised Learning (SSL), and Reinforcement Learning (RL).

These ML techniques can assist designers and engineers in product design processes from the identification of product features, determination of engineering specifications, conceptualized design solutions to final detail design, such as using big sale data-driven approach to obtain customer preferences of product features for new product development [15]. With the proposed framework and associated method, product features' characterization and customer preference can be predicted based on online product purchase data and ML techniques. Specifications and components of products are analyzed, and relationships between product specifications and components can then be established for features characterization. Customers' preferred specifications, features, and their combinations are predicted for development of new products.

As design changes often occur during the product development process and some design changes may require significant modifications of product specifications and design parameters, lifecycle performances of the product may be affected [16]. Due to complexities and possibly unknown relationships among specifications and parameters of products, the design change decisions must be carefully evaluated. A Machine Learning-based regression method can be applied to model relationships of product specifications and design parameters. The relationships are preliminary detected via the correlation coefficients-based clustering. A regression learner is trained using an input vector represented by design parameters and an output vector represented by product specifications data contained in the cluster. The trained model is tested using a dataset of new critical design parameters and related product specifications with the least differences from the target specifications to obtain an estimation of best critical design parameter combinations.

It becomes possible to generate design solutions automatically by computers along with the progress of general AI, especially, the recent development of ChatGPT.

(5) Digital Twin

In the product design and manufacturing context, a digital twin is a dynamic virtual copy or a digital representation of a physical object (a product, process, or system). It not only looks like, but behaves identically to the real-world counterpart. The digital twins connect real-world objects with real-world data so that one can better visualize the objects. Digital twins enable crossfunctional teams to collaboratively design, realize, and manipulate complex objects in an immersive and interactive manner.

For a non-existing object in product design, its digital twin can be created using CAD systems. For an existing object without a digital representation, the 3D model of the object is first established. With a digital representation of the real-world object, visualization and analysis can be carried out. When a digital twin is powered by real-time 3D and computer graphics systems, it can organize and present multiple sources of data to support real-time interactive visualization.

Digital twin can also represent dynamic properties of objects such as movements, forces, and interactions in the real world. Developers and users can engage with dynamic contents in virtual environments. In the virtual environment, they can effectively simulate real-world conditions, what-if scenarios, and any circumstance imaginable and visualize the outcomes instantly on any platform, including mobile devices, computers, and augmented, mixed, and virtual reality devices. Therefore, an evolving set of data relies on the physical object, which requires dynamically updating and adjusting the model based on the data [7, 8, 14].

(6) Design in the Future Metaverse and Industrial Metaverse Environment.

Introduction to Metaverse

The metaverse is considered as a new vision of the next or third generation of Internet. It is a shared, immersive, persistent, 3D virtual space for users to experience life in ways beyond the physical world. The term of metaverse became a household word when Facebook changed its corporate identity to Meta in October 2021 and announced plans to invest at least $10 billion in the concept that year. In addition to Meta, tech giants including Google, Microsoft, NVIDIA, and Qualcomm are also investing billions of dollars in the metaverse development. As the metaverse progresses, various industries will become important sectors [5].

Although the concept of metaverse is not new, it has still generated significant interests widely in the general public. There are mainly two types of reactions to the metaverse as follows:

(1) It is a new generation of the Internet technology after mobile networks.
(2) There is too much hype, nobody knows what exactly metaverse is.

The metaverse can be considered as a virtual environment where physical and digital worlds are blended with comprehensive social and economic systems. The word "metaverse" describes a fully realized digital world that exists beyond where we live [10]. A framework of seven rules was proposed for metaverse [3].

There are seven main features of a metaverse, including blockchain, VR/AR, digital twins, user-generated content (UGC) creator, economics, and AI [1]. The metaverse will eventually integrate the virtual and physical worlds and accomplish more than what can be done in the real world alone.

Industrial Metaverse for AD

The current discussion on the metaverse is mostly in the areas of gaming, online retailing, sports, and social media. Although these applications have attracted most of the interests from public and markets, the potential very big market for applications of metaverse is in fact in industries, which are the backbone of economies of developed and developing countries. The applications of metaverse technologies in industry constitute industrial metaverse. The industrial metaverse, although unexplored but emerging, will transform the way we build and operate physical systems such as machines, airplanes, robots, and vehicles. Although very little has been accomplished during the book writing, it can be expected that the industrial metaverse will provide new approaches, methods, and tools for carrying out product design in industries.

Boeing wishes to build new models of airplanes in the metaverse to avoid very costly mistakes in design and building processes. If the company can build and link virtual 3D digital twin replicas of the jet and the production system to run simulations, most of the potential design changes and subsequent manufacturing, assembly, operation, maintenances, and end-of-life treatments of the product can be simulated and tested with models and data before going to real environments of development. In industrial metaverse, digital twins (DTs) will play a pivotal role in the industrial metaverse, providing virtual and behaviorally accurate representations of physical systems, to simulate complex operations before taking real-world actions.

It has been claimed that NVIDIA Omniverse can be used to build future factories, autonomous vehicles, and robotics. Unity has extended its efforts in gaming industry to manufacturing. We can expect that more efforts from academia, research community, and industries will focus on both fundamentals and applications of industrial metaverse.

Based on the adaptable design discussed in this book, AD in the industrial metaverse environment will help in the following aspects.

- Rapid design of new engineering systems.
- With DT technologies and real-time data feedback from physical products to dynamic updating of digital twins in an AD system, a new design can be created with the updated existing design. This design can be simulated realistically to determine the subsequent processes from manufacturing to retirement of the entire product lifecycle. When the design is released to production, more smoother processes can be expected as potential problems have been resolved during the simulation stage.
- Major service operations such as maintenance and upgrading.
- One of the major advantages of industrial metaverse is that major service operations, in addition to manufacturing processes, can also be simulated with full details. To upgrade large infrastructure facility and equipment such as power station and heavy-duty gas turbine, the simulation of the complete operation processes can be conducted in an industrial metaverse with both real-life data and visualization of operation details. The potential issues can be resolved before realizing the operations in the real world. The training of service personnel can also be carried out to ensure smooth operations.

7.3 Summary

This chapter provides concluding remarks and future trends. As the technology is rapidly developing, it can be expected that new technologies will have profound impact on the future design capabilities. These developments and technologies include the impact and implications of adaptable product architectures, evolution mechanism and adaptability evaluation to provide a theoretical foundation for adaptable design, and introduction of artificial intelligence applications in AD. In addition, the research on metaverse is discussed for industrial metaverse and digital twins (DTs). It is certainly expected that industrial metaverse with associated technologies including digital twins will attract significant interests in research and applications of product design and manufacturing.

References

1. H. Duan, J. Li, S. Fan, Z. Lin, X. Wu, W. Cai, Metaverse for social good: a university campus prototype, in *Proceedings of the 29th ACM International Conference on Multimedia (MM'21), October 20--24, 2021*, Virtual Event, China (2021). https://doi.org/10.1145/3474085.3479238
2. https://www.arm.com/blogs/blueprint/xr-ar-vr-mr-difference
3. https://medium.com/meta-verses/the-seven-rules-of-the-metaverse-7d4e06fa864c (medium.com., 2023)
4. https://medium.com/@northof41/what-really-is-the-difference-between-ar-mr-vr-xr-35bed1da1a4e
5. metaverse-devices. https://metamandrill.com/metaverse-devices (2023)

6. https://www.techtarget.com/whatis/feature/The-metaverse-explained-Everything-you-need-to-know
7. W. Hu, T. Zhang, X. Deng, Z. Liu, J. Tan, Digital twin: a state-of-the-art review of its enabling technologies, applications and challenges. J. Intell. Manuf. Spec. Equip. **2**(1), 1–34 (2021)
8. K.Y.H. Lim, P. Zheng, C.H. Chen, A state-of-the-art survey of Digital Twin: techniques, engineering product lifecycle management and business innovation perspectives. J. Intell. Manuf. **31**, 1317–1337 (2020)
9. I.K. Nti, A.F. Adekoya, B.A. Weyori, O. Nyarko-Boatend, Applications of artificial intelligence in engineering and manufacturing: a systematic review. J. Intell. Manuf. **33**, 1581–1601 (2022)
10. N. Stephenson, *Snow Crash* (Bantam Books, 1992)
11. N.P. Suh, *The Principles of Design* (Oxford University Press, Oxford, UK, 1990)
12. N.P. Suh, *Axiomatic Design: Advances and Applications* (Oxford University Press, Oxford, UK, 2001)
13. Z. Sun, K. Wang, Y. Chen, D. Xue, P. Gu, Information entropy method for adaptable design evaluation. Chin. J. Eng. Des. **28**(1), 1–13 (2021)
14. F. Tao, H. Zhang, A. Liu, A.Y.C. Nee, Digital twin in industry: State-of-the-art. IEEE Trans. Ind. Inf. **15**(4), 2405–2415 (2018)
15. J. Zhang, A. Simeone, P. Gu, B. Hong, Product features characterization and customers preferences prediction based on purchasing data. Ann. CIRP **67**(1), 149–152 (2018)
16. J. Zhang, A. Simeone, Q. Peng, P. Gu, Dependency and correlation analysis of specifications and parameters of products for supporting design decisions. Ann. CIRP **69**(1), 133–136 (2020)

SPRINGER NATURE

GPSR Compliance

The European Union's (EU) General Product Safety Regulation (GPSR) is a set of rules that requires consumer products to be safe and our obligations to ensure this.

If you have any concerns about our products, you can contact us on ProductSafety@springernature.com

In case Publisher is established outside the EU, the EU authorized representative is:

Springer Nature Customer Service Center GmbH
Europaplatz 3
69115 Heidelberg, Germany

The manufacturer's authorised representative in the EU is Springer Nature Customer Service Centre GmbH, Europaplatz 3, 69115 Heidelberg, Germany. If you have any concerns regarding our products, please contact ProductSafety@springernature.com

Printed and bound by CPI Group (UK) Ltd, Croydon, CR0 4YY
01/10/2026
02225900-0008